New Discoveries in Agrochemicals

ACS SYMPOSIUM SERIES **892**

New Discoveries in Agrochemicals

J. Marshall Clark, Editor
University of Massachusetts

Hideo Ohkawa, Editor
Fukuyama University

Sponsored by the
ACS Division of Agrochemicals

American Chemical Society, Washington, DC

Library of Congress Cataloging-in-Publication Data

New Discoveries in Agrochemicals / J. Marshall Clark, editor, Hideo Ohkawa, editor.

    p. cm.—(ACS symposium series ; 892)

    Developed from a symposium held at the 3rd Pan-Pacific Conference on Pesticide Science, Honolulu, Hawaii, June 1–4, 2003.

    "Sponsored by the ACS Division of Agrochemicals."

    Includes bibliographical references and index.

    ISBN 0–8412–3903–7 (alk. paper)

    1. Pesticides—Congresses. 2. Agricultural chemicals—Congresses.

    I. Clark, J. Marshall (John Marshall), 1949- II. Ohkawa, Hideo., III. American Chemical Society. Division of Agrochemicals. IV. Pan-Pacific Conference on Pesticide Science (3rd : Honolulu, Hawaii) V. Series.

SB950.93.N48   2004
668.6—dc22
                                                                                            2004053129

The paper used in this publication meets the minimum requirements of American National Standard for Information Sciences—Permanence of Paper for Printed Library Materials, ANSI Z39.48–1984.

Copyright © 2005 American Chemical Society

Distributed by Oxford University Press

All Rights Reserved. Reprographic copying beyond that permitted by Sections 107 or 108 of the U.S. Copyright Act is allowed for internal use only, provided that a per-chapter fee of $30.00 plus $0.75 per page is paid to the Copyright Clearance Center, Inc., 222 Rosewood Drive, Danvers, MA 01923, USA. Republication or reproduction for sale of pages in this book is permitted only under license from ACS. Direct these and other permission requests to ACS Copyright Office, Publications Division, 1155 16th Street, N.W., Washington, DC 20036.

The citation of trade names and/or names of manufacturers in this publication is not to be construed as an endorsement or as approval by ACS of the commercial products or services referenced herein; nor should the mere reference herein to any drawing, specification, chemical process, or other data be regarded as a license or as a conveyance of any right or permission to the holder, reader, or any other person or corporation, to manufacture, reproduce, use, or sell any patented invention or copyrighted work that may in any way be related thereto. Registered names, trademarks, etc., used in this publication, even without specific indication thereof, are not to be considered unprotected by law.

PRINTED IN THE UNITED STATES OF AMERICA

Foreword

The ACS Symposium Series was first published in 1974 to provide a mechanism for publishing symposia quickly in book form. The purpose of the series is to publish timely, comprehensive books developed from ACS sponsored symposia based on current scientific research. Occasionally, books are developed from symposia sponsored by other organizations when the topic is of keen interest to the chemistry audience.

Before agreeing to publish a book, the proposed table of contents is reviewed for appropriate and comprehensive coverage and for interest to the audience. Some papers may be excluded to better focus the book; others may be added to provide comprehensiveness. When appropriate, overview or introductory chapters are added. Drafts of chapters are peer-reviewed prior to final acceptance or rejection, and manuscripts are prepared in camera-ready format.

As a rule, only original research papers and original review papers are included in the volumes. Verbatim reproductions of previously published papers are not accepted.

ACS Books Department

Contents

Preface xiii

Overview

1. Changes and Challenges Facing Modern Crop Protection 2
   Harry Strang

Biopesticides and Transgenic Crops

2. Evaluation of Herbicide Metabolism in Transgenic Rice
   Plants Expressing *CYP1A1* and *CYP2B6* 18
   Hiroyuki Kawahigashi, Sakiko Hirose, Hideo Ohkawa,
   and Yasunobu Ohkawa

3. Role of Composition and Animal Feeding Studies in the
   Safety Assessment of Biotech Crops 28
   William P. Ridley, Gary F. Hartnell, and Bruce G. Hammond

4. Monitoring of Endocrine Disruptors in Transgenic Plants
   Carrying Aryl Hydrocarbon Receptor and Estrogen
   Receptor Genes 40
   Hideyuki Inui, Hideaki Sasaki, Susumu Kodama,
   Nam-Hai Chua, and Hideo Ohkawa

5. Impact of *Bacillus thuringiensis* Corn Pollen on Monarch
   Butterfly Populations 48
   Mark K. Sears

6. Allelopathic Activities in Litters of Mushrooms 63
   Hiroshi Araya

Combinatorial Chemistry

7. CombiChem at Bayer CropScience: What We Have Learned, Exemplified by Recent Chemistries 74
   Mazen Es-Sayed, Michael Beck, Stefan Bräse, Armin de Meijere, Christian Funke, Kristian Kather, Michael Limbach, Matthias E. P. Lormann, Christian Paulitz, Heiner Wroblowsky, and Viktor Zimmermann

8. Synthesis Based on Affinity Separation: A New Methodology for High-Throughput Synthesis Using Affinity Tags 87
   Koichi Fukase, San-Qi Zhang, Yoshiyuki Fukase, Naomi Umesako, and Shoichi Kusumoto

9. Computer-Aided Library Design for Agrochemical Lead Generation: Design and Synthesis of Ring-Fused 2-Pyridinone Esters 99
   James M. Ruiz and Beth A. Lorsbach

10. Parallel Synthesis Technologies in Lead Discovery and Optimization: Strategies and Applications 109
    Robert J. Pasteris

11. A Combinatorial Synthesis Approach for Agrochemical Lead Discovery 119
    James A. Turner, Michael R. Dick, Thomas M. Bargar, Gail M. Garvin, and Thomas L. Siddall

Mode of Action

12. Fungal Site of Action Determination: Integration with High-Volume Screening and Lead Progression 132
    Steven Gutteridge, Steve O. Pember, LiHong Wu, Yong Tao, and Mike Walker

13. Three-Dimensional Modeling of Cytochrome P450 14α-Demethylase (*CYP51*) and Interaction of Azole Fungicide Metconazole with *CYP51* 142
    Atsushi Ito, Keiichi Sudo, Satoru Kumazawa, Mami Kikuchi, and Hiroshi Chuman

14. New Herbicide Target Sites from Natural Compounds 151
    Stephen O. Duke, Franck E. Dayan, Isabelle A. Kagan, and Scott R. Baerson

15. Mode of Action of Pyrazole Herbicides Pyrazolate and
    Pyrazoxyfen: HPPD Inhibition by the Common Metabolite 161
    Hiroshi Matsumoto

16. Mechanism of Selective Actions of Neonicotinoids on Insect
    Nicotinic Acetylcholine Receptors 172
    Kazuhiko Matsuda and David B. Sattelle

17. Expression of a *Bombyx mori* Tyramine Receptor in HEK-293
    Cells and Action of a Formamidine Insecticide 183
    Yoshihisa Ozoe, Hiroto Ohta, Idumi Nagai, and Toshihiko Utsumi

18. Measurement of Receptor-Binding Activity of Non-Steroidal
    Ecdysone Agonists Using in vitro Expressed Receptor Proteins
    (EcR/USP complex) of Chilo *suppressalis*
    and *Drosophila melanogaster* 191
    Chieka Minakuchi, Yoshiaki Nakagawa, Manabu Kamimura,
    and Hisashi Miyagawa

Natural Products

19. Secondary Metabolites with Diverse Activities toward
    Phytopathogenic Zoospores of *Aphanomyces cochlioides*
    in Host and Nonhost Plants 202
    Satoshi Tahara and Md. Tofazzal Islam

20. Nematicidal Compounds from the Fungi 216
    Yasuo Kimura, Miyako Kusano, and Satoshi Nakahara

21. Bioorganic Chemistry on Sex Pheromones Secreted by
    Lepidopteran Insects and Their Application for
    Plant Protection 226
    Tetsu Ando

22. Phytotoxin Produced by *Streptomyces cheloniumii* Causing
    Potato Russet Scab 239
    Masahiro Natsume, Mayumi Komiya, Fumie Koyanagi,
    Hiroshi Kawaide, Nobuya Tashiro, and Hiroshi Abe

23. Synthesis and Biological Evaluation of Abscisic Acid,
    Jasmonic Acid, and Its Analogs 246
    Hiromasa Kiyota, T. Oritani, and S. Kuwahara

New Chemistry–Green Chemistry

24. Discovery of Pyridalyl: A Novel Compound for Lepidopterous Pest Control 256
    Noriyasu Sakamoto, Shigeru Saito, Taro Hirose, Masaya Suzuki, Sanshiro Matsuo, Keiichi Izumi, Toshio Nagatomi, Hiroshi Ikegami, Kimitoshi Umeda, Kazunori Tsushima, and Noritada Matsuo

25. Synthetic Study on Macrocyclic Musks, Mints, and Jasmine Perfumes Utilizing Ti-Claisen and Aldol Reactions 267
    Yoo Tanabe

26. Oxadiazole Derivatives as Novel Insect-Growth Regulators: Synthesis and Structure–Bioactivity Relationship 273
    Xuhong Qian, Song Cao, Zhong Li, Gonghua Song, and Qingchun Huang

27. Natural Products as Green Pesticides 283
    Denise C. Manker

28. ELISA and Liquid Chromatography/Mass Spectrometry/Mass Spectrometry Methods for Sulfentrazone and Its Acid Metabolite in Groundwater Samples 295
    Audrey W. Chen

29. Synthesis and Acaricidal Activity of Novel 2-Substituted-3-trifluoromethylquinoxalines 304
    Yoshitaka Fukushima, Naoki Ishii, Tetsuya Imai, Makio Usui, and Noriharu Ken Umetsu

Human Vector Control

30. Emerging Vectorborne Diseases and Their Control 314
    John D. Edman

31. Olyset Net, a Long Lasting Insecticidal Net for Malaria Control 326
    Takaaki Itoh

32. Process of Action of Dipteran-Specific Insecticidal Crystal Proteins from *Bacillus thuringiensis* subsp. *Israelensis* 334
    Hiroshi Sakai and Masashi Yamagiwa

33. Development of Vaccines for the Control of Blood-Feeding Arthropods: The Combined Use of Proteomic and Genomic Strategies 348
   Stephen K. Wikel

34. Control and Resistance Management of Human Pediculosis 383
   Si Hyeock Lee, Kyong Sup Yoon, Jian-Rong Gao, Young-Joon Ahn, and J. Marshall Clark

Indexes

Author Index 397

Subject Index 399

Preface

Pesticide science provides human society with the crop and disease vector protectorant products necessary to sustain the food, fiber, and heath that it requires in an environmentally safe, sustainable, and affordable manner. This critical task has been made even more daunting by the rising costs of product registration and increasing environmental concerns. Nevertheless, recent technologies and approaches have fundamentally altered how we approach pest management and the control of vectors of communicable diseases. In the forefront of all new agrochemical developments are the criteria of human safety, environmental stewardship, and resistance management.

With these goals in mind, approximately 200 pesticide scientists convened the 3rd Pan-Pacific Conference on Pesticide Science, which was jointly hosted by the Pesticide Science Society of Japan and the American Chemical Society (ACS) Division of Agrochemicals, on June 1–4, 2003 in Honolulu, Hawaii. Researchers from 14 countries (Australia, Canada, China, Germany, India, Iran, Japan, Korea, New Zealand, Nigeria, Taiwan, Thailand, United Kingdom, and the United States) presented 70 invited papers, 68 posters, and 4 panel discussions that dealt with two main topics "New Discoveries in Agrochemicals" and "Environmental Fate and Safety Management of Agrochemicals".

This ACS Symposium Series Book deals with the first topic "New Discoveries in Agrochemicals". Thirty-four invited and peer-reviewed chapters from internationally recognized pesticide experts are divided into six sections: Biopesticides and Transgenic Crops; Combinatorial Chemistry; Mode of Action; Natural Products; New Chemistry/Green Chemistry; and Control Agents for Vectors of Communicable Diseases.

Acknowledgments

We thank all the authors for their presentations at the Conference and for their contributed chapters. Special thanks goes to our keynote speakers (H. Strang and S. Oba), the conference co-chairs (N. K. Umetsu and B. Cross), topic organizers (Topic A-H. Miyagawa, N. K. Umetsu, and R. M. Hollingworth), and session organizers (H. Ohkawa, W. P. Ridley, K. Wada, J. A. Turner, Y. Ozoe, J. Y. Pyon, J. Lyga, T. Ando, J. R. Coats, M. Sasaki, K. Wing, M. Hirano, and J. M. Clark). In particular, we extend our deepest appreciation to the many expert colleagues who provided helpful and necessary critical reviews. We thank Stacy VanDerWall and Bob Hauserman in acquisitions and Margaret Brown in editing/production of the ACS Books Department for all their help, suggestions, and encouragement. Lastly, we thank the Pesticide Science Society of Japan and the ACS Division of Agrochemicals and their benefactors, contributors, and donors whose financial support made this book possible.

J. Marshall Clark
Department of Veterinary and Animal Science
University of Massachusetts
Amherst, MA 01003

Hideo Ohkawa
Research Center for Environmental Genomics
Kobe University
Kobe, Hyogo 657-8501
Japan

Overview

Chapter 1

Changes and Challenges Facing Modern Crop Protection

Harry Strang

Bayer CropScience, 17745 South Metcalf Avenue, Stilwell, KS 66085

Significant increases in the costs of researching, developing and maintaining crop protection chemicals in the market place have contributed to major consolidation in the crop protection industry. The June 2002 acquisition of Aventis CropScience by Bayer to form Bayer CropScience is a recent example of this industry trend. Advances in agriculture continue to make tremendous contributions to improving human health, nutrition and quality of life around the world. Modern plant breeding, improved agronomy and the development of inorganic fertilizers and synthetic pesticides lead these advances. Pesticides used in public health programs have also saved millions of human lives from disease by controlling insects or rodent vectors and intermediate hosts. While 24 million American workers (17% of workforce) are involved in the production, processing and selling of the nation's food and fiber, high technology agriculture has led to significantly fewer workers being actively engaged in farming. Thus many consumers have lost touch with the complexity of farming and see little benefit from the use of pesticides. Activist groups and some sections of the media increasingly advocate banning or removing even the slightest traces of synthetic chemicals, such as approved pesticides or food additives. These scientifically and socially unwarranted efforts for a pesticide-free environment divert consumer attention away from the very real public-health goal of increasing consumption of fruits and vegetables. Such distortion of health risks threaten innovation, jobs and our nation's enviable high

© 2005 American Chemical Society

standard of living, and often divert critical resources needed to address proven health risks. The scientific community and agricultural industry must actively explore ways to partner with grower groups, other members of the food chain and recognized medical experts to better educate consumers and policy makers on distinguishing between known and hypothetical risks.

Market Overview

In 2001, the world market for agricultural chemicals exceeded US $28 billion, with NAFTA (US, Canada, Mexico) and Europe accounting for over 50% of this total *(Fig. 1)*. Herbicides accounted for almost half of all sales, followed by insecticides (26%) and fungicides (21%).

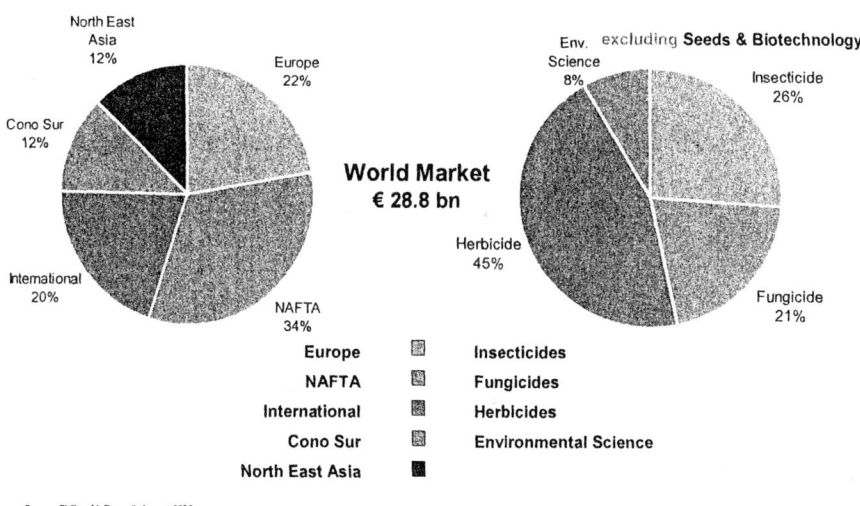

Figure 1. Agrochemical World Market 2001

Industry Consolidation

Although the world market, value-wise, has remained relatively stable over the past decade, significant increases in the costs of researching, developing and maintaining crop protection chemicals in the market place have contributed to major consolidation in the crop protection industry. As shown in Figure 2, only nine companies now account for almost 90% of the world market. The two largest companies, Bayer CropScience and Syngenta, collectively represent the consolidation of approximately fifty entities which once existed as separate companies.

Benefits of Pesticides

The essential role played by crop protection chemicals in modern agriculture and the positive impact they have on the world economy, the preservation of natural resources and the quality of life around the world, is often not well understood or recognized.

Over the past forty years, advances in agriculture such as modern plant breeding, improved agronomy and the development of inorganic fertilizers and synthetic pesticides have made tremendous contributions to improving human health and nutrition. Between 1961 and 2000 food production doubled in developed countries and increased three-fold in developing countries. Wheat yields in England climbed from 2 to 6 metric tons per hectare although it had previously taken nearly 1,000 years for yields to increase from 0.5 to 2 metric tons per hectare. In Asia, cereal production more than doubled between 1970 and 1995, although the total land area cultivated with cereals increased by only 4% *(1)*. The world population increased from 2.5 billion in 1950 to approximately 6 billion in 2000, but the area of arable land in production remained remarkably constant at approximately 1.5 billion hectares *(Fig. 3)*.

Advances in modern agriculture have successfully supported an increasing world population and have resulted in sustained food surpluses in most industrialized countries during the latter half of the 20th century. These advances have also resulted in the preservation of billions of hectares of the world's forests and grasslands. Without such productivity increases in yield per hectare, it is estimated that two to three times more land would have had to have been brought into production to feed the world's population *(Fig. 4)*. However, new agricultural advances and technologies, such as improved pesticides, fertilizers and biotechnology, must continue to be implemented in order to meet the increasing world population, which is expected to peak at ca. 10 billion by the middle of the 21st century.

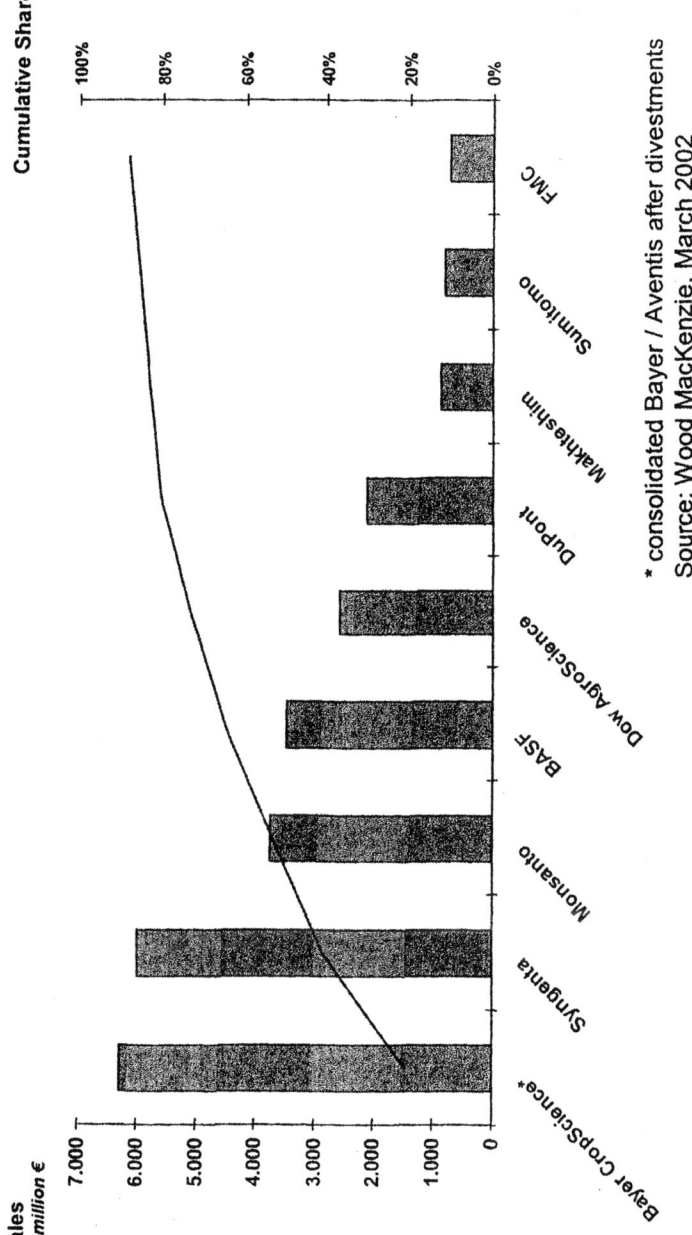

Figure 2. Agrochemical World Market 2001-2002

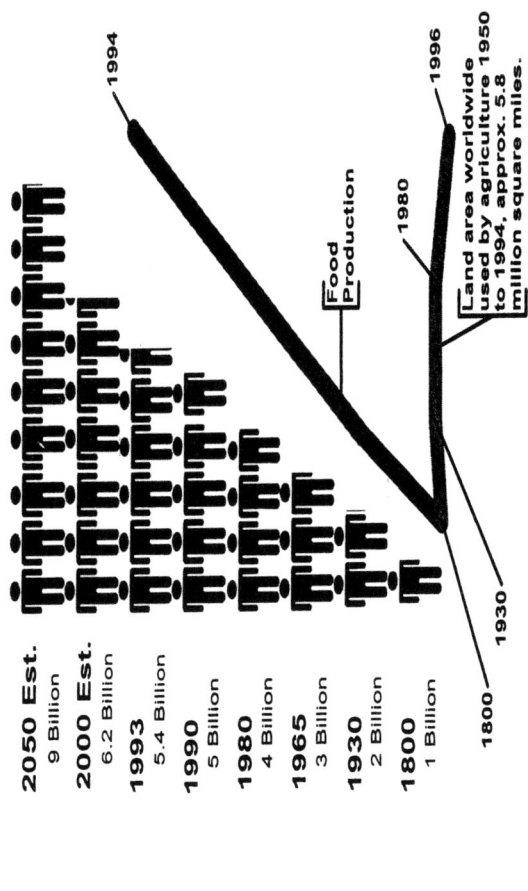

Figure 3. Relation of Arable Land and Population

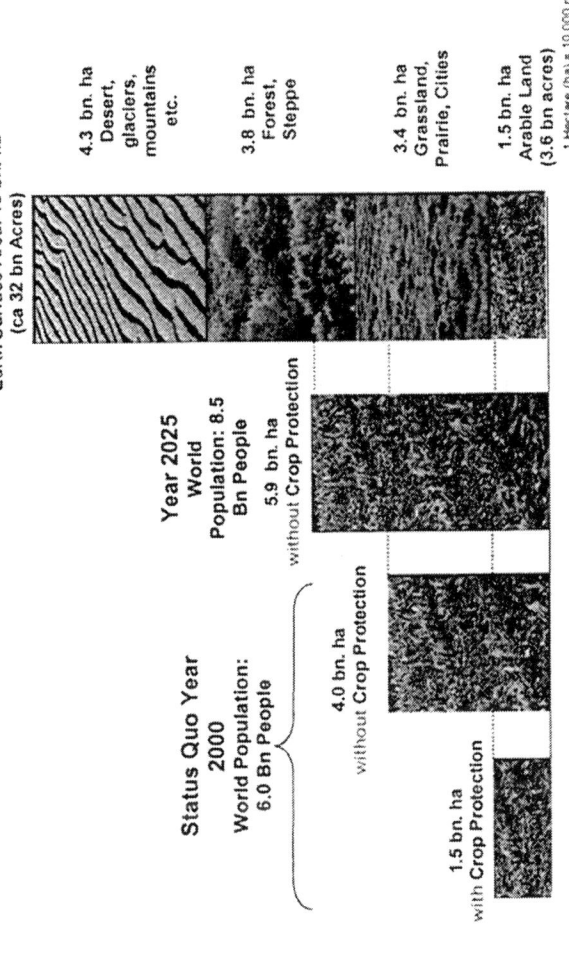

Figure 4. Agricultural Area Required To Produce Sufficient Foodstuffs With & Without Crop Protection

Increased use of pesticides since the 1950's is one of the major reasons for these productivity increases. Research indicates that without pesticides yields of many crops would drop 50-100% *(2)*. A recent report by the National Center for Food and Agricultural Policy (NCFAC) *(3)* showed that replacing herbicides with hand labor would still result in yield losses of 30-50% for most crops *(Fig. 5)*.

Figure 5. The Value of Herbicides in U.S. Crop Production

The NCFAP report further estimated that to prevent any yield loss in comparison to herbicides an estimated 70 million workers, approximately 25% of the US population, would need to be employed as hand weeders. This estimate is supported by the fact that from 25-75% of the population of many developing countries toil daily weeding their crops.

America's farmers are among the most productive in the world, producing over 40% of the world's soybeans and corn, 20% of the world's cotton and 24% of the world's beef and veal. America's agricultural production exceeds US $220 billion annually with over $50 billion worth of agricultural products exported around the world in 2001 *(4)*.

This abundance has resulted in Americans paying less for their food, 7% of income, than any other country in the world. The citizens of most developed countries spend between 10 and 25% of their income for food compared with 50 to 75% of income spent in the developing world *(Fig. 6)*.

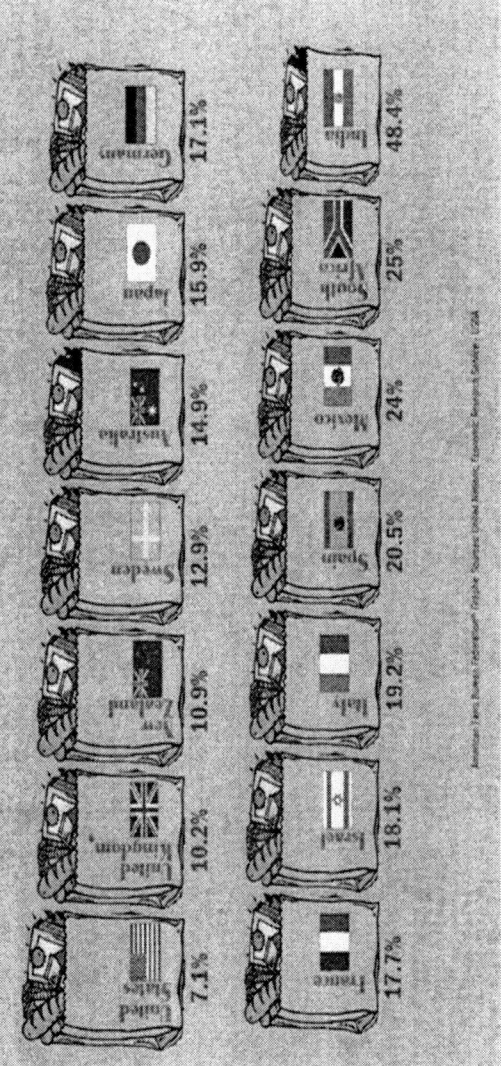

Figure 6. Food Is Most Affordable in the United States

While 24 million American workers (17% of workforce) are involved in the production, processing and selling of the nation's food and fiber, high technology agriculture has led to significantly fewer workers being actively engaged in farming. Today almost 150 fellow citizens are fed by one American farmer, compared with less than 50 in 1960 *(Fig. 7)*.

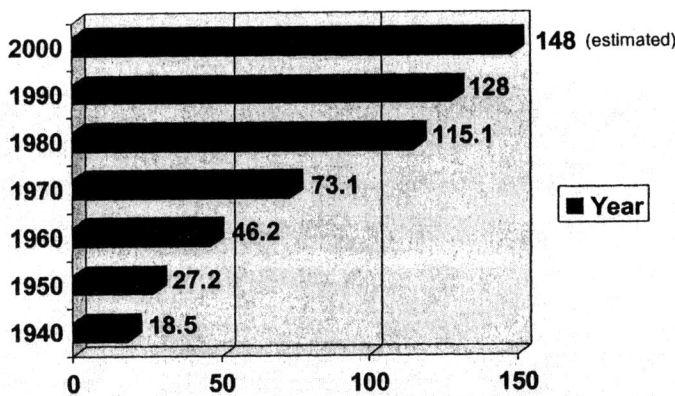

Source: Reproduced with permission from *J. Biol. Chem.* **2003**, *278(7)*, 4369–4380. Copyright 2003 The American Society for Biochemistry and Molecular Biology.

Figure 7. Number of People Fed by One Farmer

Thus, over 98% of America's citizens today work in occupations outside of farming while enjoying the most abundant and economic food supply in the world. Unfortunately, many consumers have lost touch with the complexity of farming, do not fully appreciate the multiple disease, insect and weed pressures that can devastate entire crops, and see little benefit from the use of pesticides. Many consumers in the developed world are unaware of, or choose to ignore, the fact that >99.99% (by weight) of the pesticides in their diet are natural chemicals, which plants produce to defend themselves (Table I).

Table I. Natural Chemicals, The Forgotten Control

Natural Chemicals, The Forgotten Control
Recent comments by Dr. Bruce Ames: Leading Cancer Researcher, winner of President Clinton's 1999 National Medal of Science.

- "99.9% of all chemical exposure is from ingesting natural chemicals in food."
- "99.99% of exposure to pesticides is from ingesting natural pesticides produced by plants. Only 52 natural pesticides have been tested in high-dose animal cancer tests, and about half (27) are rodent carcinogens."
- "The possible cancer hazard of traces of synthetic chemicals such as pesticides are tiny compared with natural chemicals in the diet."
- "There are over 1000 chemicals in coffee: 28 have been tested and 19 are rodent carcinogens."
- "Spending time debunking the dubious assumptions behind environmentalist fervor against traces of industrial chemicals does not prevent cancer."
- "diversion of resources and attention from programs that focus on major risks to those that focus on hypothetical risks might hurt public health."

Source: Ames, B. N. 2003; An Enthusiasm for Metabolism; J. Biol. Chem., 278(7): 4369-4380

Activist groups and some sections of the media intentionally exaggerate the often hypothetical risks of pesticide use for their own gain while totally ignoring the numerous benefits. In a recent report by the American Council on Science and Health, Dr. D. Juberg states that activist groups use alleged risks from chemicals and other environmental fears to manipulate parent's very legitimate concerns for their children's health. Most often, these campaigns are an effort to promote legislation, regulation and litigation that is based not in science, but rather in a political agenda opposed to technology, free markets and scientific progress (5). These groups increasingly advocate banning or removing even the slightest traces of synthetic chemicals, such as approved pesticides or food additives.

These scientifically and socially unwarranted efforts for a pesticide-free environment divert consumer attention away from the very real public-health goal of increasing consumption of fruits and vegetables in order to counteract the rapidly increasing obesity threat in America. The Center for Disease Control and Prevention has recently declared that obesity in America is reaching epidemic proportions with two-thirds of Americans being seriously over weight and over 25% are obese (> 30 lbs over a healthy weight). This health epidemic contributes to over 300,000 premature deaths annually in the U.S., second only to tobacco related deaths (6). A major contributing factor is that Americans consume excessive amounts of sugars and fats and less than 25% of the population is eating the minimum five daily servings of fruits and vegetables essential for good health and strongly promoted by the dietary guidelines of the U.S. Department of Agriculture (USDA), U. S. Department of Health and

Human Services (DHHS), National Cancer Institute (NCI), American Dietetic Association (ADA), Produce for Better Health Foundation (PHB), etc.

Unfortunately, what often receives the most media coverage are the activists' unfounded and biased assertions that the foods most frequently eaten by children, such as apples, are contaminated with toxic pesticides at levels that pose a significant risk to infants and children. A disturbing statistic reported by the International Food Information Council is that journalists provide adequate context in only 6% of their stories on diet, nutrition and food safety. Internet sites and publications from highly reputable science based organizations, such as The Center for Disease Control (CDC), National Cancer Institute (NCI), American Medical Association (AMA), Food and Drug Administration (FDA), American Council on Science and Health (ACSH), etc., provide needed balance but cannot compete in getting the general public's attention.

The real facts are that billions of dollars of research and reports by the world's leading health authorities confirm the negligible risk of pesticide residues in food *(Fig. 8)*.

A 650 page report reviewing over 4,500 studies which investigated the effects of foods on the development of cancer, concluded that:

"There is no convincing evidence that any food contaminant (including pesticides) modifies the risk of cancer, nor is there any evidence of a causal relationship. Indeed there is little epidemiological evidence that chemical contamination (pesticides) of food and drink, resulting from properly regulated use, significantly affects cancer risk."

World Cancer Research Foundation, American Institute for Cancer Research, World Health Organization, National Cancer Institute, Food and Agriculture Organization, United Nations, International Agency for Research on Cancer

Figure 8. Food, Nutrition and the Prevention of Cancer:
A Global Perspective, 1997

Dr. Bruce Ames, one of the world's leading cancer experts, has stated that the quarter of the population eating the fewest fruits and vegetables has double the cancer risk for most types of cancer than the quarter eating the most. He further concludes that attempting to reduce hypothetical risks has other costs as well. If reducing synthetic pesticides makes fruits and vegetables more expensive, thereby decreasing consumption, then the cancer rate will increase, especially for the poor *(7)*. Recent studies support the protective role of fruits and vegetables against many forms of cancer and other diseases (Table II).

Table II. Recent Studies (1999-2001) Support Protective Role of Fruits and Vegetables Against Many Forms of Cancer and Other Diseases

Form of Cancer	% Reduction*	Type of Fruit/Vegetable
Lung	21-32%	Cruciferous, citrus, carrots
Breast (post-menopausal women)	40-50%	Cruciferous
Prostate	35-41%	Cruciferous, tomatoes
Ovarian	40-54%	Carrots, tomatoes
Larynx	65-68%	Lettuce, tomatoes, oranges
Bladder	38-46%	Cruciferous, green-yellow
Cardiovascular Disease		
Coronary Disease	20-38%	Green vegetables, citrus
Stroke	26-39%	Cruciferous, green, citrus
Diabetes	26-39%	Fruits and vegetables

*Reduction in disease rate associated with high intake of fruits and vegetables (5-9 per day) vs. low intake (0-2 servings per day).

Source: Hyson, D. 2002. "*The Health Benefits of Fruits and Vegetables: A Scientific Overview for Health Professionals.*" Produce for Better Health® Foundation

Many health professionals and other experts have expressed concern that disproportionate resources are often spent on mitigating hypothetical risks while real risks remain almost unattended *(8)*. A Harvard University Center for Risk Analysis survey of cost estimates of regulatory "interventions that have as the primary stated political goal to save human lives" illustrates the disproportionate cost of environmental regulations in comparison to those in other federal regulatory agencies *(Fig. 9)* and have led some experts to conclude that when we ignore the cost of our environmental decisions on the lesser regulations or attention in other areas, we are in reality putting more lives at risk *(9)*.

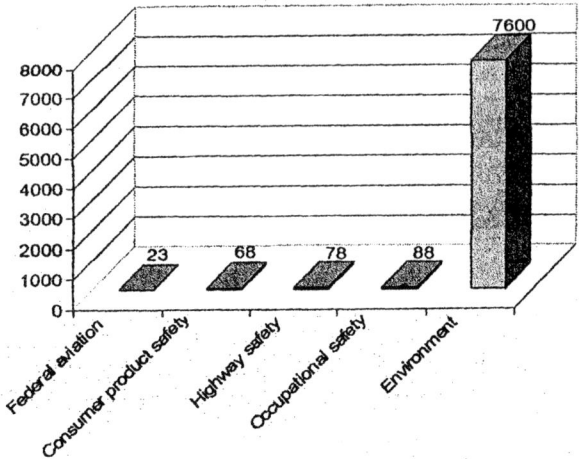

Sources: Tengs et al. 1995 Risk Analysis 15(3):369-90. Lomborg, B., "The Skeptical Environmentalist";Cambridge University Press, New York, 2001

Figure 9. Median Cost Per Life-Year Saved For Different Government Sectors (Reproduced with permission from Risk Anal. *1995,* 15(3), *369–380. Copyright 1995 Blackwell Publishing Ltd.)*

In the USA, The Center for Disease Control and Prevention (CDC) reports that whereas in the early 1900s the average life span was about 50 years, today Americans are living well into their 70's. Of the thirty years of increased life expectancy, only five can be attributed to curative medicine; the remaining 25 years are due to advances in public health and improved nutrition *(10)*. Pesticides used in food production and public health programs have saved millions of human lives by reducing mycotoxin levels in food, protecting foods from insect and fungal spoilage, filth, mold, etc. and reducing the risk of disease spread by cockroaches, flies, ticks, rodents and mosquitoes.

It is important to recognize that pesticides are among the most thoroughly tested chemicals used by man. Each new pesticide must pass up to 120 EPA required health and environmental tests before being registered. These tests take an average of 9 years to complete at an average cost of over $180 million per compound. Only 1 compound out of 139,000 tested will reach the market. Despite this thorough testing of pesticides and the assurances of health authorities that the food supply is safe, a large percentage of the general public still express concern about the safety of their food (Table III).

For many of the reasons documented in this paper, the public's understanding and acceptance of essential farming practices is disappointingly low (Table IV).

Table III. How Much Do Consumers Worry That Foods They Buy May Not Be Safe To Eat?

Year	Great Deal	Some	Little	Not at All
2002	21%	34%	30%	15%

Source: Reproduced with permission from *Examining Consumer Expectations of Agriculture;* 1999–2002. Copyright 2002 American Farm Bureau Association.

Table IV. Public Acceptance of Farming Practices

How Do Consumers Respond When Asked About The Acceptability of Farming Methods?*			
When are these methods acceptable?	Always acceptable	Sometimes acceptable	Never acceptable
Pesticides and herbicides to increase production	12%	49%	32%
Pesticides to improve appearance	5%	31%	57%
Chemical fertilizers to increase crop growth	15%	47%	29%
Natural fertilizers to increase crop growth	58%	30%	6%
Biotechnology to increase food production	18%	46%	20%

*Numbers may not total 100% due to "no response" not included.

Source: Reproduced with permission from *Examininng Consumer Expectations of Agriculture;* 1999–2002. Copyright 2002 American Farm Bureau Association.

Conclusion

It is essential for the scientific community and the agricultural industry to actively explore ways to partner with grower groups, other members of the food chain, and with recognized medical and public health experts to better educate consumers, politicians and policy makers on distinguishing between known and hypothetical risks and to appreciate the tremendous benefits provided by modern agriculture.

References

1. International Food Policy Research Institute, 2002, www.ifpri.org
2. Knutson, R. D. et. al., *Economic Impacts of Reduced Pesticide Use on Fruits and Vegetables,* American Farm Bureau Research Foundation, Chicago, IL, September, 1993
3. Gianessi, L. P.; Sankula, S.: *The Value of Herbicides in U.S. Crop Production,* April 2003. NCFAP; www.ncfap.org
4. Farm Facts 2002. American Farm Bureau Federation. www.fb.org
5. Juberg, D. R.; *Are Children More Vulnerable to Environmental Chemicals?* American Council on Science and Health, 2003 pp 5-6.
6. Center for Disease Control (CDC) website; www.cdc.gov/nccdphp/dnpa/press/archive/obesity_climbs.htm; accessed 4/9/03.
7. Ames. B.; Gold, L.: *Environmental Pollution, Pesticides and the Prevention of Cancer: Misconceptions.* 1997. FASEB, J. 11(14): 1330.
8. Koop, C. E.: Foreword in *Are Children More Vulnerable to Environmental Chemicals?* American Council on Science and Health, 2003, pp 236.
9. Lomberg, D. R.: *The Skeptical Environmentalist,* Cambridge University Press, 2001, pp 341-342.
10. Center for Disease Control and Prevention (CDC), U. S. Department of Health and Human Services. *Physical Activity and Good Nutrition: Essential Elements for Good Health, At-a-Glance 2000.* Atlanta: CDC, 2000

Biopesticides and Transgenic Crops

Chapter 2

Evaluation of Herbicide Metabolism in Transgenic Rice Plants Expressing *CYP1A1* and *CYP2B6*

Hiroyuki Kawahigashi[1], Sakiko Hirose[1], Hideo Ohkawa[2], and Yasunobu Ohkawa[1]

[1]Plant Biotechnology Department, National Institute of Agrobiological Sciences, 2–1–2 Kannondai, Tsukuba, Ibaraki 305–8602, Japan
[2]Research Center for Environmental Genomics, Kobe University, Rokkodaicho 1–1, Nada, Kobe, Hyogo 657–8501, Japan

We introduced human *CYP1A1* and *CYP2B6* cDNAs into a japonica rice *(Oryza sativa* cv. Nipponbare), to establish two different transgenic rice lines, because both CYP1A1 and CYP2B6 metabolize several herbicides having different chemical structures and modes of biological action. These transgenic rice plants showed broad cross-tolerance toward various herbicides and metabolized them. The introduced human P450 protein essentially just enhanced the natural metabolism of herbicides in the transgenic rice plants. These transgenic rice plants should be important not only for developing herbicide-tolerant lines of rice, but also for decreasing herbicide residues in rice. In addition, transgenic rice plants should prove useful in implementing effective, inexpensive methods of decreasing various chemicals that are widespread in agricultural environments.

The metabolism of herbicides in higher plants occurs in three main stages (*1*). First, herbicides are processed enzymatically via oxidation, reduction, or hydrolysis. Second, products of the first stage are conjugated with glutathione or sugars to become highly water soluble and less mobile in the plant. Finally, the conjugated metabolites are converted to secondary conjugates or insoluble bound residues. These metabolites are deposited in vacuoles or in the cell walls of plants.

Cytochrome P450 monooxygenase (P450) is one of the major enzymes involved in herbicide detoxification in plants. In particular, P450 monooxygenases play important roles in the first stage of herbicide metabolism, in cooperation with NADPH-cytochrome P450 oxidoreductase, to produce nonphytotoxic metabolites. The P450 systems in plants are responsible for herbicide sensitivity and tolerance (*2*). However, certain P450 systems involved in herbicide metabolism are not yet well documented.

Mammalian P450 systems, which metabolize xenobiotics in the liver, are well known to show broad, overlapping substrate specificity toward foreign lipophilic chemicals, including herbicides. Several transgenic plants have been engineered by introducing genes for mammalian P450 into them. The rat *CYP1A1* gene was introduced into tobacco (*3*), and human *CYP1A1* into potato (*4*). These transgenic plants showed remarkable tolerance to the phenylurea herbicide chlortoluron. The human *CYP2E1* gene was also introduced into plants, and the transformed plants showed enhanced metabolism of trichlorethylene (*5*).

We have produced transgenic rice plants expressing *CYP1A1* and *CYP2B6*; these transgenic rice plants showed remarkable cross-tolerance toward herbicides and enhanced herbicide detoxification, a property that should be useful in phytoremediation.

Herbicide Tolerance in CYP1A1 and CYP2B6 Rice Plants

Production of Transgenic Rice Plants Expressing Human P450 Genes

Expression plasmids, pIES1A1 and pIJ2B6, were constructed which contained human *CYP1A1* or *CYP2B6* cDNA (Fig. 1). Each constructed expression plasmid was introduced into *Agrobacterium* strain EHA101, which was subsequently used for transformation of rice plants (*Oryza sativa* cv. Nipponbare) (*6*). After selection on 50 mg/L hygromycin, regenerated plants were analyzed by PCR using human CYP1A1- or CYP2B6-specific primers. T_1 seedlings of the CYP1A1 and CYP2B6 rice plants were screened by germination tests in the presence of 100 μM chlortoluron or 2.5 μM metolachlor, respectively.

On the basis of the tolerance of the progeny toward these herbicides, homogeneous T_2 lines were selected and used for further study.

pIES1A1

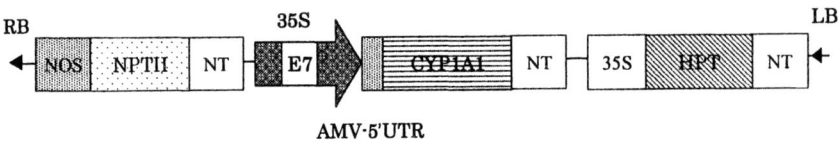

pIJ2B6

Figure 1. T-DNA region of pIES1A1 and pIJ2B6 plasmids used to express human CYP1A1 and CYP2B6 in transgenic rice plants. RB, right border; LB, left border; NOS, nopaline synthase promoter; NT, nopaline synthase terminator; NPTII, neomycin phosphotransferase II; 35S, cauliflower mosaic virus (CaMV) 35S promoter; E7, seven-enhancer region (–290 to –90) from CaMV 35S promoter; AMV-5'UTR, alfalfa mosaic virus 5'-untranslated region; HPT, hygromycin B phosphotransferase.

Germination Tests

We tested the germination of transgenic rice plants in MS agar containing various herbicides: both CYP1A1 and CYP2B6 rice plants were clearly tolerant to various herbicides having different sites of action and different chemical properties (Table I). The transgenic rice plants expressing human P450 proteins showed tolerance to most of the same herbicides as did the same P450 proteins in transgenic yeast microsome tests *in vitro* (7).

CYP1A1 rice plants were tolerant to the photosynthesis-inhibiting herbicide chlortoluron. T_2 seeds sown on MS medium containing 100 µM chlortoluron germinated and grew well, whereas untransformed Nipponbare did not grow at all (Fig. 2). CYP1A1 rice plants were tolerant to norflurazon (0.5 µM), a carotenoid-biosynthesis-inhibiting herbicide, which caused bleaching of shoots of Nipponbare. The CYP1A1 transgenic rice plants also grow well on medium

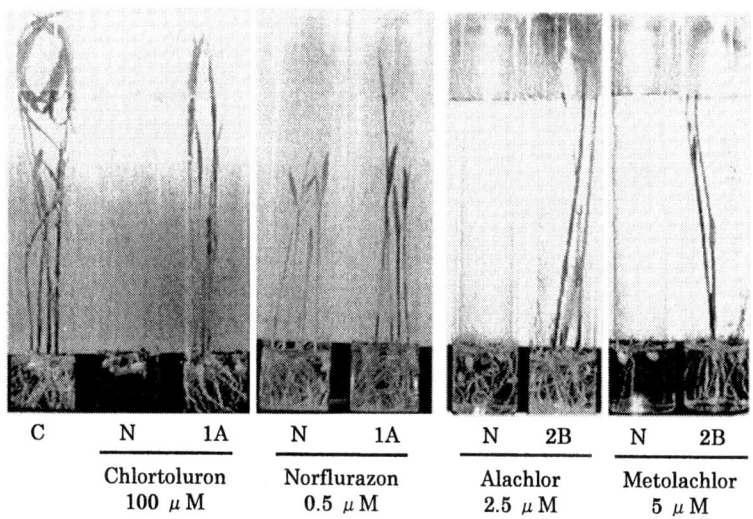

Figure 2. Phytotoxicity of various herbicides toward CYP1A1 *and* CYP2B6 *transgenic rice plants in germination tests in MS medium.* C, Nipponbare without herbicide (control); N, Nipponbare with herbicide; *1A,* CYP1A1 *plants with herbicide;* 2B, CYP2B6 *plants with herbicide.*

containing diuron, chlorpropham, mefenacet, or quizalofop-ethyl (data not shown).

Table I. Tolerance of Transgenic *CYP1A1* and *CYP2B6* Rice Plants to Various Herbicides in Germination Tests

P450	Herbicide	Conc.[1]	Mode of inhibition (Group)[2]
CYP1A1	Chlorpropham	7.5	Protein biosynthesis (Carbamate)
	Chlortoluron	100	Photosynthesis (Urea)
	Diuron	150	Photosynthesis (Urea)
	Mefenacet	2.5	Cell division (Oxyacetamide)
	Norflurazon	0.5	Carotenoid biosynthesis (Pyridazinone)
	Quizalofop-ethyl	0.2	Fatty acid biosynthesis (2-(4-Aryloxyphenoxy) propionic acid)
CYP2B6	Acetochlor	0.2	Protein biosyntehsis (Chloroacetanilide)
	Alachlor	2.5	Protein biosyntehsis (Chloroacetanilide)
	Chloridazon	300	Fatty acid biosynthesis (Pyridazinone)
	Metolachlor	5	Protein biosyntehsis (Chloroacetanilide)
	Pendimethalin	10	Cell division (Dinitroaniline)
	Pyributicarb	2.5	Sterol and triterpenoid biosynthesis
	Thenylchlor	5	Protein biosyntehsis (Chloroacetanilide)
	Trifluralin	15	Cell division (Dinitroaniline)

[1] Final concentration of herbicide in the MS agar medium
[2] As defined in (*8*)

CYP2B6 rice plants showed high tolerance toward the chloroacetanilide herbicides, which inhibit protein biosynthesis (Fig. 2). T_2 seeds on medium containing 2.5 μM alachlor or 5 μM metolachlor germinated and grew well, whereas untransformed Nipponbare did not grow at all. The CYP2B6 transgenic rice plants also grow well on medium containing acetochlor, chloridazon, pendimethalin, thenylchlor, or trifluralin (data not shown).

CYP2B6 rice plants also grew well on medium containing pyributicarb, although transgenic yeast containing the same gene did not metabolize these herbicides (*7*). It may be that the activity of human CYP2B6 is higher in rice plants than in yeast. However, it is also probable that CYP2B6 rice plants degraded the herbicides gradually during and after germination, keeping the concentration of the herbicide in plant tissues under the lethal threshold.

Metabolism of Herbicides in Transgenic Rice Plants Expressing CYP1A1 and CYP2B6 Gene

Transgenic rice plants were grown on medium containing ^{14}C-labeled herbicides to confirm the metabolism of herbicides by the introduced P450 proteins. Briefly, a six-day-old plant was transferred to an individual test tube containing 3 mL of Hyponex 5-10-5 solution containing 40 000 dpm of 10 µM ^{14}C-herbicide. Radiolabeled chemicals were extracted from the plants with 90% methanol, and 2000-dpm samples of plant extracts were applied to the origin of a silica gel 60F$_{254}$ thin-layer chromatography (TLC) plate. After development of the plates, radioactivity was measured in an FLA-2000 Bio-Imaging Analyzer.

We analyzed the metabolism of chlortoluron in CYP1A1 rice plants and control Nipponbare plants. On the 6th day of incubation, the mean amount of chlortoluron in the CYP1A1 rice plants was equivalent to only 1.5% of the added radioactivity. The total amount of metabolites, including N-demethylated (DM), ring-methyl hydroxylated (OH), N-demethylated and ring-methyl hydroxylated (DMOH), and conjugated (Ori) compounds was 24.6% added (Fig. 3A). However, in control Nipponbare plants, the amount of chlortoluron was 8.1% of the added radioactivity, and that of the metabolites was 4.5%.

Similarly, the mean amount of radioactive norflurazon in CYP1A1 rice plants decreased to 1.5% of the added radioactivity (Fig. 3B). Thus, chlortoluron and norflurazon were rapidly metabolized and detoxified in the CYP1A1 rice plants.

We also analyzed the metabolism of metolachlor in CYP2B6 rice plants. Metolachor was rapidly metabolized into its demethylated compounds in these plants and was subsequently metabolized into its conjugated compounds. These results showed that both CYP1A1 and CYP2B6 rice plants decreased the herbicide recidues in the plants more than the nontransgenic control plants.

In TLC analyses with radiolabeled herbicides, both transgenic and nontransgenic control plants metabolized these herbicides to give the same sets of metabolites. The herbicides added were metabolized more rapidly in the transgenic rice plants than in the nontransgenic control plants. Thus, the exogenous P450 species did not change the metabolic pathway of herbicides and they merely enhanced metabolism of herbicides.

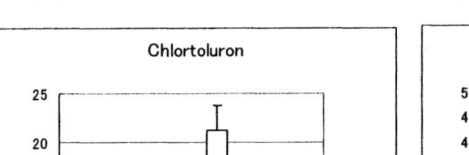
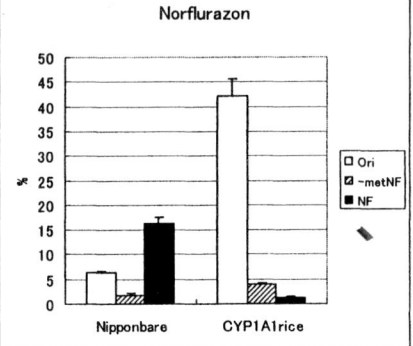

Figure 3. Metabolism of ^{14}C-labeled herbicides in *CYP1A1* rice plants analyzed by thin-layer chromatography (TLC). (A) Relative amounts of chlortoluron and its metabolites in plant extracts. (B) Relative amounts of norflurazon and its metabolites in plant extracts. CT, chlortoluron; DM, N-demethylated chlortoluron; OH, ring-methyl hydroxylated chlortoluron; DMOH, N-demethylated and ring-methyl hydroxylated chlortoluron; NF, norflurazon; -metNF, demethylated norflurazon; Ori, origin of the TLC plate (conjugated compounds). Quantity of radioactivity added to the plants was defined as 100%. Values are averages of 3 independent experiments.

Analysis of Residual Metolachlor in CYP2B6 Rice Plants

Residual metolachlor in plants and in culture medium was analyzed by GC/MS and HPLC. Ten-day-old rice seedlings were transferred into a plant-box filled with 80 mL of MS liquid medium containing 30 µM metolachlor. After 1, 3, and 6 days of incubation under light at 27 °C, the amount of metolachlor in the plants and culture medium was analyzed.

In this small-scale experiment, only a small portion of the added metolachlor was detected in the plants; most of the metolachlor remained in the culture medium. But *CYP2B6* rice plants degraded the metolachlor in the culture medium and decreased its concentration rapidly. On the sixth day of culture, the amount of metolachlor degraded by the CYP2B6 rice plants was 1.6 times that degraded by nontransgenic Nipponbare rice plants (Fig. 4).

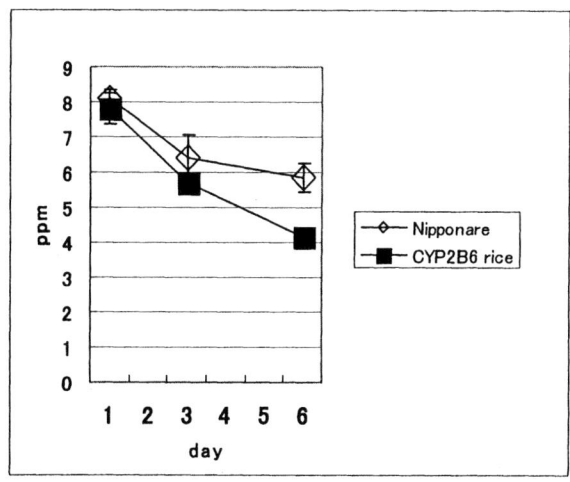

Figure 4. Analysis of metolachlor remaining in culture medium by gas chromatography. Ten-day-old rice plants were incubated in medium containing 30 μM metolachlor for 1–6 days.

To analyze residual amounts of herbicides in the culture medium more accurately, CYP2B6 and nontransgenic rice plants were cultured in 9 L of medium containing 6 μM metolachlor. Residual amounts of herbicides in the culture medium of the transgenic rice plants decreased rapidly compared with those in the culture medium of nontransgenic plants. The amount of metolachlor degraded by the CYP2B6 rice plants was about 1.4 times that by nontransgenic rice plants in both one-month-old and three-month-old plants.

The degradation per gram dry wt of biomass for metolachlor was much higher in one-month-old plants than in older plants, probably because these plants were in the vegetative stage and growing vigorously. However, the total amount of degradation of metolachlor was higher in the three-month-old plants, owing to their greater biomass.

We analyzed residual metolachlor in three-month-old plants themselves as well which were cultured with 9 L of medium containing 12 μM (3.4 ppm) metolachlor. Metolachlor was absorbed by the roots and accumulated in both CYP2B6 and nontransgenic rice plants. However, the amount of metolachlor in stems and leaves of CYP2B6 rice plants was only about one-third of that in nontransgenic Nipponbare rice plants (Table II).

These results indicate that exogenous CYP2B6 proteins enhanced the metabolism of herbicides in transgenic rice plants and had a high potential to metabolize chemical compounds *in vivo* in the xenobiotic pathway.

Table II. Degradation of Metolachlor in Three-Month-Old Rice Plants

Incubation time (day)	Concentration (ppm) [1]	
	Nipponbare	CYP2B6 rice
0	0.006	0.006
3	10.8	4.09
7	19.0	6.51

[1] concentration of the plants per wet weight

Conclusion

We introduced human *CYP1A1* and *CYP2B6* genes into rice plants, because both the CYP1A1 and CYP2B6 metabolize several herbicides having different chemical structures and biochemical effects. As a result of detoxifying herbicides by the xenobiotic pathway, the transgenic rice plants showed cross-tolerance toward several herbicides, even though most herbicide-tolerant crops show tolerance to only a particular herbicide. Therefore, we expect that transgenic plants expressing P450 components will be suitable for cropping with different herbicides in rotation.

Repeated use of one herbicide in a field tends to result in the appearance of herbicide-resistant weeds (9). The most practical approach to delaying the evolution of resistance is to rotate several herbicides in crop monocultures (10); thus, the combination of transgenic crop plants and rotation of different herbicides should prevent the appearance of mutant weeds tolerant to herbicides.

Transgenic rice plants expressing *CYP2B6* rapidly decreased the amount of metolachlor present in both the plants themselves and the culture medium. Thus, such transgenic plants should be important not only for developing herbicide-tolerant cultivars, but also for decreasing herbicide residues in rice.

The growth of industrialization and the use of chemical compounds have resulted in serious environmental pollution. Various organic contaminants, including aromatic hydrocarbons, nitroaromatic compounds and chloroaromatics persist in the environment. It is technically difficult to decontaminate these chemicals and traditional physiochemical treatments are expensive.

Phytoremediation is a relatively new approach to remove contaminants from the environment by using plants in an inexpensive way. It may be possible to use transgenic rice plants expressing these P450 systems for removing persistent xenobiotics that are widely spread in the environment. However, further studies, including tests of remediation ability in practical field and safety assessments of transgenic plants, remain to be done.

Acknowledgements

This study was supported by a program for the Promotion of Basic Research Activities for Innovative Biosciences of the Bio-oriented Technology Research Advancement Institution (BRAIN). We thank Novartis (Basel, Switzerland) for supplying the ^{14}C-chlortoluron and the Plant Biotechnology Institute (Ibaraki, Japan) for providing *Agrobacterium* EHA101.

References

1. Hatios, K.K. In *Regulation of Enzymatic Systems Detoxifying Xenobiotics in Plants*, Hatios, K.K., Ed.; Kluwer: Dordrecht, Netherlands, 1997; pp 1–5.
2. Werck-Reichhart, D.; Hehn, A.; Didierjean, L. *Trends Plant Sci.*, **2000**, *5*(3), 116–123.
3. Shiota, N.; Inui, H.; Ohkawa, H. *Pestic. Biochem. Physiol.*, **1996**, *54*, 190–198.
4. Inui, H.; Ueyama, Y.; Shiota, N.; Ohkawa, Y.; Ohkawa, H. *Pestic. Biochem. Physiol.*, **1999**, *64*, 33–46.
5. Doty, S. L.; Shang, T. Q.; Wilson, A. M.; Tangen, J.; Westergreen, A. D.; Newman, L. A.; Strand, S. E.; Gordon, M. P. *Proc. Natl. Acad. Sci. USA*, **2000**, *97*, 6287–6291.
6. Toki, S. *Plant Mol. Biol. Rep.*, 1997. *15*, 16–21.
7. Inui, H.; Shiota, N.; Motoi, Y.; Ido, Y.; Inoue, T.; Kodama, T.; Ohkawa, Y.; Ohkawa, H. *J. Pesticide Sci.*, **2001**, *26*, 28–40.
8. Tomlin, C. D. S., Ed.; *The Pesticide Manual* 12th ed.; British Crop Protection Council: Surrey, UK, 2000.
9. Benbrook, C., *Ag BioTech Infonet Technical Paper No. 1.*, 1999, http://www.biotech-info.net/RR_yield_drag_98.pdf
10. Putwain, P. D. In *Weed Control Handbook: Principles,* 8th ed., Hance, R. J.; Holly, K., Eds; Blackwell: Oxford, UK, 1990; pp 217–242.

Chapter 3

Role of Composition and Animal Feeding Studies in the Safety Assessment of Biotech Crops

William P. Ridley, Gary F. Hartnell, and Bruce G. Hammond

Product Safety Center, Monsanto Company, 800 North Lindbergh Boulevard, St. Louis, MO 63167

>The safety assessment of biotech crops is a multidisciplinary process that involves investigations of the gene, the gene product (protein), and the whole crop. Composition analyses of nutrients, anti-nutrients, and secondary metabolites in crop tissues plus the measurement of performance, carcass yield, and meat composition in broiler chickens and the measurement of growth and numerous pathology parameters in a 13 week rat safety study were conducted to investigate the safety of the whole crop for Roundup Ready® maize event NK603. The results of these studies demonstrate that NK603 is as safe and nutritious as conventional, commercial maize (*Zea mays* L.) currently in the marketplace.

The development of transformed plants during the 1980's and the product efficacy and safety testing that followed set the stage for the introduction of herbicide-tolerant soybeans and insect-protected maize during 1995-1996 (1). The increase in productivity and the decrease in the impact of agricultural practices that these new biotechnology products provided the producer led to their rapid adoption. The acreage planted with seed derived through biotechnology continues to increase and in 2003 the global area of biotech crops was 167 million acres, representing an increase of 15 percent or 22 million acres over 2002 (2).

The development of new crops containing specific agronomic traits in food and animal feed required careful evaluation to make sure there are no safety concerns. The scheme for assessing the food and feed safety of biotech products has been described previously (3, 4). Briefly, this process involves the safety evaluation of the gene and gene product (protein) responsible for the agronomical advantageous trait and a comparison of the biotechnology plant with its conventional counterpart. The comparative process using the whole plant and tissues derived from it such as forage and grain has been referred to as the evaluation of "substantial equivalence" (5, 6, 7). An essential part of the substantial equivalence process has been the analysis of composition including nutrients, anti-nutrients, and secondary metabolites, since these materials are responsible for the nutritional properties that support normal growth and development in humans and animals. Furthermore, the assessment of the equivalence of biotechnology crops compared to conventional commercial cultivars for providing nutrients to livestock and the assessment of safety in rodent feeding studies have emerged as additional studies that provide important information to support the equivalence evaluation (8, 9).

In 2001 Roundup Ready® maize event NK603 was approved for production in the United States and Canada. This product provided farmers an effective weed control option in maize since event NK603 was tolerant to glyphosate, the active ingredient in the Roundup® family of agricultural herbicides. Glyphosate acts by the inhibition of a key enzyme, 5-enoylpyruvylshikimate-3-phosphate synthase (EPSPS) in the shikimate pathway in plants, some bacteria and fungi, but not in mammals, birds, or fish (10). The introduction of a glyphosate tolerant enzyme, CP4 EPSPS, into maize event NK603 plants conferred tolerance to the herbicide. Prior to the introduction of Roundup Ready maize event NK603, extensive evaluation of composition, livestock performance, and rat performance were conducted to assess the safety of the product compared to its conventional counterpart and for evaluating any unintended effects. This report will describe the results of representative studies conducted for the assessment of plants containing Roundup Ready maize event NK603.

® Roundup and Roundup Ready are registered trademarks of Monsanto Technology LLC.

Materials and Methods

The assessment of compositional equivalence, livestock nutritional performance, and safety in rodent feeding studies share many common features. All these studies involve test, control, and reference groups, a specified experimental design that involves randomization and replication, followed by data collection (measurements) and lastly, evaluation (statistics). The study plan for the compositional analysis of maize has been described in detail previously (11) and is summarized in Table 1. The maize test line (event NK603), the control line (a near isogenic parental hybrid), and reference lines (19 conventional, non-transgenic commercial hybrids) obtained from controlled field trials were included in the study. Field sites were located in two different geographic areas, the United States (U.S.) and the European Union (E.U.), over

Table 1. Study Plan for Compositional Analysis of Maize Event NK603

Study Items	*Description*
Variables	
Test Line	Maize Event NK603
Control Line	Near isogenic hybrid (LH82 x B73)
Reference Lines	19 commercial hybrids
Groups	
Field Sites U.S. - 1998	Seven non-replicated, two replicated
Field Sites E.U. - 1999	Four replicated
Design	
Replication	Non-replicated, four replications
Randomization	Randomized complete block
Measurements	
Forage	Protein, fat, ash, moisture, fiber, carbohydrate by calculation
Grain	Protein, fat, ash, moisture, fiber, carbohydrate by calc., amino acid profile, fatty acid profile, minerals, vitamin E, trypsin inhibitor, ferulic and p-coumaric acids, raffinose, 2-furaldeyde
Evaluation	
Within sites and combination of all sites	Mixed model analysis of variance, significance: $p < 0.05$
Biological relevance	99% Tolerance Interval

a two-year period using replicated and non-replicated randomized designs resulting in a broad assessment of different growing conditions. An extensive list of nutrients, anti-nutrients, and secondary metabolites were measured in the forage and grain, followed by evaluation using a mixed model analysis of variance to assess differences between the test and control with significance assigned at the $p<0.05$ level (see Table 1). In addition, the reference hybrid data was used to develop a 99% tolerance interval for each analyte describing the boundaries of the population of conventional values. The tolerance interval was then used to assess the biological relevance of any differences detected in the initial ANOVA evaluation of the test versus the control (11).

The study plan for the evaluation of the nutritional performance of maize

Table 2. Study Plan for Poultry Nutritional Evaluation for Maize Event NK603

Study Items	*Description*
Treatments	
Test Line	Maize Event NK603
Control Line	Near isogenic hybrid (LH82 x B73HT)
Reference Lines	Five commercial hybrids
Groups	
Male broilers	10 birds/pen
Female broilers	10 birds/pen
Design	
Replication	5 pens/sex/treatment
Randomization	Randomized complete block
Measurements	
Performance	Live wt., feed intake, feed conversion
Carcass yield	Live wt., chill wt., fat pad wt., breast meat wt., thigh wt., drum wt., wing wt.
Meat composition	Moisture, protein, fat
Evaluation	
Comparison of test/control/reference	Least significant difference; significance: $p<0.05$
Comparison of test/non-transgenic	Population statistics

event NK603 and for unintended effects in poultry has been described in detail previously (15) and is summarized in Table 2. Once again the six week study utilized test, control, and reference groups in which groups of broiler chickens were fed diets containing maize (55% wt/wt for days 1 to 20; 60% wt/wt for days 20 to 42) from event NK603, a near isogenic hybrid or conventional, commercial reference hybrids, respectively. There were five pens/sex/treatment with 10 birds/pen in a randomized complete block design. The data collected in the experiment reflected the parameters typically used to assess nutritional equivalence in broiler chickens – performance (live weight, feed intake, and feed conversion), carcass yield, and meat composition. Evaluation of the data included the comparison of event NK603 to the control and each of the reference hybrids as well as all other pair wise comparisons. In addition, maize event NK603 was compared to the population of non-transgenic hybrids.

Table 3. Study Plan for Rat Safety Assurance Evaluation for Event NK603

Study Items	*Description*
Treatments	
Test Line	Maize Event NK603
Control Line	Near isogenic hybrid
Reference Lines	Six commercial hybrids
Groups	
Male rats	Housed individually
Female rats	Housed individually
Design	
Replication	Twenty rats/sex/treatment
Randomization	Stratified randomized assignment to treatment group
Measurements	
Clinical observations	Weights for seven major organs
Clinical pathology	Microscopic evaluation of 17 tissues from major organ systems
Hematology	Twelve hematology evaluations
Serum chemistry	Seventeen blood chemistry analyses
Urine chemistry	Eleven urine chemistry analyses
Evaluation	
Comparison of test/control/reference	Least significant diff.; $p<0.05$

Table 3 contains a summary of the study plan for the rat safety assurance study of maize event NK603 that has been described in detail previously (16). Diets were formulated containing event NK603 (test) and near isogenic hybrid (control) at two dietary levels, 11% and 33% or reference hybrids (six conventional, non-transgenic commercial hybrids) at a dietary level of 33%. Twenty rats/sex/treatment were housed individually following a stratified randomization assignment to treatment group and fed the formulated diets for a total of 13 weeks. A comprehensive array of clinical observations, pathology evaluations as well as hematology, serum chemistry, and urine chemistry analyses was conducted. Finally, the data from the treated group were compared to the control group and the population of non-transgenic, reference hybrids.

Results and Discussion

Composition Evaluation

The nutrients, anti-nutrients, and secondary metabolites listed in Table 1 were assessed as part of the composition evaluation of maize event NK603 (11). A total of 105 comparisons between NK603 and its non-transgenic control were conducted for U.S. sites in 1998 and the E. U. sites in 1999. The values for the biochemical components for corn event NK603 were similar to the control or were within the published range observed for non-transgenic commercial corn hybrids.

Twelve of the comparisons were shown to be statistically significantly different. However, the magnitude of the differences was small (1.3-15.56%) and none of the statistically significant differences for an individual analyte occurred in more than one year. The biological relevance of the small statistically significant differences were examined by calculating a 99% tolerance interval for the population of the commercial hybrids and then determining if the range of the values for maize event NK603 fell within the interval. An example of this process is illustrated in Table 4 for the amino acid, histidine. Histidine accounted for 2.65% of the total amino acids in NK603 and 2.77% in the control. This small difference was shown to be statistically significantly different from the control ($p = 0.003$). However, an examination of the range of values for the test (2.56-2.74%) indicated that they were contained within the tolerance interval (2.34%, 3.36%) and therefore, within the population of commercial hybrids. The same evaluation was conducted for total protein, acid detergent fiber (ADF), potassium, phenylalanine, 18:2 linoleic acid, and phytic acid as shown in Table 4. It was concluded that NK603 is compositionally equivalent to and as nutritious as commercial hybrids.

Table 4. Selected Results for Grain from NK603 Composition Study in 1999[a]

Component (Units)	NK603 Mean (Range)	Control Mean (Range)	Commercial Tolerance Interval. (Range)	Literature (Range)
Total Protein	12.07	11.34	6.84, 14.57	
(% dw)	(10.23-13.92)	10.13-13.05)	(7.77-12.99)	(6.0-12.0)[b]
ADF	3.21	3.03	1.96, 4.71	
(% dw)	(2.63-3.87)	(2.30-3.68)	(2.46-6.33)	(3.3-4.3)[b]
Potassium	0.36*	0.38	0.31, 0.45	
(% dw)	(0.34-0.38)	(0.36-0.39)	(0.32-0.45)	(0.32-0.72)[b]
Phenylalanine	5.28	5.25	4.59, 5.61	
(% total AA)	(5.13-5.46)	(5.20-5.29)	(4.85-5.54)	(2.9-5.7)[c]
Histidine	2.65*	2.77	2.34, 3.36	
(% total AA)	(2.56-2.74)	(2.69-2.85)	(2.58-3.15)	(2.0-2.8)[c]
18:2 Linoleic	63.73	63.15	44.59, 73.50	
(% total FA)	(61.94-65.25)	(61.63-64.04)	(49.72-65.98)	(35-70)[c]
Phytic acid	0.79	0.70	0.32, 1.18	
(% dw)	(0.51-0.89)	(0.55-0.77)	(0.48-1.12)	up to 0.9[d]

*Statistically different from the control (p<0.05)
[a]Adapted from Ridley et al. (11).
[b]Watson, S. A. (12), [c]Watson, S. A.(13), [d]Watson, S. A.(14).

Poultry Performance Evaluation

A 42-day experiment comparing grain from maize event NK603 to its non-transgenic control and commercial reference maize when fed to broiler chickens in their diet has been conducted (15). Rapidly growing broilers are considered to be a useful model for evaluating the wholesomeness of transgenic maize compared to conventional maize because of their sensitivity to changes in nutritional quality and because maize is a major ingredient in broiler diets. Representative data for performance, carcass yield, and meat composition are shown in Table 5. There were no statistically significant differences between maize event NK603 and its control or the references hybrids for the mean terminal body weight, feed conversion (total feed consumed divided by the total body weight of broilers), and the thigh and breast weight. As noted in Table 5, the natural variability of broilers to diets containing different conventional hybrids was observed, for example, in statistically significant differences in breast meat weight for Reference RX826 compared with References MON 847 and RX770, and with the control. It was concluded that diets formulated with maize event NK603 were as wholesome as those containing the non-transgenic control or conventional commercially available maize because they supported the growth of broiler chickens consistent with the industry norm.

Table 5. Performance of Poultry Fed NK603 Grain[a]

Test Group Males/Females	Mean Terminal Body Wt. (kg)	Feed Conversion (kg/kg)	Thigh Meat Wt. (kg)	Breast Meat Wt. (kg)
Event NK603	2.30	1.54	0.28	0.41[bcde]
Control	2.31	1.56	0.28	0.39[e]
Reference RX826	2.34	1.58	0.28	0.42[b]
Reference LH235 x LH185	2.35	1.58	0.28	0.42[bc]
Reference DK493	2.33	1.59	0.27	0.41[bcd]
Reference MON 847	2.32	1.59	0.28	0.40[cde]
Reference RX770	2.25	1.57	0.27	0.39[de]
LSD 5.0%[f]	0.07	0.03	0.01	0.02

[a] Adapted from Taylor et al. (15)
[b-e] Individual treatment means with the same superscript letter in the same column are not statistically different ($p>0.05$).
[f] Least significant difference between two means ($p<0.05$).

Rat Safety Assurance Evaluation

In addition to the evaluation of compositional equivalence of maize event NK603 and its nutritional wholesomeness in livestock feed studies compared with conventional hybrids, a 90 day or 13 week safety assurance study was conducted to assess the potential for any safety issues associated with the human consumption of NK603 grain. The laboratory rat is considered an acceptable model for assessing human food safety and these studies can be used to establish an adequate margin of safety since food consumption for rodents is several fold higher on a body weight basis than for humans. Diets were formulated to

Table 6. Body Weights, Kidney/Body Weight, and Liver/Body Weight in Sprague-Dawley Rats following 13 Weeks Exposure to Event NK603 in the Diet[a]

Parameter	N	11% Control Mean ±SD[b]	33% Control Mean ±SD[b]	11% NK603 Mean ±SD[b]	33% NK603 Mean ±SD[b]	N	Ref. Controls
Males							
Body wt (g) (week 13)	20	574 ±71	566 ±45	580 ±53	588 ±70	119	572 ±51
Kidney/ body wt	19-20	0.81 ±0.06	0.77 ±0.06	0.79 ±0.07	0.79 ±0.05	116	0.78 ±0.06
Liver/ body wt	19-20	2.85 ±0.21	2.82 ±0.19	2.87 ±0.20	2.96 ±0.17	116	2.86 ±0.21
Females							
Body wt (g) (week 13)	20	300 ±20	307 ±22	297 ±22	319 ±27	120	309 ±25
Kidney/ Body wt	20	0.77 ±0.06	0.79 ±0.06	0.78 ±0.07	0.78 ±0.05	120	0.79 ±0.06
Liver/ Body wt	20	2.99 ±0.40	2.94 ±0.22	2.92 ±0.30	2.89 ±0.35	120	3.04 ±0.36

[a]Adapted from Hammond et al. (16)
[b]SD = Standard Deviation

contain test, control, and reference grain at levels of approximately 33% weight/weight, the standard inclusion level for Certified Rodent LabDiet 5002 as well as an inclusion level of 11% to assess any potential dose-response effects. Representative results for the 13 week study for body weights, and liver and kidney/body weight ratios are shown in Table 6. There were no statistically significant differences between event NK603 grain and its control or reference hybrids in either male or female rats in the parameters presented in Table 6 as well as the other parameters evaluated in this study (16). These results were consistent with the composition and livestock feed studies in establishing the safety of maize event NK603.

Conclusions

The safety assessment of food and feed crops derived through biotechnology presents unique challenges due to the complex nature of the materials studied. The process known as "substantial equivalence" has provided a framework for the safety assessment that focuses on relative or comparable safety rather than absolute safety. The comparison of a biotech crop with its near isogenic control and conventional, commercial cultivars have proven to be a robust and scientifically rigorous way of assessing similarities and any significant differences. Testing of the whole crop tissue such as forage or grain also addresses the issue of unexpected or pleitrophic effects (17) since the entire genome, proteome, and metabolome are represented in the test substance. Compositional analyses, nutritional performance in livestock species such as the broiler chicken, and a rat safety assurance study were utilized to assess the safety of Roundup Ready maize event NK603. The results of this multidisciplinary approach have established that NK603 is as safe and nutritious as conventional, commercial maize currently in the marketplace.

Acknowledgements

The authors are grateful to Drs. Tracey L. Reynolds, William Heydens and Mark Holland for critically reviewing the manuscript.

References

1. Ridley, W. P. Introduction to agricultural biotechnology: challenges and prospects. In *Agricultural Biotechnology: Challenges and Prospects*; Bhalgat, M.K., Ridley, W. P., Felsot, A. S., Seiber, J. N., Editors; ACS Symposium Series 866; American Chemical Society, Washington, DC; 2004; pp. 3-18.

2. James, C. Global status of commercialized transgenic crops: 2003, International Service for the Acquisition of Agri-Biotech Applications, *ISAA Briefs 30*, Ithaca, NY, 2003.
3. Astwood, J. D., Fuchs, R. L. Status and safety of biotech crops. In *Agrochemical Discovery – Insect, Weed and Fungal Control*. Baker, D. R. , Umetsu, N. K., Editors; ACS Symposium Series 774, Amercian Chemcial Society, Washington, DC, 2001; pp. 152-164.
4. Ridley, W. P., Sidhu, R. S., Astwood, J. D., Fuchs, R. L. Role of compositional analyses in the evaluation of substantial equivalence for biotechnology corps. In *Agricultural Biotechnology: Challenges and Prospects*; Bhalgat, M.K., Ridley, W. P., Felsot, A. S., Seiber, J. N., Editors; ACS Symposium Series 866; American Chemical Society, Washington, DC; 2004; pp. 165-176.
5. FAO. Biotechnology and food safety. Report of a joint FAO/WHO consultation. In *Food and Nutrition Paper 61*, Food and Agriculture Organization of the United Nations, Rome, Italy; 1996.
6. OECD. *Safety Evaluation of Foods Produced by Modern Biotechnology: Concept and Principles*. Organization for Economic Cooperation and Development, Paris, France; 1993.
7. WHO. Application of the principles of substantial equivalence to the safety evaluation of foods and food components from plant derived by modern biotechnology. In Report of a WHO Workshop No. WHO/FNU/FOS/95.1, World Health Organization. Geneva, Switerland; 1995.
8. Clark, J. H., Ipharraguerre, I.R. Biotech crops as feeds for livestock. In *Agricultural Biotechnology: Challenges and Prospects*; Bhalgat, M.K., Ridley, W. P., Felsot, A. S., Seiber, J. N., Editors; ACS Symposium Series 866; American Chemical Society, Washington, DC; 2004.
9. Hammond, B.G., Rogers, S. G., Fuchs, R. L. Limitations of whole food feeding studies in food safety assessment. In Food Safety Evaluation. Organization for Economic Cooperation and Development, Paris, France; 1993; pp. 85-97.
10. Pagette, S., Re, D., Eicholtz, D., Delanney, X., Fuchs, R. L., Kishore, G., Fraley, R. New weed control opportunities: development of soybeans with a Roundup Ready gene. In Herbicide Resistant Crops; Duke, S. O., Editor; CRC: Boca Raton, FL, 1996; pp. 53-84.
11. Ridley, W. P., Sidhu, R. S., Pyla, P. D., Nemeth, M. A., Breeze, M. L., Astwood, J. D., A comparison of the nutritional profile of glyphosate-tolerant corn hybrid NK603 to that of commercial corn (Zea mays L.). *J. Agric. Food Chem.* **2002**, 50, 7235-7243.

12. Watson, S. A. Structure and composition. In *Corn: Chemistry and Technology*; Watson, S. A., Ransted, P. E., Editors; American Association of Cereal Chemists; Minneapolis, MN, 1987; pp. 53-82.
13. Watson, S. A. Amazing maize. General properties. In CRC Handbook of Processing and Utilization in Agriculture. Vol. II: Part 1, Plant Products; Wolff, I. A., Editor; CRC Press; Boca Raton, FL, 1982; pp. 3-29.
14. Watson, S. A. Structure and composition. In *Corn: Chemistry and Technology*; Watson, S. A., Ransted, P. E., Editors; American Association of Cereal Chemists; Minneapolis, MN, 1987; pp. 455.
15. Taylor, M. L., Hartnell, G. F., Riordan, S. G., Nemeth, M. A., Karunanandaa, K., George, B., Astwood, J. D. Comparison of broiler performance when fed diets containing grain from Roundup Ready (NK603), YieldGard x Roundup Ready (MON 810 x NK603), non-transgenic control, or commercial corn. *Poult. Sci.* **2003**, 82, 443-453.
16. Hammond, B., Dudek, R., Lemen, J., Nemeth, M. Results of a 13 week safety assurance study with rats fed Roundup Ready corn grain. *Food Chem. Tox.* **2004** (in press).
17. Kuiper, H.A., Kleter, G. A., Noteborn, H. P. J. M., Kok, E. J. Assessment of the food safety issues related to genetically modified foods. *Plant J.* **2001**, 27, 503-528.

Chapter 4

Monitoring of Endocrine Disruptors in Transgenic Plants Carrying Aryl Hydrocarbon Receptor and Estrogen Receptor Genes

Hideyuki Inui[1], Hideaki Sasaki[1], Susumu Kodama[1], Nam-Hai Chua[2], and Hideo Ohkawa[1]

[1]Kobe University, Rokkodai-cho 1–1, Nada-ku Kobe,
Hyogo 657–8501, Japan
[2]Rockefeller University, 1230 York Avenue, New York, NY 10021

Transgenic plants carrying mammalian hormone receptor genes were examined for monitoring of environmental chemicals including dioxins and endocrine disruptors. The transgenic tobacco plants expressing mouse arylhydrocarbon receptor gene detected 1ppb of the dioxin-like compound 20-methylcholanthrene in the cultured plants. In addition, the transgenic *Arabidopsis* plants expressing human estrogen receptor gene detected 1ppt of 17ß-estradiol and ppb levels of the other endocrine disruptors such as bisphenol A, 4-t-octylphenol, and others in potted plants. These transgenic plants appear to be useful for on site monitoring of dioxins and estrogenic contaminants in the environment.

Introduction

Contamination of the environment and agricultural products with endocrine disruptors(EDs) including dioxins, industrial chemicals and certain pesticide residues is a serious problem. These chemicals were found in the environment and agricultural products at nano-level concentrations, and are suspected of affecting ecosystems and human health. The Ministry of the Environment of Japan surveyed contamination of EDs in rivers, showing that bisphenol A, 17ß-estradiol and nonylphenol were detected in 50.3%, 40.4% and 31.0% of the water samples, respectively(Table 1).

Table I. Contamination of environmental waters with endocrine disruptors in Japan

Endocrine disruptor	Water sample[a]		RBA(%)[b]
	Detection point(%)	Maximum conc.(ppb)	
Bisphenol A	50.3	0.56	0.01
17ß-Estradiol	40.4	0.0072	100
Nonylphenol	31.0	5.9	0.05

[a]: Ministry of the Environment of Japan(Oct. 2002)(*1*)

[b]: The relative binding affinity(RBA) of each chemical was calculated as the ratio of concentrations of 17ß-estradiol and chemical required to reduce the specific radioligand binding by 50%. RBA value for 17ß-estradiol was arbitrarily set at 100(*2*).

Contamination level of these compounds was ppb to ppt levels. Thus, it is important to develop novel technologies to monitor contamination on site, since most conventional technologies are high cost, labor intensive and take a long time to obtain results. Particularly, development of novel technologies based on biological functions appears to be needed, because of the low risk potential for secondary contamination with chemicals.

Certain plant species are grounding and have deep roots. These plants absorb and accumulate nano-level concentrations of EDs from a wide area through their root systems. On the other hand, it was found that certain biological systems containing an aryl hydrocarbon receptor(AhR) and an estrogen receptor(ER) specifically bind dioxins and estrogenic compounds at low concentrations, respectively. Binding of the complex of dioxin-bound AhR and AhR nuclear translocator(Arnt) onto the xenobiotic responsive element(XRE) in a 5'-upstream region induces expression of CYP1A1 gene in mammals. Introduction of this system into plants seems to be possible for monitoring dioxins. It was also found that an estrogenic compound bound to ER induced expression of a number of genes in mammals. It was reported that the transgenic *Arabidopsis* plants with the recombinant transcription factors consisting of ligand binding domain of ER, LexA DNA binding domain(LexA) and virus VP16 transactivation domain(VP16) expressed a reporter gene by treatment with 17ß-estradiol(E_2)(*3*). This system appears to be useful for monitoring estrogenic compounds.

We attempted to produce genetically engineered plants carrying AhR and ER genes for monitoring nano-level concentrations of dioxins and estrogenic compounds on site, respectively. Mouse AhR and Arnt cDNAs were expressed in transgenic tobacco plants and examined for monitoring dioxin-like compounds. Also, the transgenic *Arabidopsis* plants carrying the LexA-VP16-ER recombinant gene were evaluated for monitoring estrogenic compounds. These transgenic plants have displayed no abnormal agronomic or physiological properties.

Transgenic tobacco plants carrying a mouse AhR gene

The plasmid pSKAVAtG was constructed for dioxin-inducible expression system of GUS(ß-glucuronidase) reporter gene and transformed into tobacco plants by *Agrobacterium tumefaciens*(Figure 1)(*4*). The recombinant AhR bound to VP16 instead of transactivation domain of native AhR was constitutively expressed in transgenic tobacco plants(AhRV-GUS plant). The transcription factor bound dioxin may make a complex with Arnt in the nucleus and its heterodimer binds XRE upstream of GUS reporter gene, resulting in expression of GUS gene. After selection of transformants with kanamycin resistance, transgenic tobacco plants were incubated on a medium containing the dioxin-like compound 20-methylcholanthrene(MC) for two weeks. The transgenic plants showed MC-dependent inducible expression of GUS gene, suggesting that transactivation of VP16 had efficiently promoted GUS gene transcription. Dose-dependent expression of GUS gene was observed in AhRV-GUS plants treated with MC. The minimum concentration of

MC enough to detect GUS activity was 5nM(approx. 1ppb) in the cultured plants(Figure 2). Histochemical staining of transgenic tobacco plants showed MC-dependent expression of GUS gene(Figure 3). These results indicated that transgenic tobacco plants absorbed MC through their roots and expressed GUS gene through AhR expression. Thus, the transgenic tobacco plants appear to be useful for monitoring dioxins in the environment.

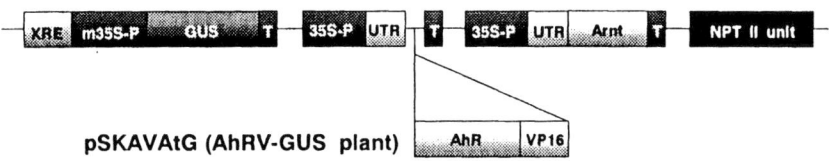

Figure 1 The expression plasmid for arylhydrocarbon receptor-dependent inducible expression system for GUS reporter gene
m35S-P, T, 35S-P and NPTII represent minimal cauliflower mosaic virus(CaMV) 35S promoter, terminator, CaMV 35S promoter and kanamycin resistance gene, respectively.

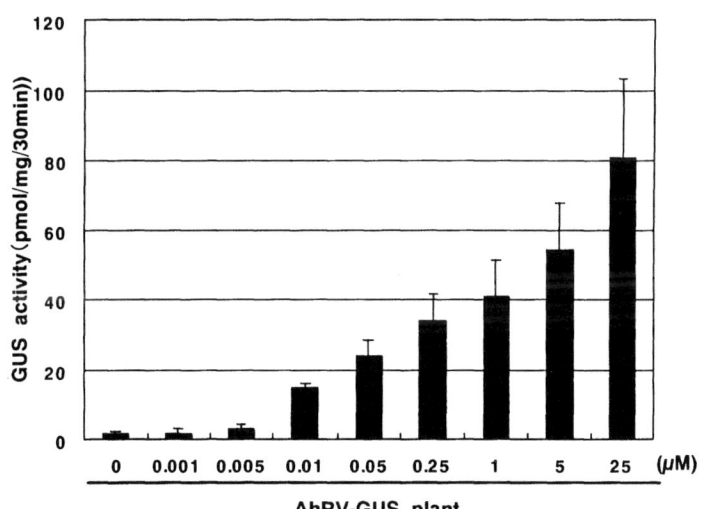

Figure 2 Dose-dependent GUS activity in the transgenic tobacco plants AhRV-GUS plant incubate in the culture mediad with MC

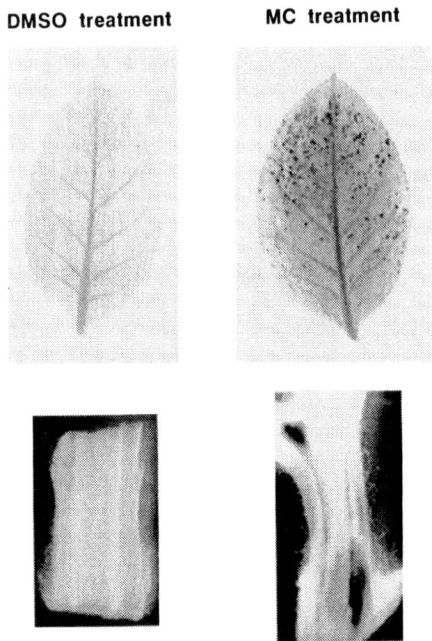

Figure 3 Histochemical staining of GUS activity in transgenic tobacco plants AhRV-GUS treated with MC

Transgenic *Arabidopsis* plants carrying a recombinant ER gene

We constructed the recombinant transcription factor based on a ligand binding domain of ER, LexA and VP16. Two types of the expression plasmids pER8-GFP and pX6-GFP were constructed to express GFP(green fluorescence protein) reporter gene for the recombinant transcription factor LexA-VP16-ER in plants(Figure 4). *Arabidopsis thaliana* was transformed by *Agrobacterium*. The transgenic *Arabidopsis* plants transformed with pER8-GFP(XVE; LexA-VP16-ER) and pX6-GFP(CLX; Cre/loxP DNA excision system) show transient inducible expression and constitutive expression of GFP reporter gene, respectively. Namely, the *Arabidopsis* plants absorbed EDs through their roots, and then EDs bind the LexA-VP16-ER transcription factor to change structural conformation. Then, the ED-transcription factor complex binds LexA operator region upstream of GFP gene, and induce transcription and translation of GFP gene transiently in XVE plants(3). In case of CLX plants, ED-transcription factor complex binds upstream of Cre recombinase gene, and then Cre recombinase expressed truncates *loxP* sites located at downstream region of G_{10-90} promoter and terminator for Cre recombinase. After truncation, G_{10-90} promoter binds upstream of GFP gene, and GFP gene expresses constitutively(5). This system also brings us marker-free transgenic plants.

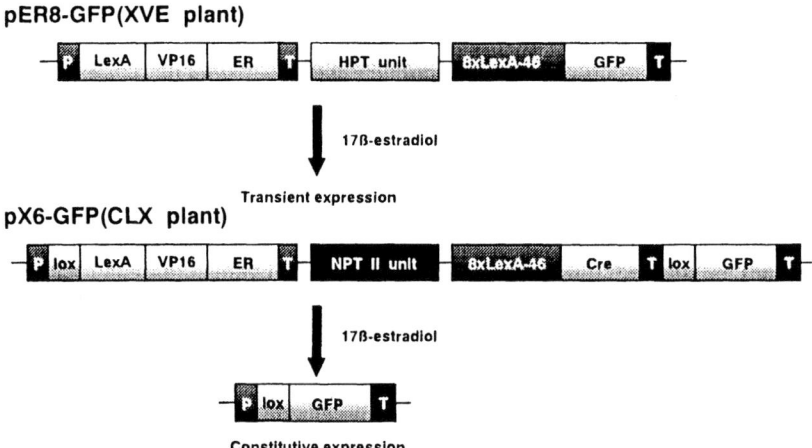

Figure 4 Expression plasmids for estrogen receptor-dependent inducible expression system for GFP reporter gene
P, NPTII, 8xLexA-46 and T represent promoter, kanamycin resistance gene, 8 copies of LexA binding domain and terminator, respectively.

Dose-dependent transcription of GFP gene was observed by RT-PCR when 0.0001ppb to 10ppb of E_2 were treated. Minimum concentration of E_2 detected was 0.1ppb and 1ppb of E_2 in XVE and CLX plants, respectively, when plants were incubated for 7 days on a E_2-containing medium(data not shown)(6). Dose-dependent fluorescence of GFP was also observed(Figure 5). Furthermore, 11 of 19 chemicals tested induced transcription of GFP gene. Of these, estrone, atrazine and 4-*t*-octylphenol were detected at 1ppb, 0.1ppm and 0.1ppm, respectively. These results revealed that XVE plants showed higher sensitivity than CLX plants, because transcription and translation of Cre recombinase gene are not required to induce GFP gene expression in XVE plants. Six EDs did not activate GFP gene transcription, suggesting that affinity of these EDs towards ER or uptake efficiency of these EDs into plants are low. ^{14}C-labeled E_2 and nonylphenol were taken into apical part of plants within 48 hours(Figure 6). Moreover, XVE and CLX plants were found to detect 0.001ppb and 0.1ppb of E_2 in the potted *Arabidopsis* plants.

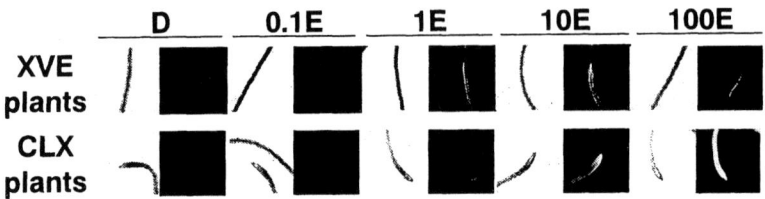

Figure 5 Dose-dependent expression of GFP gene in transgenic Arabidopsis plants treated with 17β-estradiol
Right pictures of each column were taken by fluorescence microscopy. D, 0.1E~100E represent treatment of 0.1% dimethyl sulfoxide, 0.1ppb~100ppb of 17β-estradiol, respectively.

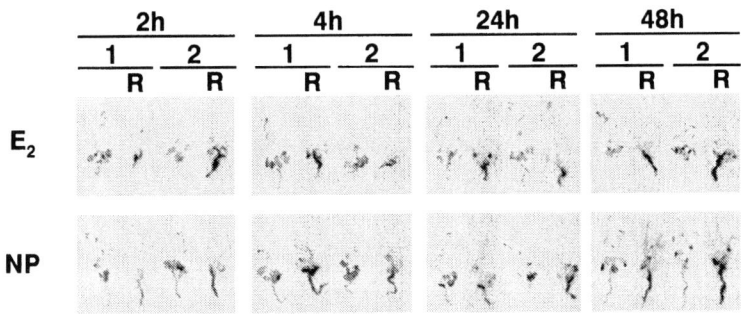

Figure 6 Absorption of ^{14}C-labeled 17β-estradiol and nonylphenol in transgenic Arabidopsis plants
E_2 and NP represent treatment of 17β-estradiol and nonylphenol, respectively. R represent autoradiogram of plants treated with ^{14}C-labeled compounds.

Concluding remarks

The transgenic plants expressing AhR and ER detected 1ppb of MC under cultured conditions and 1ppt of E_2 under potted conditions, respectively. Basic technology was established for monitoring dioxins and EDs in the transgenic plants with the receptor genes. However, increase of the sensitivity towards other EDs with low affinity towards receptors is necessary. In order to overcome this problem, (1)sensitive receptors may be useful instead of mouse AhR and human ER. It is known that fish ERs were more sensitive to alkylphenols such as nonylphenol and octylphenol than mammalian ones. (2)Increase of uptake efficiency of dioxins and EDs is another way to increase sensitivity. Zucchini plants are known to have efficient translocation of polychlorinated dibenzo-*p*-dioxins and dibenzofurans(7). Furthermore, (3)genes synthesizing flower colors as reporter genes may be useful for visible monitoring of contaminations on site.

References

1. Ministry of the Environment of Japan, (2002)
2. Kuiper, G. G. J. M.; Lemmen, J. G.; Carlsson, B.; Corton, J. C.; Safe, S. H.; van der Saag, P. T.; van der Burg, B.; Gustafsson, J-A.; *Endocrinology*, **1998**, *139(10)*, 4252-4263
3. Zuo, J.; Niu, Q-W; Chua, N-H.; *The Plant Journal*, **2000**, 24(2), 265-273
4. Kodama, S.; Okada, K.; Akimoto, K.; Inui, H.; Ohkawa, H.; **2003**, submitted
5. Zuo, J.; Niu, Q-W; Moller, S. G.; Chua, N-H.; *Nature Biotechnology*, **2001**, 19, 157-161
6. Inui, H.; Sasaki, H.; Chua, N-H.; Ohkawa, H.; **2003**, submitted
7. Hülster, A.; Müller, J. F.; Marschner, H.; *Environmental Science and Technology*, **1994**, 28, 1110-1115

Chapter 5

Impact of *Bacillus thuringiensis* Corn Pollen on Monarch Butterfly Populations

Mark K. Sears

Department of Environmental Biology, University of Guelph, Guelph, Ontario N1G 2W1, Canada

Nearly four years ago, in 1999, a note submitted to the editor of *Nature* from scientists at Cornell University claimed that pollen from a *Bt* corn hybrid harmed larvae of the monarch butterfly (1). They had sprinkled an unspecified amount of pollen from a Bt11 corn hybrid on milkweed leaves that they fed to small larvae of monarchs. The ensuing international reaction from anti-biotechnology activists created the impression that this pollen and its underlying technology might cause significant damage to populations of non-target insects such as the monarch as well as detrimental effects to the environment. A number of scientists in Canada and the United States who had evaluated *Bt* corn for efficacy, economic viability and its impact on non-targets in the agricultural environment gathered to initiate a collaborative investigation into the impact of *Bt* corn pollen on monarchs. As a group, we felt that, due to the naïve methodology and inconclusive results of the reported laboratory study and the unjustified statements from activist groups, further appropriate and scientifically sound investigations were warranted. We approached this process with the objective of obtaining information on toxicity and exposure of monarch larvae to *Bt* corn pollen and to develop a classical risk assessment using this information.

We performed a series of replicated bioassays using both purified toxins and pollen samples from a number of corn hybrids created with different transformation events. In these, we measured the effect of increasing doses of Bt pollen on feeding behavior and weight gain of first-instar monarch larvae. We established a no-observable-effect level (NOEL) from these results and compared that dose with the density of pollen that we estimated to occur on milkweed leaves in and around corn fields to determine the likelihood of encounter with an effective dose. To validate our laboratory observations and estimates of impact, we performed a series of field experiments over two years in three different locations within the Corn Belt. Monarch larvae were either caged on milkweed plants or fed milkweed leaves from plants in and around corn fields that had previously received corn pollen during pollen shed. Larvae were allowed to feed on the milkweed for the period of their development and their weight gain and survival monitored. Treatments were carried out on milkweed plants placed in fields of *Bt* corn, in conventional corn fields and in areas removed from corn altogether. Concurrently, we examined populations of monarchs in several locations within the Corn Belt to determine the synchrony of larval development in relation to pollen shed activity in corn fields. In addition, we attempted to estimate for the entire Corn Belt the amount of milkweed found in corn fields, other agricultural fields and in non-agricultural land to determine the proportion of monarch larvae that could receive significant amounts of pollen in their diets.

Response of monarch larvae to *Cry* proteins produced from various events used to create commercial corn hybrids was variable. *Bt* corn pollen from commercial hybrids at levels exceeding 1000 pollen grains/cm^2 of leaf area did not cause measurable impact to monarch larvae in bioassay trials. Only one type of pollen produced any impact, from Event Bt176 corn, and hybrids using that event never exceeded 2% of the production area and no longer are registered for use in North America (2). In every field trial, no impact was measured to larvae exposed to pollen from any Bt corn hybrid, except those developed from Event 176 technology. Even in subsequent trials, no monarch larvae have shown a measurable reaction to *Bt* corn pollen at field densities. Average densities of corn pollen on milkweed leaves in and around corn fields were less than 200 pollen grains/cm^2, far less than the 1000 grains/cm^2 that we estimated to be the NOEL. Fewer than 1% of larvae within corn fields would encounter densities of corn pollen that exceeded the NOEL. Only a small fraction of monarch larvae in each region of the Corn Belt are exposed to any pollen from corn plants due to asynchrony with pollen shed, their geographical separation from corn fields and the proportion of *Bt* corn grown in the area. We estimated that fewer than 1/100 of 1% of larvae would be exposed to *Bt* corn pollen to a degree that could be measured and concluded that the impact of *Bt* corn on populations of monarch butterflies in the Corn Belt was negligible (3). In comparison, mortality from sudden frosts while butterflies overwinter in Mexico have caused significant losses and the impact of insecticide use in and around corn fields causes significant loss of butterflies or their larvae.

References:
(1) Losey J.E. et al, Nature, **399**, 214, 1999; (2) Helmich R.L. et al, Proc. Natl. Acad. Sci., **98**, 1128-33, 2001; (3) Sears, M.K. et al. Proc. Natl. Acad. Sci., **98**, 1152-57, 2001.

In May of 1999, a note submitted to the editor of *Nature* from scientists at Cornell University purported to demonstrate that pollen from a Bt corn hybrid caused harm to monarch butterfly larvae that consumed it with their diet of milkweed leaves [1]. This led to the formation of a collaborative group of scientists from Canada and the United States to investigate the nature of pollen from Bt corn hybrids and the potential of pollen from these commercial hybrids to impact populations of monarch butterflies in the Corn Belt.

Introduction

Transgenic crops expressing Bt proteins have been grown commercially in North America for the past eight years. Questions have been raised about their impact on non-target organisms and other possible effects on the environment such as genetic contamination and increased selection pressure on the target organisms. A particularly alarming report on the impact of Bt corn pollen on monarch butterfly larvae, *Danaus plexippus* L. (Lepidoptera: Danaidae) feeding on milkweeds in and near cornfields caused an uproar of controversy in Europe and subsequently in North America (*1*). In following months, a group of researchers with USDA and several Universities across the Corn Belt collaborated on research agendas and procedures to resolve the scientific concerns resulting from the publication of the short note cited previously. The research contributions reported here represent a collaborative effort established to specifically address the question of risk associated with *Bt* corn pollen to the monarch butterfly. In December 1999, the EPA issued a data-call-in requesting industry, researchers and all interested parties to submit information and comments by March 2001 for use in evaluation and potential re-registration of corn hybrids containing Cry proteins to which the effort reported here was a contribution.

In this paper, a weight-of-evidence approach is described for the risk of exposure of monarch larvae to *Bt* corn pollen and the impact of such exposure on populations of the monarch butterfly in eastern North America (*2*). Our conclusions are based on collaborative research by scientists in the United States and Canada (*3-6*). This approach to risk assessment has been performed for many non-target species in relation to pesticides (*7-10*), industrial by-products (*11-13*) and other potential toxicants found in the environment (*14*). The approach to this process is consistent, well documented and standardized. It requires consideration of both the expression of a toxicant and the likelihood of exposure to the toxicant as the basic components for a risk assessment procedure *(15)*.

Materials and Methods

Toxicity of purified *Bt* proteins to larval stages of butterflies and moths is well known *(16,17)*. Studies conducted on the use of *Bt* sprays in forests for gypsy moth control have shown that Cry proteins can adversely affect non-target Lepidoptera *(18,19)*. Field data from these studies indicated a temporary reduction in lepidopteran populations during prolonged *Bt* use, although widespread irreversible harm was not apparent *(20)*. Lepidopteran-active *Bt* protein expressed in pollen of *Bt* corn hybrids may pose a risk to sensitive species, such as monarch butterflies, in or near cornfields during anthesis *(1,21,22)*. Milkweeds, *Asclepias* spp., and especially common milkweed, *A. syriaca* (L.), are the sole larval food source for monarch butterfly larvae and are abundant throughout the corn growing regions of North America *(23)*. As such, hazard from *Bt* corn pollen deposited on milkweed leaves warrants consideration of its ecological risk to monarch populations.

Risk assessment requires knowledge of four essential components: 1) hazard identification, 2) nature of dose-response to a toxin, 3) probability of exposure to an effective dose, and 4) characterization of risk *(24)*. Components of a risk assessment approach as applied to the case of *Bt* corn and monarch butterfly are depicted in Figure 1. *Bt* proteins expressed in corn plant tissues can bring about specific reactions in the gut of lepidopteran larvae *(25)* including non-target caterpillars that consume *Bt* corn pollen. The magnitude of the reaction will depend on the type of protein produced by various events of hybrid *Bt* corn, the amount of protein expressed in pollen grains from different events, the amount of pollen consumed by larvae of different developmental stages and the susceptibility of larvae to the *Bt* protein. That a hazard may exist was suggested by Losey *et al.* *(1)*. Characterization of toxic effects is necessary to establish the first component of risk. Laboratory and field assays of lethal and sublethal toxicity resulting from exposure to doses of *Bt* pollen likely to be encountered are required to establish an effective environmental concentration (EEC), the toxicity threshold of *Bt* pollen in the environment. The EEC will vary based on expression levels for individual *Bt* corn events in conjunction with environmental factors determining ecological exposure.

Consideration of risk as a function of exposure and effect requires that lines of evidence be established in four areas of inquiry: 1) Is there some density of *Bt* pollen on milkweed leaves that represents a lethal or sublethal threat to monarch larvae or later stages of development? 2) What proportion of *Bt* pollen deposited on milkweed leaves in and around cornfields exceeds the EEC for larvae of monarchs? 3) What proportion of monarch populations utilize milkweed in and near cornfields? 4) What is the degree of overlap between the phenological stages of monarch larvae and corn anthesis over the shared range of these species?

Results

Cry1A proteins expressed in most commercial *Bt* corn hybrids are toxic to the monarch butterfly (Hellmich & Siegfried, 2001). Mortality, expressed as LD_{50}, was estimated at 3.3 ng protein/ml diet, while growth inhibition (EC_{50}) was estimated to be 0.76 ng/ml². However, the expression of Cry1Ab endotoxin within pollen of various events varies considerably depending on the promoter gene involved (Christensen *et al.*, 1992). Expression is greatest in event 176 Bt corn (1.1 to 7.1 μg/gm pollen), a line that is being phased out. This event exceeds, by nearly two orders of magnitude, protein expression in events Bt11 and Mon810 (0.09 μg/gm pollen)), which is near the current level of detection by immunoassay.

Laboratory bioassays of pollen fed to neonate monarchs on leaf disks or whole, detached leaves of common milkweed, *A. syriaca*, indicate that pollen from event 176 *Bt* corn causes mortality and sublethal effects, such as growth inhibition, at very low concentrations (*3*). Growth inhibition, a more sensitive measure of protein intoxication, could be detected at <10 grains/cm². Pollen from all other events, including Mon810 and Bt11 corn hybrids as well as events not presently grown, such as Dbt418, Cbh351 and Tc1507 (expressing Cry1Ac, Cry9C, and Cry1F proteins, respectively), did not demonstrate any lethal or sublethal effects, even at densities above 1000 pollen grains/cm² (*3*). These data were used to establish a no-observable-effect-level (NOEL) for growth inhibition of larvae for event 176 pollen and for Bt11 and Mon810 pollen.

Five field bioassays were undertaken to determine the outcome of exposure of larvae under field conditions on milkweed plants growing or placed in the field. In Iowa, reduced weight gain was noted for larvae exposed to event 176 pollen on milkweeds within cornfields at densities of 20-25 pollen grains/cm². Both survival and weight gain were affected in Maryland, where a series of assays using milkweed leaves collected from plants in an event 176 cornfield were carried out over the pollen-shed period (*6*).

In a separate field trial in Maryland, the effects were evaluated on survival and growth of monarch neonates on leaves of milkweed within a field of a sweet corn hybrid expressing Bt11 endotoxin and compared with the effects of residues following applications of a pyrethroid insecticide. Survival of larvae that fed on insecticide-treated milkweed leaves from within the cornfield was low (0-10%). Survival also was influenced significantly (65-79%) by insecticide that drifted onto milkweeds leaves 3 m outside the field. In contrast, survival of larvae exposed to leaves taken from within both *Bt* and non-*Bt* corn plots ranged from 80-93%, and there were no significant differences in larval survival between these two plots (*6*).

Exposure to Bt corn pollen depends on 1) the phenological overlap between monarch populations and corn anthesis, 2) the spatial overlap between milkweeds used by monarchs and cornfields, and 3) the pollen densities encountered on leaves of milkweed plants in and near cornfields. Pollen from corn plants within a particular field is shed over a period of 7-15 days during the season, while larvae develop over a more prolonged period. Potential for exposure of susceptible stages of monarch larvae to corn pollen depends on synchrony of their development with pollen shed of corn plants. Locations in Iowa, Maryland, Minnesota/Wisconsin and Ontario were monitored for phenological development of monarch populations and anthesis (*4*). Overlap of the more susceptible stages of monarchs, primarily 1st and 2nd instars, with pollen shed was considered for purposes of risk assessment.

Presence of susceptible larvae at the time of corn anthesis varied considerably across the regions studied (*4*). In the more northern locations (MN/WI and ON), about 40 and 62% of the larvae overlapped with pollen shed, respectively, while in areas further south (IA and MD), about 15 and 20% of the larval stages overlapped, respectively. Data from a computer simulation of monarch phenology and corn development support the general observation that overlap increases at higher latitudes across the Corn Belt[1].

Density of milkweed stands in cornfields compared with non-agricultural lands and data on the proportion of the landscape in corn and non-agricultural lands provided a basis on which to determine the proportion of the milkweed population that was in cornfields (*4*). In all locations, densities were higher in non-agricultural lands than in cornfields, but the range of difference was considerable. In Minnesota/Wisconsin and in Iowa, the density of milkweed was approximately 4-7 times greater in non-agricultural fields than in cornfields, while in Ontario the density was up to 115 times greater. In areas where corn is more intensively cultivated, as in Iowa and southern Minnesota/Wisconsin, less non-agricultural land exists and the overall proportion of milkweed on a

---

[1] Pers. comm., D. Calvin, Department of Entomology, Pennsylvania State University, University Park, PA 16802.

landscape basis is higher in cornfields and other crop lands than in non-agricultural land. In regions of the corn growing area where mixed habitats are more common, such as in Maryland and Ontario, milkweeds are more abundant in the non-agricultural landscape and provide proportionately greater habitat than corn (4).

Dispersal of corn pollen was described by Raynor et al. (26), who demonstrated deposition of pollen as much as 60 m from field edges. During periods of pollen shed, samples of pollen were collected on sticky trap surfaces and on milkweed leaves at various distances within and beyond the margins of cornfields to estimate the concentration of pollen that could be encountered by monarch larvae (5). Data from three locations, Iowa, Maryland and Ontario demonstrated a 5-fold reduction in concentration of pollen from just within the edge of the cornfield to about 2-3m distant. Within-field densities across the different studies averaged between 65-425 pollen grains/cm^2 on milkweed leaves at the peak of corn anthesis with an average of 171 grains/cm^2.

To determine risks to monarch larvae associated with *Bt* corn pollen, two components of greatest significance are: 1) the frequency with which effective environmental concentrations (EEC) exceed the thresholds for mortality or sublethal effects, such as growth inhibition, of each *Bt* pollen type, and 2) the proportion of monarch larval populations in eastern North America that are exposed to toxic levels of *Bt* pollen.

It is clear from both laboratory and field-based studies (3,6) that pollen from the dominant commercial *Bt* corn hybrids (Mon810 and Bt11) does not express Cry1Ab protein to a level that will impact monarch populations to any significant degree. Hellmich et al. (3) suggested a conservative LOEC be established for these hybrids at 1000 pollen grains/cm^2 of milkweed leaf surface based on a combined analysis of laboratory bioassays exposing larvae to 1000-1600 of pollen grains/cm^2. Growth inhibition was evident for larvae exposed to event 176 pollen at 5-10 grains/cm^2, the lowest dose where activity was noted by Hellmich et al. (3), therefore the EEC for event 176 corn pollen will frequently exceed this threshold in fields where it is planted.

Probabilities of toxicity for events 176, Bt11 and Mon810 pollen are depicted in Figure 1 as a dose-effect relationship for exposure of larvae to pollen plotted on log-probability scales following methods accepted by EPA. Growth inhibition of monarch neonates in response to increasing concentrations of event 176 pollen, as reported by Hellmich et al. (3), is illustrated, with a no-observable-effect-level (NOEL) at 5-10 pollen grains/cm^2. In comparison, a hypothetical response curve for Bt11 and MON810 pollen is depicted using the same slope parameter for the event 176 response (the Cry1Ab protein is identical in each event), and with a lowest-observable-effect-concentration (LOEC) established, for sake of argument, as a range between 1000-4000 grains/cm^2.

Pollen deposition on milkweed leaves during 1999-2000 (5) is represented on a separate scale in a cumulative frequency occurrence curve.

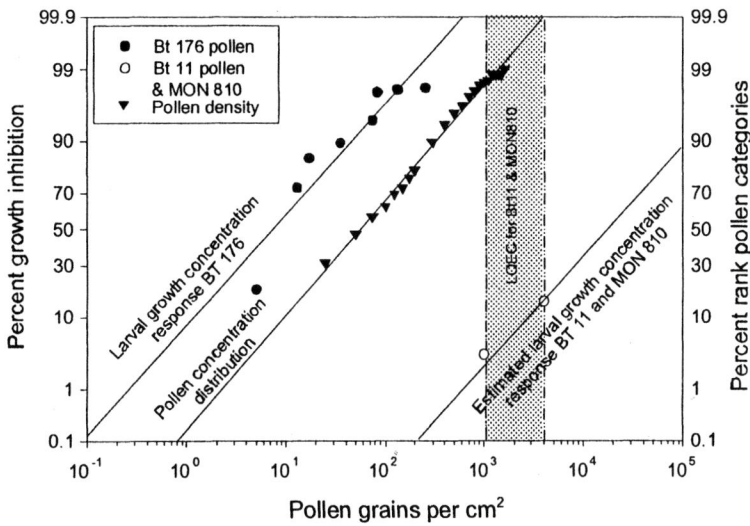

Figure 1. Percent growth inhibition of monarch larvae

It is apparent from Figure 1 that significant overlap of commonly encountered concentrations of pollen from event 176 hybrids with doses necessary for growth inhibition (and mortality) would likely occur within or near cornfields. Ninety percent of the samples of milkweed leaves examined in our study of pollen deposition had a density at or above this level. By contrast, the LOEC range for Bt11 and Mon810 of 1000-4000 pollen grains/cm^2 would be encountered by larvae in only 0.7-0.1% of natural in-field situations. Pollen deposition estimated from any single sample over two seasons never exceeded 1600 grains/cm (3,5). Even if the LOEC for Bt11 and Mon810 were established at the conservative concentration of 1000 grains/cm^2, 99.3% of encounters by monarch larvae of pollen would be below this concentration (5).

Milkweeds exist in cornfields across most of the North American corn-growing regions and monarch butterflies utilize this resource as a host for their offspring during the period of pollen shed (*4*). Quantification of the proportion of monarch populations in each region potentially exposed to *Bt* corn pollen is difficult to ascertain. Wassenaar and Hobson (*27*), using isotope analysis of overwintering monarchs in Mexico, estimated that 50% of the monarch population originates within all or part of 15 states and one province that represent the central core of the North American Corn Belt. More than 93% of North American corn is grown in this area that extends from eastern Kansas/Nebraska to western New York. Using USDA statistics, about 28% of crop and pasture land within this area, which together constitute the monarch breeding habitat, consists of corn. Adoption of *Bt* corn across this area encompassing 50% of the monarch-breeding habitat was about 19% of the corn crop in 2000 (*28*).

Distribution of milkweeds within and around cornfields is variable across the Corn Belt. In Iowa and southern Minnesota, milkweed in cornfields represents a large proportion of the total abundance in those areas, but in Ontario and Maryland, where corn represents a smaller proportion of the landscape, milkweed in cornfields constitutes only a small proportion of its overall abundance (*4*). These data provide a variable picture, but by using data from Iowa a suitable estimate of exposure to *Bt* corn pollen by monarchs can be obtained. In addition to the fact that a significant proportion of land in Iowa is devoted to corn production, a milkweed plant in cornfields was 1.7 times more likely to receive a monarch egg than a milkweed plant in non-agricultural land (*4*). Even though milkweed densities are approximately 7 times higher in non-agricultural land than in cornfields, the proportion of monarch larvae contributed by milkweed from within cornfields is 45 times greater than that of non-agricultural land. If all of the landscape within Iowa that represents breeding habitat for monarchs is coupled with the higher per-plant egg densities on milkweeds within cornfields, the relative density of milkweeds in different habitats and the predominance of corn as a proportion of the total habitat, 56% of monarchs in Iowa are estimated to originate from within cornfields (*4*). This value probably differs considerably from region to region, but comparable data for other locations are not available, although the situation in Iowa likely represents the upper end of the exposure scale.

Temporal overlap of pollen shed with the presence of sensitive larvae in Iowa was 15% (*3*). An estimate of the probability of exposure (P_e) to *Bt* corn pollen by larvae of the final monarch generation arising within Iowa can be expressed as:

$$P_e = l \times o \times a = 0.56 \times 0.15 \times 0.25 = 0.021, \tag{1}$$

or **2.1%** of the population, where P_e = probability of exposure, l = proportion of monarchs from corn, o = overlap of pollen shed with susceptible larval stages, and **a** = adoption rate of *Bt* corn. Because this represents an estimate of the potential exposure for Iowa, similar data would be required to estimate the probability of exposure for other locations in the Corn Belt. We do not have complete data for monarch and milkweed densities across the Corn Belt, and cannot assume that relative productiveness of crop and non-agricultural habitats is the same in other states. For Iowa, the proportion of the monarch population estimated to come from corn, 56%, was roughly similar to the proportion of the breeding habitat in Iowa that is corn. If we assume that this same relationship holds in other areas, we can use the proportion of corn grown relative to the total breeding habitat in other states as an estimate of relative monarch production.

Estimates of these three exposure factors and the estimated contribution of each state and province to approximately 50% of the eastern North American monarch population arising from the portion of the Corn Belt, as indicated by Wassenaar and Hobson (*27*), provides a broad view of potential exposure (Table 1). Our estimates of overlap of the pollen-shed period in each location with the presence of monarch larvae are based partly on the projections of the simulation model described previously and partly on our own observations. In this instance, our estimate for the exposure of monarchs in the Corn Belt states and Ontario is 1.6%. Since monarchs in the Corn Belt represents 50% of the total monarch population, the exposure for the entire monarch population would be 0.8%.

The proportion of the population of monarchs in Iowa that would be exposed to pollen levels that exceed the NOEL for each event (P_t) can be derived from data presented in Figure 1. For event 176 pollen, monarch larvae would likely encounter pollen densities equal to or exceeding the LOEC in 90% of field situations during anthesis, while this would be true in only 0.7% or less of field situations for Bt11 and Mon810 pollen. Overall risk **(R)** is the combined probability of exposure and toxic effect or:

$$R = P_e \times P_t. \tag{2}$$

If we assume that event 176 comprised 5% (**a** from equation **(1)** = 0.05) of planted corn acres in Iowa (or 20% of planted *Bt* corn acres in 2000), an extreme upper bound estimate based on historical marketing data, the risk of impact **(R)** to monarch populations exposed to effects from event 176 pollen is:

$$R = P_e \times P_t = 0.0042 \times 0.9 = 0.0038, \tag{3}$$

or **0.4%** of the population.

The LOEC of pollen for all other events (Bt11 and Mon810 comprise the remaining 20% of total area planted; $a = 0.20$) equals or exceeds 0.1% of expected pollen densities, thus the proportion of the monarch population at risk of impact from effects of Cry proteins, other than event 176, in *Bt* cornfields in Iowa is:

$$R = P_e \times P_t = 0.0168 \times 0.007 = 0.00012, \qquad (4)$$

or **0.012%** of the population. The combined risk estimate for monarchs in Iowa is the sum of these two values, or 0.41%.

Following the same logic as above and assuming that 1) the adoption rate of *Bt* corn reached its maximum limit of 80% ($a = 0.80$) in Iowa, and 2) pollen from current and future *Bt* corn events will pose a hazard less than or equal to that established here for Bt11 and Mon810, the proportion of the monarch population in Iowa that would be at risk with market saturation is:

$$R = P_e \times P_t = 0.067 \times 0.007 = 0.00047, \qquad (5)$$

or **0.05%** of the Iowa population. If, instead, only event 176 hybrids were grown to the maximum extent in Iowa, 6.1% of the monarch population would be at risk. Using this format and data from Table 1, risk of exposure and toxicity from *Bt* corn can be applied to each of the states and provinces in which monarch breeding and corn production overlap.

Discussion

Previous reports (*1,21,22*) indicating the hazard of *Bt* corn pollen to monarch butterfly are inadequate to assess risk, since assigning risk can only be accomplished when both the toxicity of a potential hazard can be properly expressed and the likelihood of exposure is estimated through appropriate observations. We have utilized a comprehensive set of new data and a formalized approach to risk assessment that integrates aspects of exposure to characterize the risk posed to monarch from *Bt* corn pollen. Characterization of toxic effects alone indicates that the potential for hazard to monarchs is currently restricted to event 176 hybrids, which express Cry1Ab protein in pollen at a level sufficient to show measurable effects. Other events either express negligible Cry1Ab protein in corn pollen (Mon810 and Bt11) or express Cry protein of significantly less toxicity to monarch (Dbt418, Cbh351 and Tc1507 expressing Cry1Ac, Cry9c, and Cry1F proteins, respectively). Event 176 hybrids

Table I. Parameter estimates for probability of exposure (P_e) of monarch larvae to *Bt* corn pollen

State	m[3]	l[1]	P_e	o²[2]	a[1]
IA	0.192	0.560	0.0040	0.15*	0.25
IL	0.177	0.423	0.0026	0.25	0.14
IN	0.101	0.376	0.0007	0.25	0.07
KS	0.038	0.119	0.0002	0.15	0.26
KY	0.024	0.132	<0.0001	0.15	0.11
MI	0.40	0.211	0.08	0.035	0.0002
MN	0.083	0.287	0.0028	0.40*	0.30
MO	0.15	0.120	0.22	0.054	0.0002
NE	0.071	0.393	0.0011	0.15	0.26
NY	0.003	0.135	<0.0001	0.40	0.11
OH	0.090	0.263	0.0004	0.25	0.06
ON	0.020	0.300	0.0007	0.62*	0.20
PA	0.019	0.220	0.0005	0.25	0.11
SD	0.063	0.229	0.0021	0.40	0.37
WI	0.045	0.261	0.0007	0.40	0.14
WV	0.005	0.026	<0.0001	0.15	0.11
Avg. (Totals)	(1.020)	0.253	(0.0161)	0.28	0.17

[1] l=proportion of monarchs from corn, o=overlap of pollen shed with susceptible larval stages, a=adoption rate of *Bt* corn.

[2] values marked with a (*) were derived from field observations in 2000 (see ref. 4).

[3] m represents the proportion land area of each state or province that constitutes 50% of the breeding habitat of the eastern North American monarch population.

have always had a minor presence in the corn market and current plantings, which comprise <2% of corn acres, are rapidly declining.

Monarch populations share their habitat with corn ecosystems to a degree previously undocumented (4). Despite this, the portion of the monarch population that is potentially exposed to toxic levels of *Bt* corn pollen is negligible and declining as planting of event 176 is phased out. Because the effects portion of the risk probability equations described above (P_t) is such a small value for the dominant corn hybrids currently planted, the sensitivity of the model to factors describing ecological exposure (P_e) and for risk (**R**) will remain low.

Evidence supporting this risk conclusion has been collected over a wide geographic area and under a variety of conditions in both laboratory and field settings (2-6). Findings from studies done in multiple locations were consistent, even though methods differed from one study to another. This approach to risk characterization is consistent with accepted risk assessment procedures and shares many similarities with previous assessments over a wide range of situations describing potential risk associated with a described hazard. It is imperative that future conclusions concerning the environmental or non-target impacts of transgenic crops be based on appropriate methods of investigation and sound risk assessment procedures.

Acknowledgements

Numerous technical and summer research interns supported the research summarized here and deserve our thanks. Richard Hellmich, Diane Stanley-Horn and Galen Dively provided much of the direction for this work. Jeffrey Wolt, Keith Solomon and Anthony Shelton contributed input and critical comments and suggestions during the development of the project. This research was supported by a pooled grant provided by USDA, Agricultural Research Service and the Agricultural Biotechnology Stewardship Technical Committee, and funding from Canadian Food Inspection Agency (CFIA), Environment Canada, and the Ontario Ministry of Agriculture, Food and Rural Affairs, the Maryland Agricultural Experiment Station, the Leopold Center for Sustainable Agriculture, Ames, IA.

References

1. Losey, J. E.; Rayor, L.S., Carter, M.E. *Nature (London)* **1999**, 399, 214.
2. Sears, M.K.; Hellmich, R.L.; Stanley-Horn, D.E.; Oberhauser, K. S.; Pleasants, J.M.; Mattila, H.R.; Siegfried, B. D.; Dively, G. *Proc. Natl. Acad. Sci.* **2001**, 98, 11937-11942.
3. Hellmich, R. L.; Siegfried, B. D.; Sears, M.K.; Stanley-Horn, D.E.; Mattila, H.R.; Spencer, T.; Bidne, K. G.; Lewis, L *Proc. Natl. Acad. Sci.* **2001**, 98, 11925-11930.
4. Oberhauser, K. S.; Prysby, M.; Mattila, H.R.; Stanley-Horn, D.E.; Sears, M.K.; Dively, G.; Olson, E.; Pleasants, J.M.; Lam, W-K. F.; Hellmich, R.L. (2001) *Proc. Natl. Acad. Sci* **2001**, 98, 11913-11918.
5. Pleasants, J. M.; Hellmich, R.L.; Dively, G.; Sears, M.K.; Stanley-Horn, D.E.; Mattila, H.R.; Foster, J.E.; Clark, T.L.; Jones, G.D. *Proc. Natl. Acad. Sci.* **2001**, 98, 11919-11924.
6. Stanley-Horn, D.E.; Dively, G. P.; Hellmich, R. L.; Mattila, H. R.; Sears, M. K.; Rose, R.; Jesse, L. C. H.; Losey, J. F.; Obrycki, J. J.; Lewis, L. *Proc. Natl. Acad. Sci.* **2001**, 98, 11931-11936.
7. Giddings, J.M.; Hall Jr. L. W.; Solomon, K.R. *Risk Anal.* **2000**, 2, 545-572.
8. Giesy, J.P.; Solomon, K. R.; Coats, J. R.; Dixon, K. R.; Giddings J. M. Kenega, E. E *Rev. Environ. Contam. Toxicol.* **1999**, 160, 1-129.
9. Solomon, K.R.; Baker, D. B.; Richards, P.; Dixon, K. R.; Klaine, S. J.; La Point, T. W.; Kendall, R. J.; Weisskopf, C. P.; Giddings, J. M.; Giesy, J. P.; Hall Jr., L. W.; Williams, W. M. *Environ. Toxicol. Chem.* **1996**, 15, 31-76.
10. Solomon, K.R.; Giesey, J.P.; Kendall, R.J.; Best, L.B.; Coats, J.R.; Dixon, K.R.; Hooper, M.J.; Kenaga, E.E.; McMurry, S.T. *Environ. Toxicol. Chem.* **2001**, 7, 497-632.
11. Giesy, J.P.; Dobson, S.; Solomon, K. R. *Rev. Environ. Contam. & Toxicol.* **2000**, 167, 35-120.
12. Klaine, S.J.; Cobb, G. P.; Dickerson, R. L.; Dixon, K. R.; Kendall, R. J.; Smith E. E.;Solomon, K. R. *Environ. Toxicol. Chem.* **1996**, 15, 21-30.
13. Hall, L.W., Jr.; Giddings, J. M.; Solomon, K. R.; Balcomb, R. *Critical Rev. Toxicol.* **1999**, 29, 367-437.
14. Kendall, R.J.; Lacher Jr., T.; Bunck, E. C.; Daniel, F. B.; Driver, C.; Glue, G. E.; Leighton, F.; Stansley, W.; Watanabe, P. G.; Whitworth, M. (1996) *Environ. Toxicol Chem.* **1996**, 15, 4-20.
15. U.S. EPA (U.S. Environmental Protection Agency). Ecological Committee on FIFRA Risk Assessment Methods. (http://www.epa.gov/NCEA/ecorisk.htm) **1999**, EPA/OPP/EFED, Wahsington, D.C.

16. Kreig, A.; Langerbruch, G.A. In Microbial Control of Pests and Plant Diseases; Burges, H.D.; Ed.; Academic Press, New York, NY, 1981; pp. 837-896.
17. Peacock, J.W.; Schweitzer, D.F.; Dale, F.; Carter, J.L.; Dubois, N.R. *Environ. Entomol.* **1988**, 27, 450-457.
18. Miller, J.C. *Amer. Entomol.* **1990**, 36, 135-139.
19. Johnson, K.S.; Scriber, J.M.; Nitao, J.K.;Smitley, D.R. (1995) *Environ. Entomol.* **1995**, 24, 288-297.
20. Hall, S.P.; Sullivan, J.B.; Schweitzer, D.F.; 1999 USDA Bull. No. FHTET-98-16.
21. Jesse, L. C. H.; Obrycki, J. J. *Oecologia* **2000**, 125, 241-248.
22. Zangerl, A.R.; McKenna D.; Wraight, C.L.; Carroll M.; Ficarello P.; Warner R.; Berenbaum M.R. *Proc. Natl. Acad. Sci.* **2001**, 98, 11908-11912.
23. Malcolm, S. B., Cockrell, B. J. & Brower, L. P. In Biology and Conservation of the Monarch Butterfly; Malcolm S. B.; Brower L. P.; Eds. Natural History Museum of Los Angeles County, Los Angeles, CA 1993; pp. 253-267.
24. NRC (National Research Council) Issues in Risk Assessment; National Acad. Pr. Washington, D.C. 1993
25. Koziel, M. G.; Beland, G.L.; Bowman, C.; Carozzi, N.B.; Crenshaw, R.; Crossland, L.; Dawson, J.; Desai, N.; Hill, M.; Kadwell, S.; Launis, K.; Lewis, K.; Maddox, D.; McPherson, K.; Meghji, M.R.; Merlin, E.; Rhodes, R.; Warren, G.W.; Wright, M.; Evola, S.V. *Biotechnol.* **1993**, 11, 194-200.
26. Raynor, G. S.; Ogden, E.C.; & Hayes, J.V. *Agronomy Journal* **1972**, 64, 420-427.
29. Wassenaar, L. I.; Hobson, K.A. *Proc. Natl Acad. Sci. USA* **1998**, 95, 15436-15439.
30. U.S.D.A.-NASS (2000) U.S. Dept. of Agriculture National Agricultural Statistics Service, Census of Agriculture 2000, Vol.1, Part 57.

Chapter 6

Allelopathic Activities in Litters of Mushrooms

Hiroshi Araya

Chemical Ecology Unit, Department of Biological Safety, National Institute of Agro-Environmental Sciences, 3–1–3 Kannondai, Tsukuba, Ibaraki 305–8604, Japan

I examined the influence of the mushrooms of almost 60 fungal species on the growth of lettuce (*Lactuca sativa* L.) seedlings. Most of the species inhibited the growth of the seedlings. The results indicate that compounds produced by mushrooms are allelopathic, and suggest that chemicals derived from certain mushrooms could be used to control weeds.

Introduction

A mushroom is the fruiting body of a fungus; it typically appears above the ground and contains spores. There are at least 8000 species of mushroom in the world. Although often considered plants, fungi differ from common plants in that they lack chlorophyll and must rely on organic material produced by other organisms for nutrition (*1*). Many species influence plants directly or indirectly. Several species cause fairy rings, which form large circles in lawns (*2*). The ring encloses a zone of diseased or dead grass. Another zone of stimulated growth may occur inside the dead or dying zone. Nada reported that the stimulated growth of turf was caused by plant growth hormones produced by both *Calocybe georgii* L. and *Agaricus campester* Fr. (*3*). Other researchers have reported several plant growth inhibitors in mushrooms. These plant growth regulators can be considered allelochemicals. However, little research on mushroom allelopathy has been reported. In this chapter, I report the allelopathic activity of almost 60 mushroom species. The possibility of allelopathic activity of mushrooms for weed control is discussed.

Allelopatic activity in mushroom

In allelopathy, plant species interfere or stimulate with the germination, growth, or development of the other plant species by releasing chemical compounds. The phenomenon has been known for over 2000 years (Theophtastus, 300 B.C. (4)). The term, derived from the two Greek words *allelon* ('one another') and *pathos* ('suffering'), was coined by Prof. Hans Molisch in 1937 to refer to biochemical interactions between plants of all kinds, including organisms which placed in the plant kingdom (4). Chemicals released from plants and having allelopathic effects are termed allelochemicals (or allelochemics). Thus, exogenous plant hormones must act as allelochemicals, and endogenous hormones may affect to other plant by release as litter. The emission of allelochemicals can occur via various routes. Possible routes are shown in Figure 1.

Allelopathy plays an important role in both agroecosystem and natural ecosystem. In agroecosystems, allelopathic activities of weeds affect on crops yields. Many agronomy researchers have tried various strategies to utilize allelopathy. For example, rye (*Secale cereale* L.) produces allelochemicals and is used to some extent in weed management (5).

On the other hand, only two studies of mushroom allelopathy have been reported. In 1937, Molisch first reported the influence of volatiles from fruiting bodies of *Agaricus campestris* on vetch (*Vicia sativa*) (4). Fifty years later, Chamont and Simeray reported that water extracts of 114 species inhibited the growth of red radish (*Raphanus sativus* var. *radicicola*) seedlings (6). The possible emission routes of allelochemicals from mushrooms are proposed in Figure 2. Since the longevity of fruiting bodies is generally short, the main source of allelochemicals would be mycelia in soil.

Mushrooms must be selected for testing on the basis of observations of allelopathy in nature. Some phenomena have been reported: for example, soil sickness caused by Matsutake (*Tricholoma matsutake* (S. Ito et Imai) Sing) (7); and stimulation of bamboo growth by a fungus in the Ramariaceae (8). However, wheat (*Triticum aestivum* (L.) Thell.) could not grow well in the field fertilizing immature compost of bed log for lacquered bracket fungus (*Ganoderma lucidum*) (*9*). This may result in their pathogenicity excluding toxic compounds (*10*).

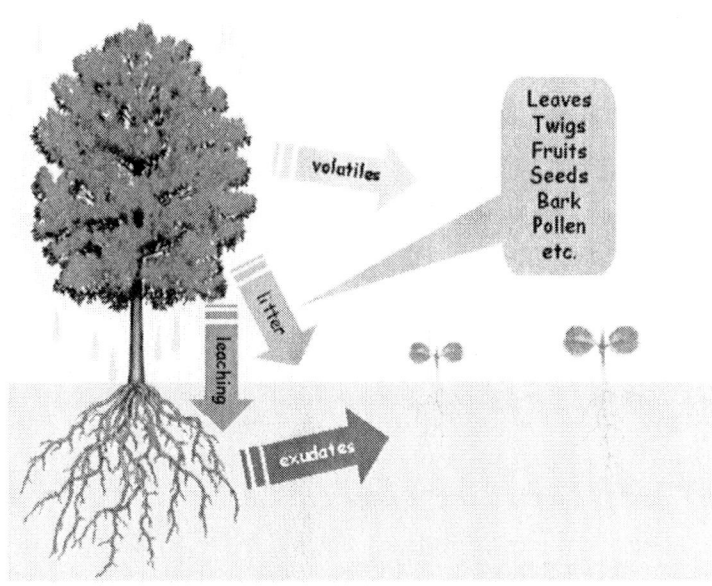

Figure 1. Possible Release Routes of Allelochemicals from Common Plants (See page 1 of color insert.)

Several compounds that inhibit plant growth have been isolated from mushrooms: *e.g.*, azetidine-2-carboxylic acid from *Clavaria miyabeana* (*11*); neogrifolin and grifolin from *Polyporus confluens* (*12*); and 2-amino-3-cyclopropyl-butanoic acid and 2-amino-5-chloro-4-pentenoic acid from *Amanita castanopsidis* (*13*). In addition, the production of plant hormones has been reported: *e.g.* indole-3-acetic acid from *Schizophyllum commune* Fries (*14*); cytokinins from *Rhizopogon roseolus* (Cda.) Th. Fr. (*15*) and *Suillus punctipes* (*16*); and ethylene from *Agaricus bisporus* (Lange) Sing. (*17*). These are candidate allelochemicals.

Allelopathy offers the opportunity for identifying new agrochemicals. Its study is therefore important.

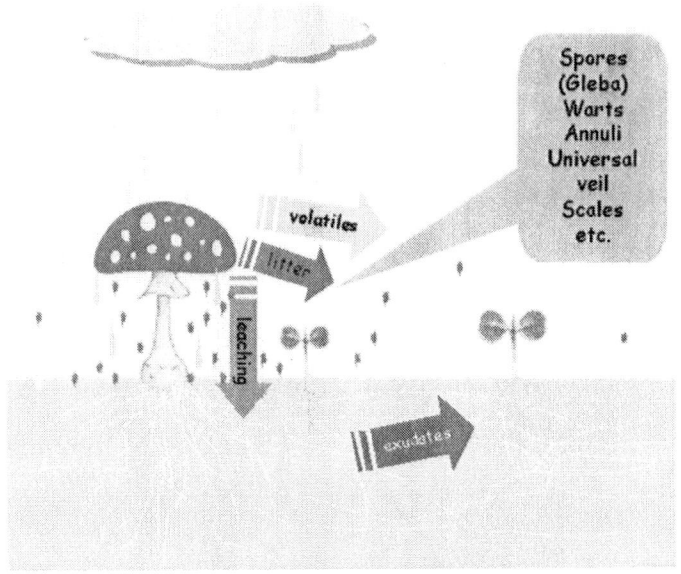

*Figure 2. Proposed Release Routes of Allelochemicals from Mushrooms
(See page 1 of color insert.)*

Estimation of Allelopathic Activity of Mushrooms

Several bioassay methods have been used to screen allelopathic plants. Fujii demonstrated that a method using two layered agar is useful for screening of the allelopathic potential of leaf litter (*18*). I have been investigating the allelopathic activities in litters of mushroom fruiting bodies by application of this method (*19–21*). The procedure and a picture as an example are shown in Figure 3. Here I present the results of about 60 mushroom species.

All the wild mushrooms were collected in Ibaraki, Tochigi, and Fukushima prefectures, Japan, except for one specimen of *Morchella esculenta*, which was collected and air dried in Canada. Cultivated mushrooms were obtained from Dr. A. Sekiya of which Forestry and Forest Products Research Institute, Japan. All mushrooms were lyophilized after removal of stray materials. Fragments of dried mushroom (10 mg and 50 mg) were bioassayed by standard procedures (Figure 3). Because of its sensitivity to various chemicals, lettuce (*Lactuca sativa* L. cv.

'Great Lakes 366') was used as a test plant. For the control treatment, mushrooms were not added. The lengths of the radicle and hypocotyl of lettuce seedlings were measured after incubation at 20 °C in the dark for 3 days. Efficacy of lettuce seedling growth was calculated from the following formula:

$$\text{Efficacy (\%)} = 100 - 100 \times \text{sample growth} / \text{control growth}.$$

100% means completely killed, -100% means two fold growth

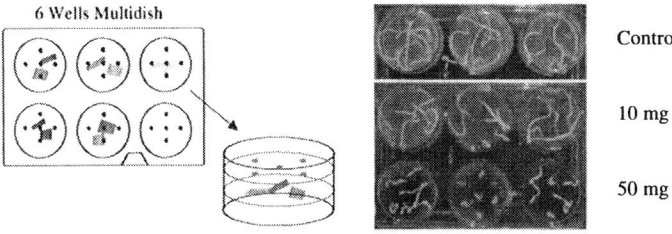

(1) add 5 ml of agar (0.75 %, 40℃)
(2) place fragments of mushroom (50 and 10 mg/well) on gelatinized agar
(3) add 5 ml of agar (0.75 %, 40℃)
(4) place lettuce seeds (Great Lakes 366) on gelatinized agar
(5) incubate at 25℃ in the dark for 3 days
(6) measure radicle length, hypocotyl length and germination %

Figure 3. Procedure and an example for allelopathy assay for mushroom litters

Results and Discussion

Most of the mushrooms inhibited the growth of lettuce seedling, as shown in Tables I and II. Several dried mushroom fruiting bodies inhibited the lettuce growth strongly: *Agaricus subrutilescens* (Kauffm.) Hotson et Stuntz, *Amanita cokeri* (Gilb. et Kühn.) Gilb. f. *roseotincta* Nagasawa et Hongo, *Amanita sinensis* ZL Yang, *Boletus fraternus* Peck, *Boletus reticulates* Schaeff., *Coprinus comatus* (Müller: Fr.) Pers., *Cortinarius violaceus* (L.: Fr.) Fr., *Gymnopilus spectabilis* (Fr.) Sing., *Lanopila nipponica* (Kawam.) Y. Kobayasi, *Pleurocybella porrigens* (Pers.: Fr.) Sing., *Ramaria formosa* (Fr.) Quél.,

Lyophyllum decastes (Fr.: Fr.) Sing., *Morchella esculenta* (L.: Fr.) Pers. var. *esculenta* (collected in Japan), *Panellus serotinus* (Pers.: Fr.) Kühn., *Pleurotus ostreatus* (Jacq.: Fr.) Kummer, and *Hypsizigus marmoreus* (Peck) Bigelow. On the other hand, *F. obliqua* stimulated lettuce seedling growth. The chemicals that diffused from these mushrooms through the agar are therefore plant growth regulators and candidate allelochemicals, which could yield new herbicides.

Differnt effect on two *M. esculenta* collected from Japan and Canada was observed. Decomposition of chemicals may have been a cause in air drying of Canadian *M. esculenta*. Further study is needed to verify the allelopathy.

Table I. Percent Inhibition of Lettuce Seedling Growth by Wild Mushrooms

Mushrooms	Radicle		Hypocotyl	
	10 mg	50 mg	10 mg	50 mg
Agaricus subrutilescens	84.81	91.31	63.99	80.98
Albatrellus confluens	40.09	70.36	3.98	23.11
Amanita cokeri	76.45	93.63	44.14	66.97
Amanita pantherina	53.61	76.68	-8.60	43.74
Amanita sinensis	88.99	95.71	70.07	86.30
Amanita vaginata	41.27	76.01	28.17	55.90
Armillariella mellea	71.64	84.22	31.08	56.18
Armillariella tabescens	35.23	78.16	-4.77	56.69
Auricularia auricula	61.70	73.30	14.52	26.89
Boletus fraternus	74.63	92.13	25.90	65.00
Boletus griseus	53.73	81.66	15.14	51.79
Boletus pseudocalopus	72.92	88.99	47.25	72.10
Boletus regius	60.88	77.40	31.96	56.97
Boletus reticulates	78.12	90.40	40.85	81.37
Boletus sensibilis	59.94	87.67	16.81	63.48
Cantharellus cibarius	19.58	61.23	4.37	23.58
Cantharellus luteocomus	38.55	53.73	3.96	8.76
Coprinus comatu	84.65	91.09	69.72	73.11
Cortinarius purpurascens	40.54	76.21	1.20	54.63
Cortinarius violaceus	87.23	96.92	67.54	81.23
Craterellus cornucopioides	53.99	76.45	0.07	25.94
Fuscoporia obliqua	-1.35	36.71	-50.73	-36.42

Table I. Continued on next page

Table I. (Continued)

Mushrooms	Radicle		Hypocotyl	
	10 mg	50 mg	10 mg	50 mg
Gymnopilus spectabilis	76.39	90.79	29.31	70.43
Hygrophorus capreolarius	70.50	79.31	19.86	36.09
Hygrophorus fagi	51.74	66.41	-15.24	22.52
Ionomidotis frondosa	28.46	71.60	-3.99	23.41
Lactarius volemus	38.36	79.53	-7.03	42.17
Lanopila nipponica	79.34	90.73	30.03	63.36
Leatiporus sulphureus	54.13	78.50	18.34	57.21
Leccinum extremioriental	73.81	87.45	34.57	63.99
Lycoperdon perlatum	92.86	94.65	58.56	71.24
Lyophyllum sykosporum	68.93	84.86	24.87	58.57
Morchella esculenta (CAN)	54.19	94.76	1.64	61.31
Morchella esculenta (JPN)	97.12	98.08	92.13	95.19
Mycena haematopoda	76.11	81.47	36.19	58.11
Naematoloma sublateritium	49.28	77.40	20.18	54.98
Pholiota lubrica	83.52	87.85	44.38	67.73
Pleurocybella porrigens	82.41	85.07	61.16	68.92
Polyporus arcularius	72.49	81.29	16.34	28.97
Ramaria formosa	81.73	91.53	50.80	66.78
Ramaria fumigata	57.39	91.36	61.79	82.64
Russula virescens	67.64	85.85	37.10	53.26
Sarcodon imbricatus	60.25	78.53	16.33	45.36
Sarcodon scabrosus	79.32	89.77	49.00	52.59
Strobilomyces seminudus	38.78	84.01	7.80	64.14
Strobilomyces strobilaceus	28.24	51.13	-30.36	6.67
Tricholoma portentosum	77.08	87.84	24.17	52.70
Xanthoconium affine	80.90	89.43	48.37	64.75

The overuse of synthetic chemicals for pest control is a threat to the environment (22–23). There is a misleading tendency that compounds produced by plants are eco-friendly. In fact, many allelopathy researchers feel that allelochemicals could be ideal agrochemicals (24–33). But we must assess the safety of these products before using allelochemicals just the same way as synthetic agrochemicals.

Table II. Percent Inhibition of Lettuce Seedling Growth by Cultivated Mushrooms

Mushrooms	Radicle 10 mg	Radicle 50 mg	Hypocotyl 10 mg	Hypocotyl 50 mg
Ganoderma lucidum	26.65	55.44	-5.98	8.37
Grifola frondosa	68.32	88.63	19.91	68.03
Hericium erinaceum	24.38	53.72	1.97	47.13
Hipsizigus marmoreus	79.44	96.05	53.63	87.30
Lentinula edodes	40.86	91.67	5.25	69.74
Lyophyllum decastes	82.39	94.36	41.31	84.84
Panellus serotinus	80.20	90.33	40.23	74.71
Pholiota nameko	58.40	73.64	11.12	55.39
Pleurotus ostreatus	77.99	94.44	59.60	89.81
Pleurotus pulmonarius	73.59	85.94	48.85	75.41
Tremella fuciformis	24.35	56.54	-11.15	-1.97

Conclusion

Based on this experiment, almost all mushrooms showed inhibitory activity on lettuce seedling growth. Consequently, broad leaf plant growth will be inhibited at the place after mushrooms grew, or mushroom was incorporated, if the soil did not degrade or strongly adsorb the allelochemicals or the chemicals are easily dissipated by leaching or photodegradation.

Acknowledgments

I would like to express my thanks to the staff of the Chemical Ecology Unit, National Institute for Agro-Environmental Sciences, Japan, for their help and useful comments on this research work. I am also grateful to Mr. Ryuichi Koshihara, Mitsui & Co., Ltd., for his kind donation of morel (*M. esculenta*). I also wish to thank Dr. Atsushi Sekiya, Forestry and Forest Products Research Institute, Japan, for generous gifts of several cultivated mushrooms.

References

1. Lincoff, G. H.: *National Audubon Society Field Guide to North American Mushrooms*: Chanticleer Press Inc., New York, NY, **1981**, p.11–13.
2. Smiley, R. W; Dernoeden, P. H; Clarke, B. B.: Fairy Rings. In *Compendium of Turfgrass Diseases* 2nd Ed.: APS Press, Minnesota, MN, **1992**, p.61–63.
3. Nada, G.: *Biol. Vestn. (Ljubljana)* **1975**, *23*, 89–96.
4. Molisch, H.: *Der Einfluß einer Pflanze auf die andere Allelopathy*: Fischer, Jena, **1937**.
5. Weston, L. A. *Agron. J.* **1996**, *88*, 860–866.
6. Chaumont, J. P.; Simeray, J. *Rev. Écol. Biol. Sol* **1985**, *22*, 331–339.
7. Ogawa, M. *Biology of Matsutake*: Tsukiji Shokan Publishing Co., Ltd., Tokyo, **1978**, p.326 (Jpn).
8. Hongo, T.: *Narrow Road to Mushrooms*: Tombow Publishing Co., Ltd., Tokyo, **2003**, p.30–31 (Jpn).
9. Araya, H., *unpublished data*, 2001.
10. Ariffin, D.; Idris, A. S.; Singh, G.: Status of *Ganoderma* in oil palm. In *Ganoderma Diseases of Perennial Crops*: CABI Publishing, Oxford, **2000**, p.49–68.
11. Ikeda, M; Naganuma, Y.; Ohta, K.; Sassa, T; Miura, Y. *Nippon Nogeikagaku Kaishi* **1977**, *51*, 519–522.
12. Ikeda, M.; Kanou, H.; Nukina, M. *Bull. Yamagata Univ. Agr. Sci.* **1989**, *10*, 849–852.
13. Yoshimura, H.; Takegami, K.; Doe, M.; Yamashita, T.; Shibata, K.; Wakabayashi, K.; Soga, K.; Kamisaka, S. *Phytochemistry* **1999**, *52*, 25–27.
14. Epstein, E.; Miles, P. G. *Plant Physiol.* **1967**, *42*, 911–914.
15. Miller, C. O. *Science* **1967**, *157*, 1055–1057.
16. Crafts, C. B.; Miller, C. O. *Plant Physiol.* **1974**, *54*, 586–588.
17. Turner, E. M.; Wright, M.; Osborne, D. J.; Self, R. *J. Gen. Microbiol.* **1975**, *91*, 167–176.
18. Fujii, Y. *Bull Natl. Inst. Agr.-Env. Sci.* **1994**, *10*, 115–218.
19. Araya, H.; Hattori, M.; Nishihara, E.; Hiradate, S.; Fujii, Y.; Sekiya, A. Fruiting bodies of mushroom as allelopathic plants. In *Third World Congress on Allelopathy – Challenge for the New Millennium, Abstracts Book* **2002**, *8*, 244.
20. Araya, H. Fruiting bodies of mushrooms as allelopathic plants In *Allelopathy: Challenge for the New Millennium*: Science Publisher, Tokyo, **2004** (in preparation).
21. Araya, H; Hiradate, S.; Fujii, Y. *J. Weed Sci. Tech.* **2003**, *48 (sup.)*, 198–199.

22. Chou, C. H.: Allelopathy and sustainable agriculture. In *Allelopathy: Organisms, Processes and Applications*: ACS Symposium Series 582, American Chemical Society, Washington DC, **1995**, p.211–223.
23. Cutler, H. G.; Cutler, S. J.: Agrochemicals and pharmaceuticals. In *Biologically Active Natural Products: Agrochemicals*: CRC Press, Boca Raton, FL, USA, **1999**, p.1–14.
24. Cutler, H. G. *Weed Tech.* **1988**, *2*, 525–532.
25. Cutler, H. G. Potentially useful natural product herbicides from microorganisms. In *Principles and Practices in Plant Ecology: Allelochemical Interactions*, CRC Press, Boca Raton, FL, USA, **1999**, p.497–516.
26. Dayan, F.; Romagni, J; Tellez, M; Rimando, A.; Duke, S. *Pesticide Outlook* **1999**, *Oct.* 185–188.
27. Duke, S. O.; Abbas, H. K.; Duke, M. V.; Lee, H. J.; Vaughn, K. C.; Amagasa, T; Tanaka, T. *J. Environ. Sci. Health* **1996**, *B31*, 427–434.
28. Duke, S. O.; Abbas, H. K.; Amagasa, T; Tanaka, T.L Phytotoxins of microbial origin with the potential for use as herbicides. In *Crop Protection Agents from Nature: Natural Products and Analogues. Critical Reviews on Applied Chemistry, Vol. 35*, Society of Chemical Industry, Cambridge **1996**, p.82–113.
29. Duke, S. O.; Dayan, F. E.; Romagni, J. G.; Rimando, A. M. *Weed Res.* **2000**, *40*, 99–111.
30. Hoagland, R. E.: Microbes and microbial products as herbicides: an overview. In *Microbes and Microbial Products as Herbicides, ACS Symposium Series 439*, American Chemical Society, Washington DC, **1990**, p.2–52.
31. Hoagland, R. E.: Biochemical interactions of the microbial phytotoxin phosphinothricin and analogs with plants and animals. In *Biologically Active Natural Products: Agrochemicals*, CRC Press, Boca Raton, LFL, USA, **1999**, p.107–125.
32. Macías, F. A.; Molinillo, J. M. G.; Galindo, J. C. G.; Varela, R. M.; Torres, A.; Simonet, A. M.: Terpenoids with potential use as natural herbicide templates. In *Biologically Active Natural Products: Agrochemicals*, CRC Press, Boca Raton, FL, USA, **1999**, p.15–31.
33. Putnum, A. R. *Weed Tech.* **1998**, *2*, 510–518.

Combinatorial Chemistry

Chapter 7

CombiChem at Bayer CropScience: What We Have Learned, Exemplified by Recent Chemistries

Mazen Es-Sayed[1], Michael Beck[1], Stefan Bräse[2], Armin de Meijere[3], Christian Funke[1], Kristian Kather[1], Michael Limbach[3], Matthias E. P. Lormann[2], Christian Paulitz[1], Heiner Wroblowsky[1], and Viktor Zimmermann[2]

[1]Bayer CropScience AG, BCS-R LG, Alfred-Nobel-Strasse 50, 40789 Monheim, Germany
[2]Department of Chemistry and Biochemistry, University of Bonn, Gerhard-Domagk-Strasse 1, 53121 Bonn, Germany
[3]Department of Chemistry, University of Göttingen, Tammannstrasse 4, 37077 Göttingen, Germany

At Bayer CropScience combinatorial chemistry has been used to significantly increase and diversify the compound collection for high throughput *in vitro* and *in vivo* screening. Quality aspects were seen critical along this process, from the selection of novel chemotypes with a known biological background to the purity of the test compounds. This will be exemplified by a benzotriazole library obtained by a polymer-bound version of the Sanger reagent attached *via* a traceless triazene linker (T1). After identification of initial fungicidal activity a thorough analysis in all relevant internal chemical and biological databases revealed the structural requirements for 2nd validation libraries, increasing the success likelihood to find potential leads worth further optimization. In a second combinatorial approach peptides with an agro-relevant mode of action will be used to derive new active ingredients. Based on cyclopropylideneacetate chemistry, a structurally extremely flexible peptidomimetic system has been developed for systematic modification of biologically important peptide motifs in potentially optimized conformations. The next step will be the identification of analogues based on this structural informations.

Introduction

The introduction of miniaturized high-throughput *in vivo* and *in vitro* screening has significantly reduced the compound quantities required and allowed one to test more compounds much faster. Both aspects have helped to better position combinatorial chemistry in agro-research, by applying this technology to significantly enlarge and diversify the existing compound repository. Over the last 5 years more than half a million compounds comprising over 345 novel chemotypes have been added to the screening pool. During this process quality criteria had been considered carefully. Most of the selected chemotypes were not only novel but reported with interesting biological activities. All libraries were decorated with agrophilic side chains and carefully profiled (clogP and MW). Except for a little percentage, only those compounds were accepted for screening with a calculated clogP below 6 and a molecular weight below 600, with an average per library way below those figures. All compounds were provided in sufficient quantities for secondary screening. Additionally, compound purity was continuously improved to average 85% (by LC-MS) per library. Today, combinatorial chemistry resources are shifted from expanding the repository to validation and optimization of screening hits to lead structures. This process will be exemplified by a solid phase benzotriazole library developed in cooperation with Prof. Bräse et al.

An Efficient Solid-phase Benzotriazole Synthesis

Benzotriazoles are often described in the literature with interesting biological activities, ranging from antiemtic (*1*) to CNS activity (*2*). Many existing methods for this class of substances are low yielding, accompanied with poor regioselectivity for N-alkylated derivatives and in general with limited structural diversity (*3*). For these reasons benzotriazoles were considered an interesting motif for a hit-finding library. To increase the optimization potential some of the existing limitations will be also addressed. Starting point is 2-fluoro-4-nitroaniline (**1**), which was first transformed into the diazonium salt **2**, followed by reaction with benzylamino-substituted Merrifield resin **3**, to generate a triazene-linked nitro-fluorobenzene **4** (Figure 1). The two electron-withdrawing substituents on the fluorobenzene activate the fluorine for nucleophilic substitution just like two nitro groups, making **4** a solid-phase analogue of the Sanger reagent (4). With the exception of anilines the fluorine is easily substituted by a structurally diverse set of amines (5).

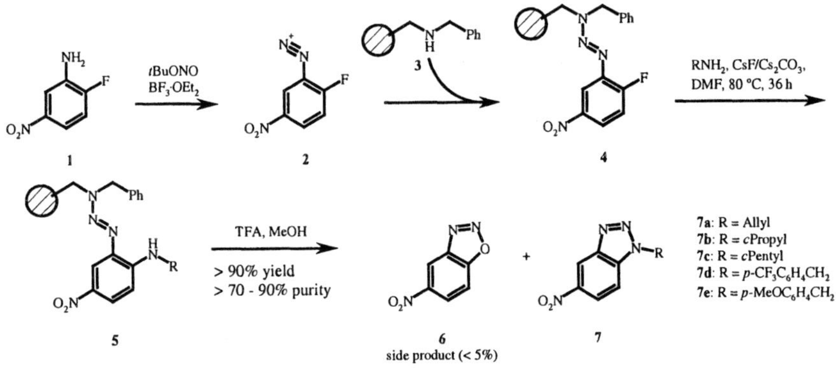

Figure 1. Traceless T1 linked Sanger reagent.

The benzotriazole is formed upon treatment of the triazene **5** with acid to regenerate the *ortho*-amino-substituted diazonium salt, which immediately cyclizes in good yields (>90%) and good purity (70 – 90% by LC-MS) to the benzotriazole **7**. The only side product observed is the benzoxadiazole **6** resulting from traces of water, often carried in with the reagents. A small hit-finding library with 130 compounds has been synthesised. With the clogP covering 3 orders of magnitude and centred around 3, the molecular weight below 400 and the polar surface area in the range of 78, the library was perfectly profiled by the rules of Briggs (6). Screening revealed weak fungicidal activity. Typically at this point, all relevant internal chemical and biological databases are searched for analogues or activities found in other *in vitro* or *in vivo* assays. This step is critical to precisely define the structural requirements for a follow-up library. The search engine for this is an in-house developed web-application called Pythia, which uses an intuitive Isis Draw® based query language. Besides accessing all relevant databases via one interface, it also calculates physico-chemical properties, like clogP or polar surface area etc. and activity indices. The last one is a rule-based prediction of *in vivo* activity, based on up to 30 molecular descriptors, derived from analysis of *in vivo* data from over 300T compounds tested at Bayer CropScience. The results are visualised in a user friendly Excel® format (Figure 2). This comprehensive analysis supports the identification of all biological data reported for the class of substances in question, hints toward the relevant physico-chemical profile of actives and helps to focus synthesis on the most promising compounds.

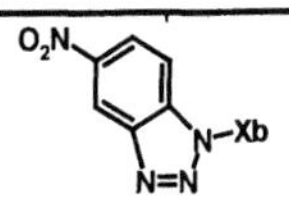	Xb(molw	F-Index	PHY	VTR	PMM	H-Index	I-Index	logP_pH	PSA	F_ALL
16024-2-1	Clc1c(c(c	139,606	143	143	5	0	0	0	3,2	75,1875	0,56
16024-1-1	Clc1c(c(c	139,606	143	143	5	0	0	0	3,2	75,1875	0,56
16193-1-1	C1(CCC(	97,1817	91	81	36	22	28	6	3,5	77,5	0,44
16192-1-1	C1CCCC	111,209	83	81	17	2	28	0	3,8	76,875	0,56
15882-1-1	Clc1c(cc	139,606	76	0	1	76	0	0	3,3	78,0625	0,67
15903-1-1	Clc1c(cc	139,606	62	0	8	61	0	0	3,3	78,5	0,67
16204-1-1	C(C)C[X	43,0892	48	45	16	0	0	0	2,4	78,3125	0,11
15884-1-1	Clc1ccc(	139,606	47	45	11	8	0	0	3	78,4375	0,56
15888-1-1	C1CC1[X	41,0733	45	0	0	45	0	0	1,9	79,0625	0,33
15902-1-1	FC(F)(F)	159,132	38	0	37	8	0	0	3,2	77,75	0,56
15893-1-1	C1(CC1)(	55,1004	38	29	2	25	0	0	2,8	78,4375	0,33
15886-1-1	Clc1c(cc	160,024	37	0	15	34	0	0	3,3	78,3125	0,56
16171-1-1	C(CCC)(	99,1976	36	35	7	0	0	0	3,8	78,75	0,56
16174-1-1	C(CCC[X	85,1705	36	35	10	2	0	0	3,5	78,375	0,44
15885-1-1	FC(F)(F)(	151,153	35	0	8	34	10	0	3,5	77,5625	0,67
15883-1-1	FC(F)(F)(	173,159	35	0	9	34	0	0	3,5	77,8125	0,67
15899-1-1	o1c(ccc1	81,095	34	0	5	34	0	0	2,1	89,9375	0,33
15898-1-1	C1(CCC(	97,1817	33	0	26	20	22	0	3,4	77,625	0,44

Figure 2. Pythia results for benzotriazole library in Excel® format.

As a consequence, the diversity in the substitution pattern of the benzotriazole ring and in the amine-set had to be increased to include also anilines and polar functionalised amines. This was accomplished by switching from solid-phase based Sanger-type to the Buchwald-Hartwig reaction (7). Various substituted chloro- and bromobenzenes linked *via* the triazene are readily available (8). The palladium catalyzed substitution works well using BINAP as a ligand, but this catalysts suffer from air sensitivity. Higher stability of the catalyst was achieved with the (biphenyl)dicyclohexylphosphine ligand (9) (Figure 3). Along this route, anilines and amines with polar side chains were incorporated and delivered, after acid catalyzed cleavage, benzotriazoles in high purity (73 – 99% by GC). Due to the long reaction times and high temperature, the yields were only in the range of low to medium (10 – 40%).

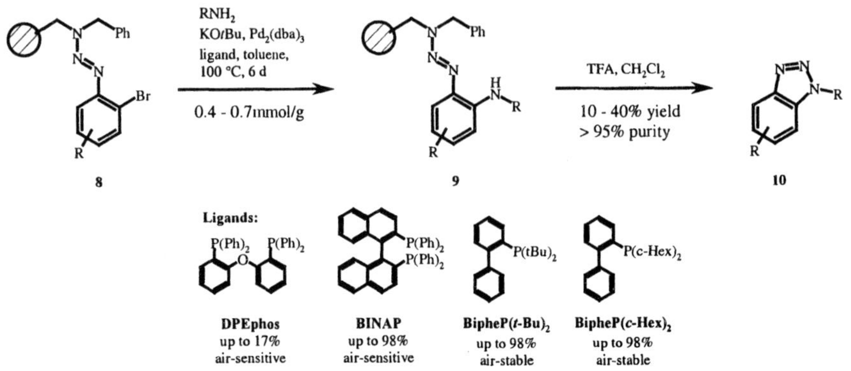

Figure 3. Buchwald-Hartwig reaction on triazene-linked phenyl bromides.

Benzylamines and the chloro or bromo substituted anilines cannot be applied in the Buchwald-Hartwig reaction and this is another limitation. These limitations can be overcome by triazene-linked anilines. On one hand, the amine could be benzylated by reductive alkylation with benzaldehydes, or on the other hand, be used for an inverse Buchwald-Hartwig reaction, e.g. with dichlorobenzenes for the assembly of chloroanilines.

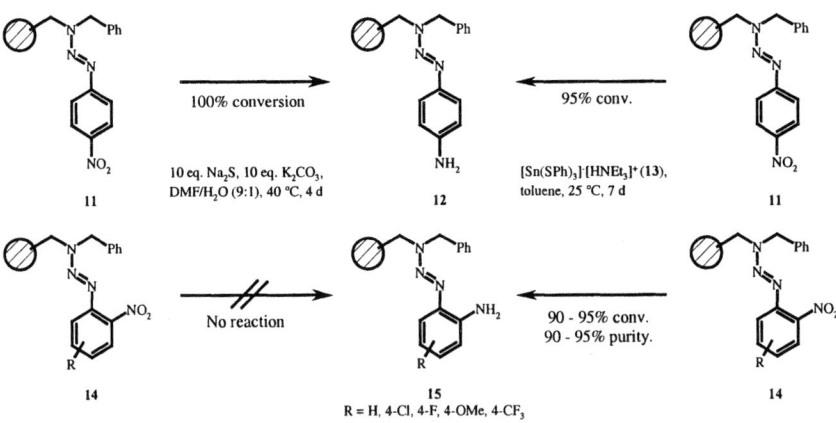

Figure 4. Synthesis of the triazene-linked anilines.

The formation of triazene-linked anilines proved to be rather difficult and succeeded, when the Bartra reagent (**13**) (*10*) was used to reduce the nitro group (Figure 4). With the resin-bound T1-linked aniline in hand, the reductive alkylation (*11*) and inverse Buchwald-Hartwig reaction worked as planned (Figure 5).

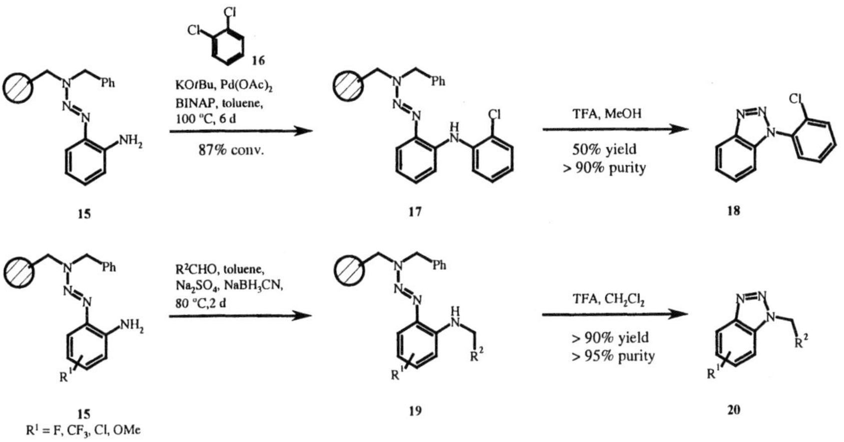

Figure 5. Inverse Buchwald-Hartwig reaction and reductive amination.

To fulfill the up-front defined structural requirements for the validation libraries, three new methods have been developed. As a general trend, the methods become more complex, requiring more demanding reaction conditions, e.g. inert gas atmosphere. The latter has been addressed by linking all available technologies, to one IT platform. Switching to the most appropriate technique, also used in gloveboxes, is made much easier that way. But this is only part of the solution for complex combinatorial chemistry, as these reactions are much more sensitive to impurities, requiring reagents of higher quality. Especially for validation purposes, SAR relevant reagents in sufficient quantities and of appropriate quality have to be available. With all those hurdles the library size typically becomes smaller and smaller, the further it moves on along the optimization process, while the efforts for methodology or synthesis of specific templates are increasing at the same time. But emphasizing quality over quantity is the only option to reduce the risk of overlooking potentially interesting novel substance classes, which are increasingly more difficult to be found.

Geometrically Defined Peptidomimetics from Cyclopropylideneacetates

Instead of searching in large compound collections for novel structures with a novel mode of action (MOA), one can use an inverse approach, i.e. search for structural analogues of biologically active compounds. In this respect, nature provides numerous structural novelties, most of them with an innovative MOA, like the fungicidal peptides Hectochlorin (*12*), Serratamolide (*13*) or Rhodopeptin (*14*). They represent interesting templates for the search of structural analogues. The successful transfer of the structural information of the cyclic peptide Rhodopeptin to a quinoline carboxamide analogue has been described (*14*). This analogue showed higher *in vitro* potency and also better physico-chemical properties. As in the case of Rhodepeptin, only parts of the peptide, but these in a defined conformation, are required for biological activity. Based on the well established multi-facetted cyclopropylideneacetate **21** (*15*) novel geometrically defined mono- **22** or bicyclic peptidomimics **23** and **24** have been developed in co-operation with the group of de Meijere at the University of Göttingen (Figure 6), to be used for the systematic search for preferred conformation of the biologically important motif of peptides.

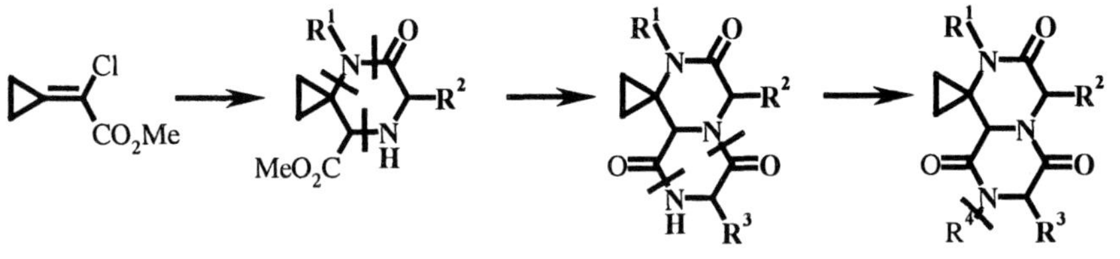

Figure 6. Geometrically defined peptidimimetics from cyclopropylideneacetates.

The advantage of the cyclopropylideneacetates is a unique combination of high reactivity and high density of functionalities (*16*), which both are needed for a high degree of structural flexibility (*17*). The synthesis starts with a Michael addition, occuring with most amines in quantitative yields (Table I), followed by coupling of the Michael adduct with a Boc-protected amino acid. Deprotection and cyclization under basic conditions yielded the piperazinone **26** (*18*). The major product even at low temperature is the seven-membered lactam **27** (Figure 7).

In this synthetic scheme, chiral amino acids with an α-substituent can only be attached onto the Michael adducts of small nucleophiles like methylamine (Figure 8). Upon cyclization to the piperazinones **29** a kinetic resolution of the epimers with respect to the chiral center in position 2 is observed (Table II). One

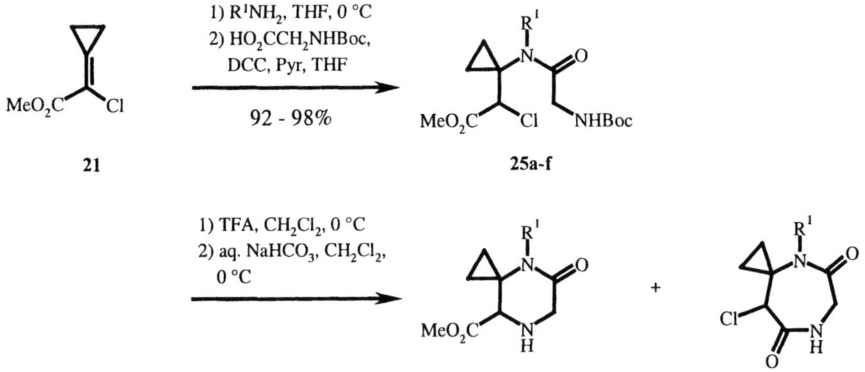

Figure 7. Synthesis of a monocyclic peptidomimetic.

Table I. Structural flexibility in the Michael addition to 2-chloro-2-cyclopropylideneacetate 21 and subsequent cyclizations

Entry	R^1	26 (%)	27 (%)
a	n-pentyl	22	60
b	Phenethyl	18	63
c	p-MeO(C$_6$H$_4$)CH$_2$	16	56
d	p-Cl(C$_6$H$_4$)CH$_2$	20	61
e	(furan-2-yl)CH$_2$	15	55
f	(indolyl-3-yl)CH$_2$CH$_2$	22	58

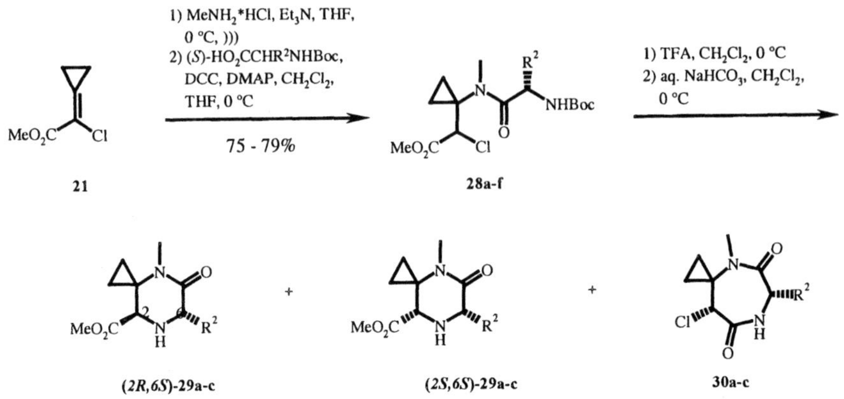

Figure 8. Kinetic resolution during ring closure.

epimer preferentially cyclizes to the piperazinone **(2R,6S)-29** and the other to the lactam **30** (*18*), making both accessible in enantiomerically pure form (Figure 8).

To overcome the selectivity problems during ring closure and the limitations for the coupling of Michael adducts with chiral amino acids, a modified approach has been developed. In this, the Michael adduct of **21** was treated with unsubstituted or α-substituted bromoacetic acid chlorides under modified Schotten-Baumann conditions (Figure 9).

Table II. Optically active piperazinones and lactams by kinetic resolution

Entry	R^2	(2R,6S)-29 Yield (%)[1]	(2S,6S)-29 Yield (%)[1]	30 Yield (%)[1]
A	*sec*-butyl	23	0	19
B	MeSCH$_2$CH$_2$	18	0	20
C	(indolyl-3-yl)CH$_2$	21	2	22

[1]overall yield (3 steps).

The bromoacetyl **31** adds another point of diversity, as the bromine can be displaced in good yields by various amines, and the resulting substitution products undergo immediate cyclization under basic conditions. When aliphatic

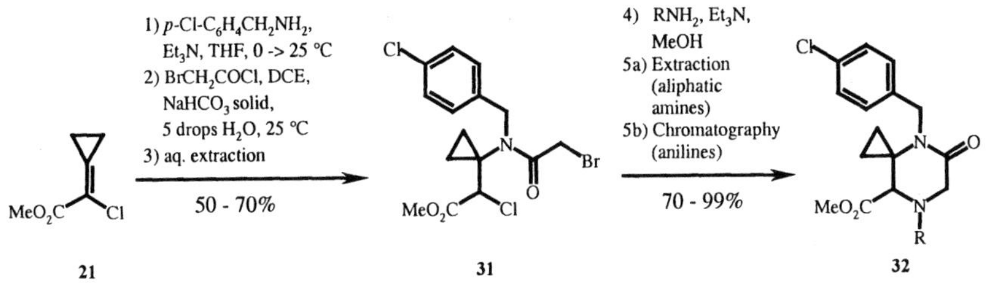

Figure 9. Piperazinone formation under modified Schotten-Baumann reaction.

amines are used in these last steps, the products are obtained in high yields (70 – 87% overall) and high purity by simple extraction. For anilines, the final product required chromatography (*19*). The flexibility at all three positions of the monocyclic peptidomimetic, high yielding steps and simple work-up made the combinatorial exploration of biologically relevant motifs of peptides possible.

The build-up of the bicylic peptidomimetics (*18*), follows the same sequence as that used for the monocyclic compounds followed by, coupling of the piperazinone **33** with a Boc-protected amino acid, deprotection and thermally induced cyclization (Figure 10). With chiral amino acids used in the second step, in most cases separation of the two diastereomers by chromatography was possible (Table III).

The described peptidomimetics with an unprecedented structural flexibility, allows one to combinatorially address up to four points of diversity for the mono- or bicyclic systems, including also the control of all chiral centers.

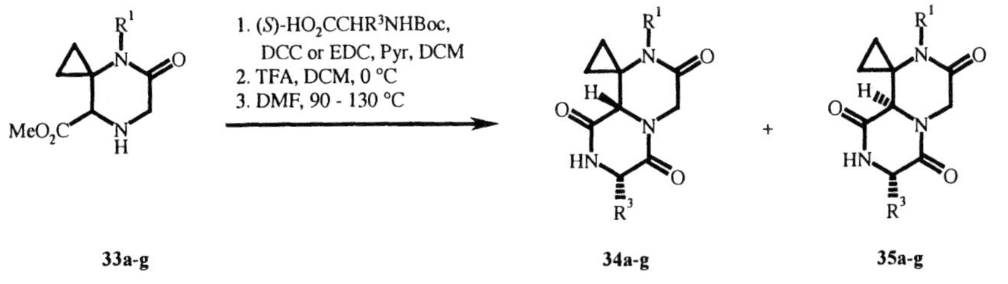

Figure 10. Sequence for the construction of bicyclic peptidomimetics.

Table III. Bicyclic peptidomimetics using chiral amines

Entry	R^1	R^3	34 + 35
A	n-pentyl	benzyl	50
B	n-pentyl	sec-butyl	38
c	PhCH$_2$CH$_2$	H	46
d	PhCH$_2$CH$_2$	(indol-3-yl)CH$_2$	56
e	(furan-2-yl)CH$_2$	CH$_2$CO$_2$Bn	18 + 24
f	4-MeO(C$_6$H$_4$)CH$_2$	CH$_2$CO$_2$Bn	23 + 25
g	4-Cl(C$_6$H$_4$)CH$_2$	CH$_2$CO$_2$Bn	24 + 29

Current work is focussed on the application of this methodology on agro-relevant peptides.

Conclusion

In the beginning, combinatorial chemistry was preferentially used for expanding the compound repository made necessary after introduction of high throughput screening. Today the focus is shifting toward hit validation and optimization. This requires increasingly more complex methodologies for precisely defined libraries. Some of them are much more sensitive to specific reaction conditions or impurities in the reagents. In addition, the demand for SAR relevant reagents in sufficient quantities and purities is growing. The increasing chemical complexity has been reflected by a higher degree of technical flexibility, e.g. by linking all available instrumentation to one IT platform. This allows the chemist to switch, whenever necessary, to the most appropriate technology. A similar approach has been followed with Pythia, which made all relevant internal chemical and biological databases accessible by one web-based interface, using a simple ISIS Draw® query language. Results, including physico-chemical properties and activity predictions, are reported in a user-friendly Excel® format. Both are critical to obtain, fast and in a simple way, the informations needed for precise library design. In a second approach the library design comes from agro-relevant peptides, incorporated in a cyclopropylideneacetate based peptidomimetic. With four points of diversity accessible by combinatorial chemistry and the control of stereochemistry, this tool will help to analyse the conformational space covered by the peptide fragment, responsible for biological activity, more systematically. Adding it all up, combinatorial chemistry becomes much more complex, emphasising quality over quantity. The challenge for the combinatorial chemist is to act faster and more specifically. But he will be rewarded with simply more active and better compounds.

Acknowledgements

We would like to express our sincere thanks to our colleagues from the engineering, the analytical and the modelling departments at Bayer CropScience for supporting the described work.

References

1. Monkovic, I.; Willner, D.; Adam, A. A.; Brown, M.; Crenshaw, R. R.; Fuller, C. E.; Juby, P. F.; Luke, G. M.; Matiskella, J. A.; Montzka, T. A. *J. Med. Chem.* **1988**, *28*, 1548–1557.
2. [a] Mokrosz, J. L.; Bojarski, A. J.; Charakchieva-Minol, S.; Duszynska, B.; Mokrosz, M. J.; Paluchowska, M. H. *Arch.Pharm.* **1995**, *328*, 604-608; [b] Augelli-Szafran, C. E.; Purchase, T. S.; Roth, B. D.; Tait, B.; Trivedi, B. K.; Wilson, M.; Suman-Chauhan, N.; Webdale, V. *Bioorg. Med. Chem. Lett.* **1997**, *7*, 2009–2014.
3. [a] Katritzky, A. R.; Lan, X. F.; Yang, J. Z.; Denisko, O. V. *Chem.Rev.* **1998**, *98*, 409–548; [b] Fries, K.; Empson, J. *Liebigs Ann.Chem.* **1912**, *389*, 345–367; [c] Schiemann, K.; Showalter, H. D. H. *J. Org. Chem.***1999**, *64*, 4972–4975.
4. Sanger, F. *Biochem. J.* **1949**, *45*, 563–574.
5. Lormann, M. E. P.; Walker, C. H.; Es-Sayed, M.; Bräse, S. *Chem. Commun.* **2002**, 1296–1297.
6. Briggs, G. G. *SCI Meeting*, **1997**, „Predicting uptake & movement of agrochemicals from physical properties".
7. [a] Guram, A. S.; Rennels, R. A.; Buchwald, S. L. *Angew. Chem. Int. Ed.* **1995**, *34*, 1348; [b] Willoughby, C. A.; Chapman, K. C. *Tetrahedron Lett.* **1996**, *37*, 7181–7184; [c] Wolfe, J. P.; Tomori, H.; Sadighi, J. P.; Yin, J.; Buchwald, S. L. *J. Org. Chem.* **2000**, *65*, 1158–1174.
8. [a] Bräse, S.; Enders, D.; Köbberling, J.; Avemaria, F. *Angew. Chem.* **1998**, *110*, 3614–3616; [b] Dahmen, S.; Es-Sayed, M. (Bayer CropScience), *Unpublished data*, **1987**.
9. Wolfe, J. P.; Buchwald, S. L. *Angew. Chem.* **1999**, *111*, 2570–2573; *Angew. Chem. Int. Ed.* **1999**, *38*, 2413–2416.
10. Bartra, M.; Romea, P.; Urpí, F.; Vilarrasa, J. *Tetrahedron* **1990**, *46*, 587–594.

11. Lane, C. F. *Synthesis* **1975**, 135.
12. [a] Luesch, H.; Yoshida, W. Y.; Moore, R. E.; Paul, V. J. *Tetrahedron* **2002**, *58*, 7959–7966; [b] Marquez, B. L.; Watts, K. S.; Yokochi, A.; Roberts, M. A.; Verdier-Pinard, P.; Jimenez, J. I.; Hamel, E.; Scheuer, P. J.; Gerwick, W. H. *J. Nat. Prod.* **2002**, *65*, 866–871; [c] Cetusic, J. R. P.; Green III, F. R.; Graupner, P. R.; Oliver, M. P. *Org. Lett.* **2002**, *4*, 1307–1310.
13. [a] Strobel, G. A.; Morrison, S. L.; Cassella, M. WO 02/091825 A2, **2002**; [b] Shemyakin, M. M.; Ovchinnikov, Y. A.; Antonov, V. K.; Kiryushkin, A. A.; Ivanov, V. T.; Shchelokov, V. I.; Shkrob, A. M. *Tetrehedron Lett.*, **1964**, *1*, 47–54.
14. [a] Chiba, H.; Agematu, H.; Sakai, K.; Dobashi, K.; Yoshioka, T. *J. Antibiotics* **1999**, *52*, 710–720; [b] Kawato, H. C.; Nakayama, K.; Inagaki, H.; Nakajima, R.; Kitamura, A.; Someya, K.; Ohta, T. *Org. Lett.* **2000**, *2*, 973–976; [c]. Kawato, C.; Nakayama, K.; Inagaki, H.; Ohta, T. *Org. Lett.* **2001**, *3*, 3451–3454.
15. [a] de Meijere, A.; Kozhushkov, S. I.; Hadjiarapoglou, L. P. *Top. Curr. Chem.* **2000**, *207*, 149–227; [b] Liese, T.; Teichmann, S.;. de Meijere, A. *Synthesis* **1988**, 25–32; [c] Seyed-Mahdavi, F.; Teichmann, S.; de Meijere, A. *Tetrahedron Lett.* **1986**, *27*, 6185–6188; [d] Nötzel, M. W.; Tamm, M.; Labahn, T.; Noltemeyer, M.; Es-Sayed, M.; de Meijere, A. *J. Org. Chem.* **2000**, *65*, 3850–3852; [e] de Meijere, A.; Teichmann, S.; Yu, D.; Kopf, J.; Oly, M.; v. Thienen, N. *Tetrahedron* **1989**, *45*, 2957–2968; [f]. Wessjohann, L.; Krass, N.; Yu, D.; de Meijere, A. *Chem. Ber.* **1992**, *125*, 867–882; [g] de Meijere, A.; von Seebach, M.; Kozhuskov, S. I.; Boese, R.; Bläser, D.; Cicchi, S.; Dimoulas, T.; Brandi, A. *Eur. J. Org. Chem.* **2001**, 3789–3795.
16. [a] Es-Sayed, M.; Heiner, T.; de Meijere, A. *Synlett* **1993**, 57–58; [b] Es-Sayed, M.; Gratkowski, C.; Krass, N.; Meyers, A. I.; de Meijere, A. *Tetrahedron Lett.* **1993**, *34*, 289–292
17. Belov, V. N.; Funke, C.; Labahn, T.; Es-Sayed, M.; de Meijere, A. *Eur. J. Org. Chem.* **1999**, 1345–1356.
18. For similar peptidomimetics: [a] Kim, H.-O.; Nakanishi, H.; Lee, M. S.; Kahn, M. *Org. Lett.* **2000**, *2*, 301–302; [b] Golebiowski, A.; Klopfenstein, S. R.; Shao, X.; Chen, J. J.; Colson, A.-O.; Grieb, A. L.; Russel, A. F. *Org. Lett.* **2000**, *2*, 2615–2617.
19. Limbach, M.; de Meijere, A.; Es-Sayed, M. (Bayer CropScience), *Unpublished results*, **2003**.

Chapter 8

Synthesis Based on Affinity Separation: A New Methodology for High-Throughput Synthesis Using Affinity Tags

Koichi Fukase, San-Qi Zhang, Yoshiyuki Fukase, Naomi Umesako, and Shoichi Kusumoto

Department of Chemistry, Graduate School of Science, Osaka University, Machikaneyama 1-1, Toyonaka, Osaka 560-0043, Japan

A new "tag" strategy, termed "synthesis based on affinity separation (SAS)", was developed for high throuput synthesis of organic compounds. In this method, the desired tagged compound was separated from the reaction mixture by solid-phase extraction using specific molecular recognition. The interaction between a crown ether (32-crown-10) and ammonium ion and the interaction between a barbituric acid derivative and its artificial receptor were used for SAS.

"Phase tagging" methodologies have been developed as a hybrid method that combines the merits of solid phase synthesis (easy separation) and solution phase synthesis (homogeneous reaction conditions). In this methodology, a compound having a tag group is easily separated from untagged molecules. Various phase tags such as soluble polymers, fluorous, hydrophobic, and basic tags have been reported (*1, 2*). We have developed a new "tag" strategy, termed "synthesis based on affinity separation (SAS)", in which the desired tagged compound was separated from the reaction mixture by solid-phase extraction using specific molecular recognition (Fig. 1).

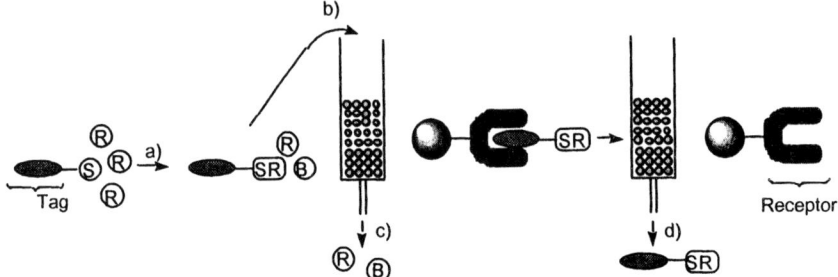

Fig. 1. Synthesis based on affinity separation. S: substrate, R: reagent, SR: product, B: byproduct. a) reaction in solution, b) adsorption of tagged product, c) elution of excess reagent and untagged byproduct, d) desorption of tagged product.

SAS Using Crown Ether (32-Crown-10) and Ammonium Ion Interaction

We first employed the interaction between a crown ether (32-crown-10) and ammonium ion for SAS (**Method A**) (*3*). Aminomethylated polystyrene resins were used as a column stationary phase. A highly cross-linked macroporous resin [ArgoPore^{TM}-NH$_2$ (amine loading: 1.39 mmol/g)] gave better separation than the usual 1% cross-linked gel-type resin. The desired compounds possessing the crown ether moiety as a tag were readily isolated from the reaction mixture by the following procedure. After each reaction cycle, the reaction mixture was applied to the aminomethylated polystyrene column [trifluoroacetic acid (TFA) form]. By using nonpolar eluents such as CH_2Cl_2 and toluene, the tagged compound was selectively adsorbed on the column, whereas other untagged impurities were washed off. Subsequent desorption by CH_2Cl_2-Et_3N or CH_2Cl_2-MeOH (1:1) afforded the desired compound with high purity.

We applied this method to the synthesis of peptides and heterocyclic compounds. After removal of the Fmoc group of **2**, reaction with benzaldehyde in 1% TFA / CH_2Cl_2 gave the β-carboline derivative **3** *(4)*. The reaction mixture was applied to an ArgoPore^{TM}-NH_3^+•$CF_3CO_2^-$ column. The desired **3** was adsorbed on the resin, whereas other byproducts and excess reagents were removed by elution with CH_2Cl_2. Compound **3** was then eluted with CH_2Cl_2-MeOH (1:1). The crown ether moiety of **3** was then removed by transesterification. After usual work-up, the mixture of **4** and the crown ether **1** in CH_2Cl_2 was applied to the ArgoPore^{TM} - NH_3^+•$CF_3CO_2^-$ column. Compound **4** passed through the column, whereas the crown ether **1** was adsorbed on the column. After evaporation, **4** was obtained in 80.2% yield. Crown ether **1** was then recovered by elution with CH_2Cl_2-MeOH (1:1).

a: 5% piperidine in DMF, 8 min; b: PhCHO (2eq), 1% TFA in CH_2Cl_2, 24 h; c: ArgoPore^{TM}- NH_3^+•$CF_3CO_2^-$ column; d: elution with CH_2Cl_2-MeOH (1:1); e: 0.2 M MeONa/MeOH, 10 min; f: 4-fluoro-3-nitrobenzoic acid (1,25eq), DIC (2.5eq), DMAP in CH_2Cl_2; g: i) N-phenylpiperazine (2 eq.) in NMP, ii) Ac_2O.

Diarylpiperazine **7** was prepared via aromatic nucleophilic substitution *(5, 6)*. 1-Methyl-2-pyrrolidone (NMP), which reduces the interaction of the crown ether and the ammonium ion, was used as a solvent at the nucleophilic substitution step. After excess *N*-phenylpiperazine was quenched with Ac$_2$O, the NMP solution was diluted with CH$_2$Cl$_2$ and then subjected to the affinity separation. Purification of diarylpiperazine **6** was thus effected in the same manner without the additional step for the removal of NMP. Transesterifaction and isolation by using ArgoPore^{TM} column gave **7** in total 88.6% yield.

Since the crown ether **1** is not commercially available, we then investigate the use of commercial available short chain polyethylene glycol (PEG) in place of the crown ether. We first employed Triton X-100 as a tag. Since Triton X-100 is a detergent having a PEG chain and hydrophobic moiety, we expected that tagged compounds show good solubility in many organic solvents. The desired compounds possessing the Triton X-100 tag were separated from the reaction mixture in a manner similar to the SAS using crown ether.

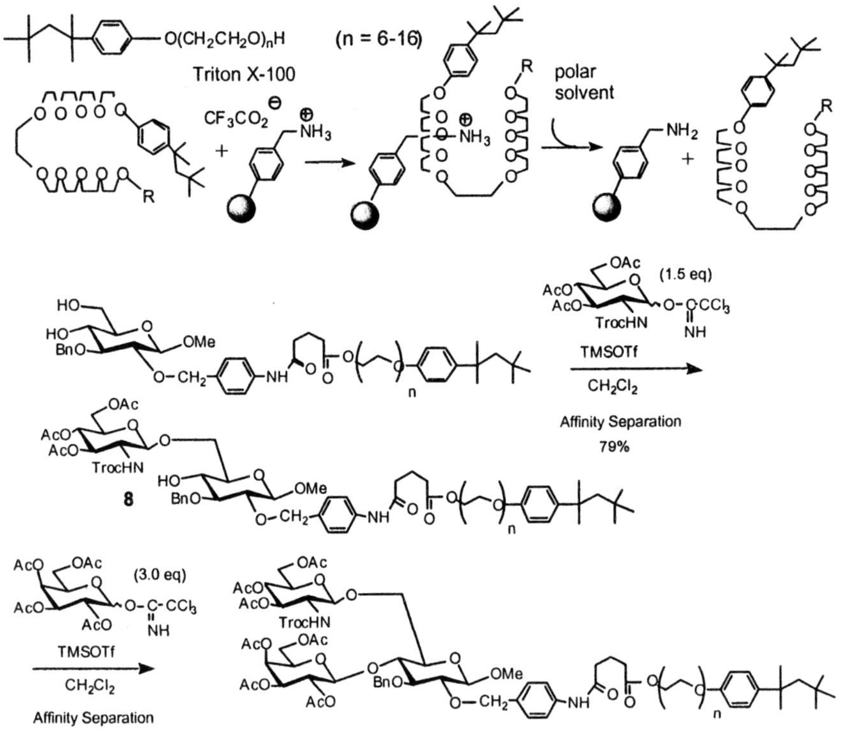

We next applied this method to the synthesis of oligosaccharides. One of the examples is shown above. The binding of Triton X-100 to the polymer-supported ammonium is somehow weaker than that of the crown ether. In addition, the chain length of Triton X-100 is heterogeneous and Triton X-100 with the shorter chain length only showed weak binding. The recoveries of the disaccharide **8** and the trisaccharide **9** were therefore not quantitative but practical enough. The present method enabled the use of ether as a reaction solvent and monitoring the reaction with TLC, both of which are difficult by the method using usual unmodified PEG tag.

SAS Using Interaction of Barbituric Acid and Its Artificial Receptor

We also found interaction between a barbituric acid derivative and its artificial receptor can be used for affinity purification (**Method B**) (*7*). By using nonpolar solvents, the compound possessing the barbituric acid tag was selectively adsorbed on the column, whereas other impurities without the tag such as excess reagents and byproducts were washed off. Subsequent desorption with CH_2Cl_2-MeOH (1:1) afforded the desired compound with high purity. This method was applied for the synthesis of peptides, heterocycles, and oligosaccharides.

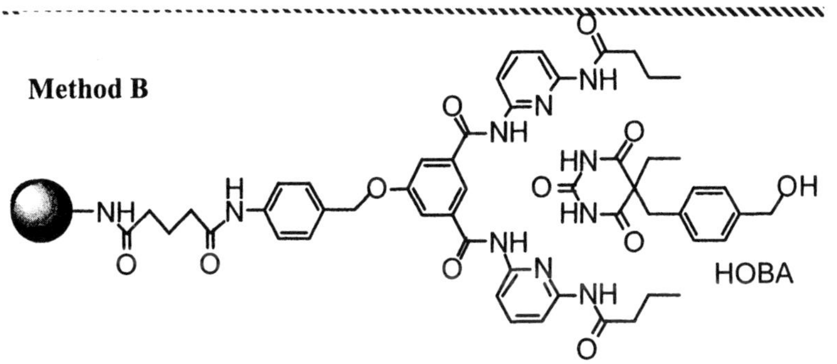

Method B

One of the examples for oligosaccharide synthesis was shown below. α-Selective glycosylation was effected by virtue of the solvent effect of ether using 2-*O*-benzylated thioglycosyl donors. Since the solubilities of glycosyl acceptors having the barbituric acid tag in ether were low, a mixture of ether-dioxane was used as the reaction solvent. The combination of PhIO and TMSOTf was employed as the promoter of thioglycoside (*8*).

a) $NiCl_2 \cdot 6H_2O$, $NaBH_4$, MeOH; b) DIC, CH_2Cl_2; c) affinity separation; d) PhIO, TMSOTf, Et_2O/dioxane; e) Zn-Cu couple, AcOH.

Glycosylation of the tagged acceptor **11** with excess phenyl 6-*O*-Troc-thioglycoside **12** (1.5 equiv.) gave disaccharide **13** (α:β =10:1). The 6-*O*-Troc group of **13** was removed and the resulting 6-*O*-free disaccharide **14** was then glycosylated with **12** to give the trisaccharide **15** (28.7% from **11**). After each reaction step, the product was rapidly separated from the reaction mixture by the affinity separation. Since the acylaminobenzyl linker was partly cleaved at the glycosylation step, the final product **15** was purified by silica-gel column chromatography after the affinity separation.

Lipid A is the hydrophobic lipid part of lipopolysaccharide (LPS), which is the immunostimulating glycoconjugate of Gram-negative bacteria. We previously clarified that the lipid A is the bioactive principle of LPS by the total synthesis of *Escherichia coli* lipid A. In order to investigate its biological function in relation to the chemical structure, we have investigated the efficient synthesis of lipid A and its analogues. A new and efficient synthesis of lipid A was hence achieved by application of the SAS concept (**Method B**) (*9*).

Glycosylation of a glycosyl acceptor **18** possessing the BA-tag with a 4'-phosphorylated *N*-Troc glucosaminyl trichloroacetimidate **20** gave the disaccharide 4'-phosphate **21**, which was purified by the affinity separation.

Acyl groups were then introduced to the disaccharide step by step. Acylation of the 3-position of **20** with (R)-3-(4-trifluoromethylbenzyloxy)tetradecanoic acid (=R^{1}COOH) proceeded smoothly by using DIC and DMAP to give **21**. Cleavage of 2'-N-Troc group of **21** with Zn–Cu couple was followed by N-acylation with (R)-3-(dodecanoyloxy)tetradecanoic acid (=R^{2}COOH) and DIC.

The third acyl group ((R)-3-(4-trifluoromethylbenzyloxy)tetradecanoic acid: R^3COOH) was introduced to the 2-amino group after cleavage of the Fmoc group of **23** followed by *N*-acylation using DIC. The 3'-*O*-Alloc group of **24** was then removed with Pd(PPh$_3$)$_4$, HCO$_2$H, and butylamine. Final acylation to the 3'-position with (R)-3-(tetradecanoyloxy)tetradecanoic acid (R^4COOH) was carried out by using DIC and DMAP to give fully acylated disaccharide **25**.

After each reaction step, the product was rapidly separated from the reaction mixture by the affinity separation. Though with less affinity than the BA-tag, DMAP was unexpectedly also retained to the column of BA-receptor so that the fractions of BA-tagged products **22** and **25** were inevitably contaminated with DMAP even after affinity separation. Successive silica-gel short column chromatography was, therefore, used for the complete removal of DMAP.

The linker moiety was then removed. We previously reported that *p*-acylaminobenzyl ethers were selectively cleaved by DDQ oxidation in a rate comparable to the cleavage of an MPM group (*10*). DDQ oxidation of **25**, however, gave complex mixture of undesired debenzylated products formed by over-oxidation of many benzyl groups. Cleavage of the BA tag moiety in **25** was thus carried out in 20% TFA in CHCl$_3$ to give the desired **26** in 15% yield. The low yield of this step can be explained as follows. Byproducts possessing the BA moiety can not be removed by the affinity separation. All byproducts accumulated during multistep of SAS were removed after cleavage of the tag. The 1-*O*-allyl group of **26** was then removed to yield 1-hydroxy disaccharide **27**. Finally, the 1-*O*-α-phosphorylation was carried out via 1-*O*-lithiation and subsequent treatment with tetrabenzyl diphosphate in THF at −78°C. The protected 1,4'-bisphosphate thus obtained was purified by silica-gel column chromatography. Subsequently, all the protecting groups were removed by hydrogenolysis (10 kg cm^{-2} of H$_2$) with Pd-black in one-step to give *E. coli* lipid A **28** in 75% yield.

We have demonstrated the importance of acyl groups to the biological activity of lipid A. For example, tetraacylated biosynthetic precursor lipid A having four (*R*)-3-hydroxytetradecanoic acid (C14 acid) shows antagonistic activity in human, whereas tetraacylated having (*R*)-3-hydroxytdecanoic acid (C10 acid) does not show the activity (*11*). NMR and molecular dynamics simulation indicated that biosynthetic precursor lipid A took a particular conformation, whereas the short chain analogue took several conformations, giving a strong evidence that the acyl moieties play an important role in holding the overall conformation of lipid A. LPS receptors including toll like receptor 4 (TLR4) might recognized the particular conformation, though the hydrophobic interaction of lipid A and the receptors should be also important for the recognition.

A : *Rubrivivax gelatinosus* -type **B** **C**

D **E** **F**

In the present study, we synthesized six analogues having different acylation distribution in order to investigate the effect of the acylation distribution to the biological activity. All the target structures possess hexaacyl groups: each has two (*R*)-3-hydroxydecanoyl groups and two (*R*)-3-(dodecanoyloxy)decanoyl groups. Since the 1-*O*-carboxymethyl (CM) analogues had proved to exhibit indistinguishable activity with lipid A but are chemically stable (*12*), six CM analogues were synthesized by using SAS in the present study.

One of the target compounds, i.e., **A** corresponds to the structure proposed for *Rubrivivax gelatinosus* lipid A. Another analogue **C** has the same acylation pattern as *E. coli* lipid A 28. Biological test of **C** is expected to give additional information on the effect of the chain length on the activity. Other four analogues (**B**, **D-F**) have unnatural distribution of acyl groups on the same disaccharide.

In the above synthesis, a trichloroethoxycarbonyl (Troc) group was used for the key protection of the 2-amino group. A considerable amount of the undesired *N*-2,2-dichloroethoxycarbonylated compound was, however, formed during the reductive cleavage of the *N*-Troc group. We hence used the propargyloxycarbonyl (Proc) group in place of the Troc group. The Proc group is stable to neat TFA but can be readily cleaved by treatment with $Co_2(CO)_8$ and TFA via alkyne-Co complex (*13*). We also found the Proc group can be cleaved by Zn-AcOH, Pd(0)-Et_3SiH, or $[Ir(cod)(MePh_2P)_2]PF_6$ (Ir-complex) (*13b*).

Glycosylation of the acceptor **29** with the trichloroacetimidate donor **30** afforded the disaccharide **31**, which was purified by affinity separation. Successive removal of protective groups and introduction of acyl groups were then effected and the synthetic intermediate at each step was purified rapidly by the affinity separation. The final deprotection and cleavage of the tag by catalytic hydrogenolysis afforded the desired CM-analogues **A – F**. Compound **A** showed weak but definite cytokine inducing activity and compound **E** showed potent antagonistic activity. Other analogues did not show the activity. The biological tests clearly demonstrated the importance of acylation distribution to the biological activity. Acylation distribution of lipid A probably influences the molecular conformation.

As described, SAS was effectively applied to the synthesis of peptides, heterocycles, oligosaccharides, and glycoconjugates. These new strategies are expected to be useful for multiple parallel synthesis and combinatorial library preparation of complex molecules.

References

1. (a) Porco, J. A., Jr. *Comb. Chem. High Throughput Screen.* **2000**, *3*, 93-102, (b) Yoshida, J.; Itami, K. *Chem. Rev.* **2002**, *102*, 3693-3716.
2. Lepore, S. D. *Tetrahedron Letters* **2001**, *42*, 6437-6439.
3. Zhang, S.-q.; Fukase, K.; Kusumoto, S. *Tetrahedron Lett.* **1999**, *40*, 7479-7483.
4. Pictet-Spengler reaction on Wang resin: Mayer, J. P.; Bankaitis-Davis, D.; Zhang, J.; Beaton, G.; Bjergarde, K.; Anderson, C. M.; Goodman, B. A.; Herrera, C. J. *Tetrahedron Lett.* **1996**, *37*, 5633-5636.
5. Solid-phase synthesis of aryl and benzylpiperazines: Dankwardt, S. M.; Newman, S. R.; Krstenansky, J. L.; *Tetrahedron Lett.* **1995**, *36*, 4923-4926. In the case of the solid-phase aromatic substitution, it took 48 h at r.t. to complete the reaction of the polymer-supported fluoronitrobenzene with 1-phenylpiperazine (10 eq.).
6. Pan, P.-C.; Sun C.-M. *Tetrahedron Lett.* **1998**, *39*, 9505-9508.
7. Zhang, S.-Q.; Fukase, K.; Izumi, M.; Fukase, Y.; Kusumoto, S.; *Synlett*, **2001**, 590-596.
8. Fukase, K.; Kinoshita, I.; Kanoh, T.; Nakai, Y.; Hasuoka, A.; Kusumoto, S. *Tetrahedron* **1996**, *52*, 3897-3904. b) Fukase, K.; Nakai, Y.; Kanoh, T.; Kusumoto, S. *Synlett* **1998**, 84-86.
9. Fukase, Y.; Zhang, S.-Q.; Iseki, K.; Oikawa, M.; Fukase, K.; Kusumoto, S.; *Synlett*, **2001**, 1693-1698.

10. (a) Fukase, K.; Nakai, Y.; Egusa, K.; Porco, J. A.; Kusumoto, S. *Synlett* **1999**, 1074-1078. (b) Fukase, K.; Tanaka, H.; Torii, S.;, S. *Tetrahedron Lett.* **1990**, *31*, 389-392.
11. Fukase, K.; Fukase, Y.; Oikawa, M.; Liu, W.-C.; Suda, Y.; Kusumoto, S. *Tetrahedron* **1998**, *54*, 4033-4050.
12. Liu, W.-C.; Oikawa, M.; Fukase, K.; Suda, Y.; Kusumoto, S. *Bull. Chem. Soc. Jpn.* **1999**, *72*, 1377-1385.
13. (a) Fukase, Y.; Fukase, K.; Kusumoto, S. *Tetrahedron Lett.* **1999**, *40*, 1169-1170. (b) Fukase, Y.; Fukase, K.; Kusumoto, S. In *Peptide Science 1999*; Fujii, N., Ed.; The Japanese Peptide Society: Osaka, 2000: pp 93-96.

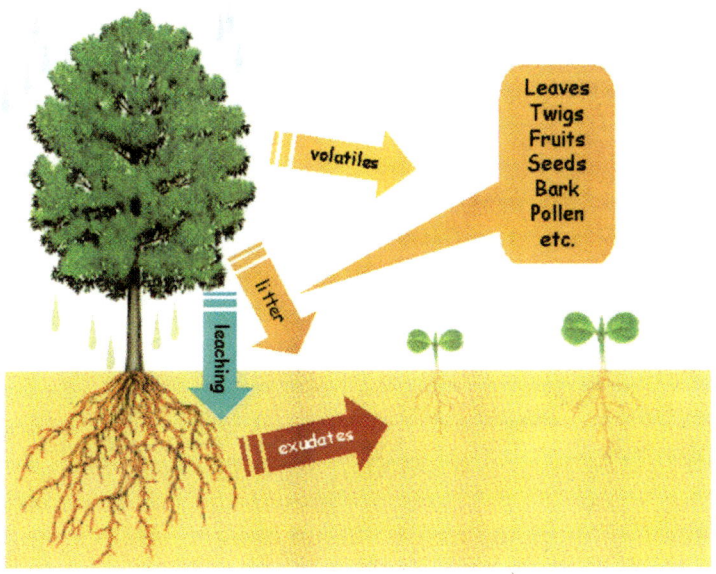

Plate 6.1. Possible Release Routes of Allelochemicals from Common Plants

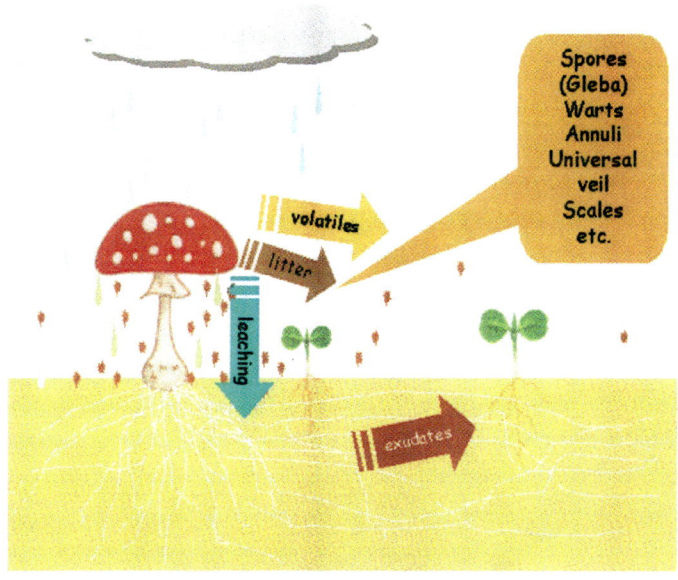

Plate 6.2. Proposed Release Routes of Allelochemicals from Mushrooms

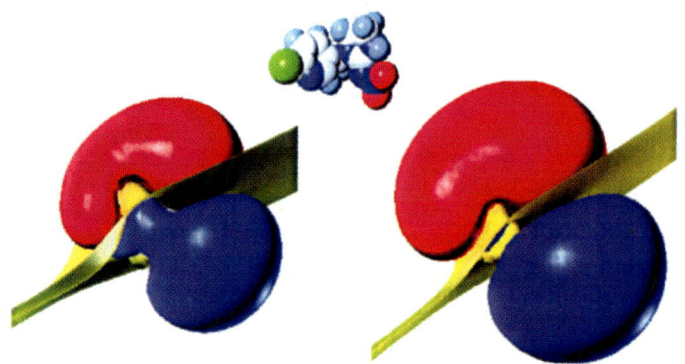

Plate 16.2. Contour plots for electrostatic potential of imidacloprid calculated by a semi-empirical molecular orbital method PM3 combined with the AMSOL programme. The plot on the left shows unpolarized imidacloprid, whereas that on the right shows the compound polarized by an ammonium located in the vicinity of the nitro group. Contours are plotted at +1 (red), zero (yellow) and -1 kcal/mol (blue). A space-filling model of imidacloprid is shown between the two plots. Reproduced from ref. 10 with permission of Elsevier.

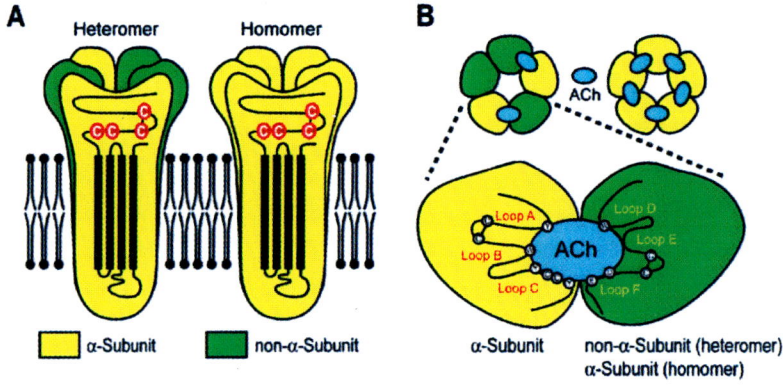

Plate 16.3. Schematic representation of nicotinic acetylcholine receptors. Side (A) and top (B) views of heteromeric and homomeric nicotinic receptors. Loops A-F forming the agonist binding site of heteromeric receptors and key amino acid residues that have been shown to contribute to the agonist binding are also illustrated in B. Abbreviation: ACh, acetylcholine. Reproduced from Ref. 10 with permission of Elsevier.

Chapter 9

Computer-Aided Library Design for Agrochemical Lead Generation: Design and Synthesis of Ring-Fused 2-Pyridinone Esters

James M. Ruiz and Beth A. Lorsbach

Discovery Research, Dow AgroSciences LLC, 9330 Zionsville Road, Indianapolis, IN 46268

Lead generation combinatorial libraries are valuable sources of new chemistry for the product pipeline. The purpose of these libraries is to produce molecules with activity and enough information to facilitate optimization. By utilizing ag-like and lead-like criteria, libraries are designed to have physical properties and structural features that favor activity and bioavailability as agrochemicals. Using this strategy, a ring-fused 2-pyridinone library was designed and synthesized for herbicide discovery. Substituents for the library were chosen based on herbicidal lead-like properties and input from scientists. Molecules with a cyclopropyl substituent produced herbicidal activity. As a lead generation effort, this library achieved its goal of producing actionable chemistry that warranted synthetic follow up.

Two of the major sources of novel and actionable chemistry for agrochemical discovery are external acquisitions from third parties and lead generation combinatorial libraries. Compounds from these sources are screened to identify those with properties and activity sufficient to serve as starting points for lead optimization. A major difference between the sources is that with the third party compounds, selections are limited to what are available. Often molecules have substructures or groups that are unsuitable for agrochemistry. With lead generation libraries, however, we can incorporate structures and properties into the molecules that we think are important for good agrochemical leads.

What are the features for agrochemical molecules? An important one is good bioavailability via absorption and distribution in the host or target organisms. Clearly there are different requirements for herbicides, insecticides and fungicides but some important physical properties affecting bioavailability include: acidity/basicity, lipophilicity, and aqueous solubility. We can predict and screen for these properties computationally, and discard molecules with poor properties before they are synthesized. For lead generation libraries, additional features are also important. Molecules must serve as starting points for optimization, so they must have positions which can be modified with a reasonable synthesis plan. Lastly, to increase the likelihood of success, molecules should also be designed based on a hypothesis that supports biological activity. Some examples of biological hypotheses include the use of privileged structures(*1*), pharmacophore matches with known target sites(*2*) and similarity matches with molecules with desired modes of action(*3,4*).

Ring-Fused 2-Pyridinone Library Scaffold

As an example of a lead generation effort, we describe the design and synthesis of a library with a ring-fused 2-pyridinone scaffold. The library was targeted primarily for Weed Management (WM) applications. The ring-fused 2-pyridinone scaffold had two desirable features for herbicides, an ester functionality that can be converted to an acid in the plant and a heterocycle with the potential for low lipophilicity. There was also precedence in the literature for molecules with this privileged scaffold to be biologically active. The 2-

pyridinone core was found in many molecules with medicinal uses such as antifungal(5), antibacterial(6,7,8), and anticancer agents(9,10). Therefore, these molecules had a good biological basis for activity and possessed desirable herbicide physical properties. Lastly, these molecules were under-represented in the company molecule collection.

A method of synthesis was recently reported for these compounds, using nitriles and Meldrum's acid derivatives as inputs(11,12). By starting from a known protocol, the time to create the library was greatly reduced. Additionally, several Meldrum's acid derivatives were available in-house from a previous combinatorial library.

Design Strategy

The strategy for designing lead generation libraries has changed recently in the pharmaceutical and agrochemical industries(13,14). Initially, large libraries were created with emphasis on maximizing the diversity of the library. The thought was that increasing the number of molecules screened would lead to more hits and leads. The results show that this strategy was not successful(15). Although many hits were generated, most failed to advance to leads because of poor bioavailability due to low aqueous solubility or membrane permeability. The poor physical properties were usually the result of molecules having relatively high molecular weights (MW) or lipophilicities (as measured by the octanol/water partition coefficient, log P)(16). Now, the focus is on smaller libraries with properties leading to good bioavailability ("drug-like" and "ag-like"). Simple rules such as Lipinski's "Rule of 5" (16), and Tice's limits for herbicides and insecticides(17) were developed to help guide the selection of better molecules and eliminate potentially problematic ones.

Another trend in the design of lead generation libraries is to make the molecules better starting points for lead optimization(18). If the initial molecules have relatively high MW and log P, it is very difficult to improve activity without adding groups that increase these properties and produce poor bioavailability. Therefore, libraries are now designed to create molecules with lower average MW and log P that allow for increases during optimization.

The goal of this library design was to define the set of reactants that would lead to ring-fused 2-pyridinones that 1) explore variations around the core scaffold ("diverse"), 2) have properties that are relevant to agrochemicals ("ag-like") and 3) serve as good starting points for lead optimization ("lead-like"). This design increased the quality of the molecules in the library by incorporating as many favorable features in the molecules as early as possible.

To choose a diverse set of molecules, we used the Cerius2 program(19) to "virtually" create all of the possible products from available reagents. For all

Table I. Herbicide Property Ranges for Commercial Products and the Ring-Fused 2-Pyridinone Ester Library

Properties	Commercial Products[a]	Lead-Like
MW	>150 & <500	>150 & <450
Alog P	≤ 5.0	≤ 4.0
Hbond donor	≤ 3	≤ 3
Hbond acceptor	≥ 2 & ≤ 12	≥ 2 & ≤ 11
Rotatable Bonds	≤ 12	≤ 11

[a]reference (17)

molecules, we calculated the default two-dimensional descriptors and properties including MW, Alog P (20), number of hydrogen bond acceptors and donors and number of rotatable bonds. A principal component analysis of the descriptors and properties was made to eliminate dependencies among the parameters. Six principal components were used for the diversity calculations. We used the "diversity-penalty" reagent selection method to pick molecules. Penalties occurred when molecules were chosen with calculated properties outside the range of lead-like herbicides. These limits were based on those found by Tice for commercial herbicides(17), but the MW and log P were reduced to foster molecules being better starting points for lead optimization (see Table I).

The lead-like limits were useful in eliminating compounds with undesirable physical properties. To select molecules with desirable agrochemical features (ag-like), we employed a computer expert system called CLASS (Chemical Library Acquisition Selection System)(21). Within this program are a set of rules developed by chemists and biologists to prioritize molecules for acquisition. For the design, we used the CLASS rules to virtually screen for molecules with the most desirable herbicidal features and properties.

The final component of the library selection was the manual review and modification of input lists by computational and WM chemists and biologist. Based on their feedback, reagents were substituted using either their suggestions, or another diversity selection. The cycle of diversity selection, CLASS screening and manual review continued until no further changes were required.

Synthesis

The synthesis begins with imino ethers, produced from nitriles, that are condensed with racemic cysteine methyl ester hydrochloride to give a set of substituted thiazolines (Scheme 1)*(11,12)*.

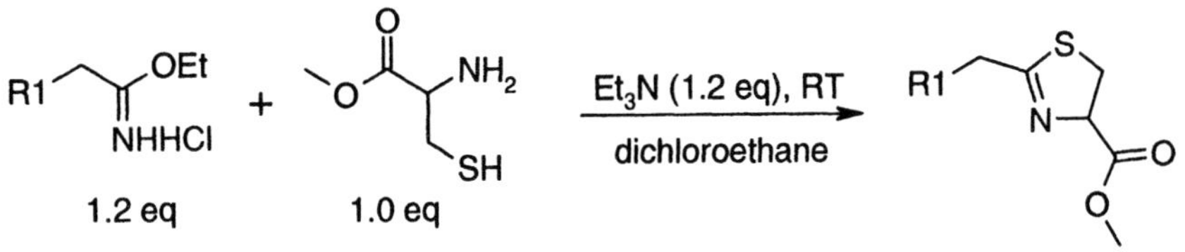

Scheme 1

The Meldrum's acid adducts (MAA) were prepared previously from a set of carboxylic acids and Meldrum's acid. The thiazolines and MAA were treated with 4 M HCl in dioxane and heated at 65 °C for 12h (Scheme 2). A major concern with any solution-phase library is purity. The major impurities in the first step were triethylamine salts, which were removed by filtering through diatomaceous earth. Analysis of the crude products from the cyclization step by NMR and LC MS confirmed product formation with little to no starting materials present. Filtration through diatomaceous earth pre-loaded with $NaHCO_3$ removed any salts and purity of final products ranged from 65-85% by LC MS. Library production required three days for each plate. The reactions were carried out in 4-dram vials in a J-Chem block on an orbital shaker which allowed for simultaneous heating and mixing.

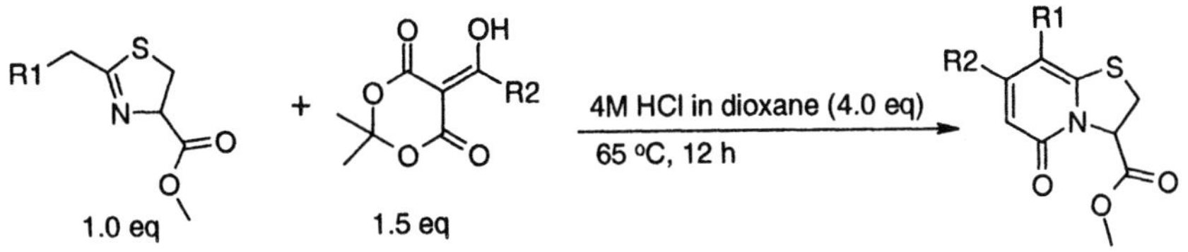

Scheme 2

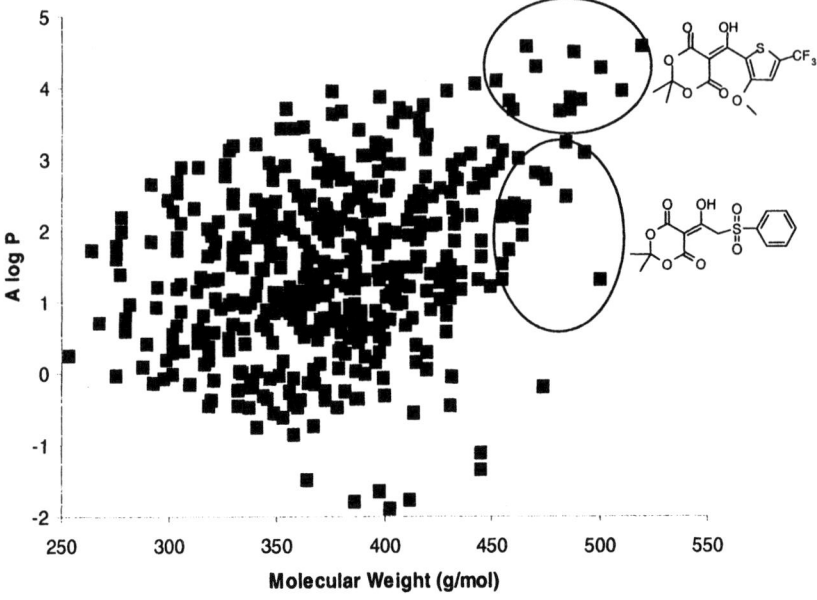

Figure 1. Calculated log P versus Molecular Weight for Candidate Virtual Library

Results and Discussion

The library inputs were selected in two steps. First, 16 MAA were chosen computationally from a list of 40 that were available in-house at Dow AgroSciences. The computer selections were reviewed by the product team biologist and chemists and a consensus list of 16 was chosen. Second, the MAA inputs were held fixed during computer diversity selection of nitrile inputs.

The candidate library was then screened against the CLASS expert system. We found the CLASS filter was useful in identifying unattractive inputs that would have been removed during the scientist reviews. For example, molecules with reactive groups or long alkyl chains received low scores and subsequently were removed from the candidate library. We made substitutions for these inputs and screened using CLASS until all molecules "passed".

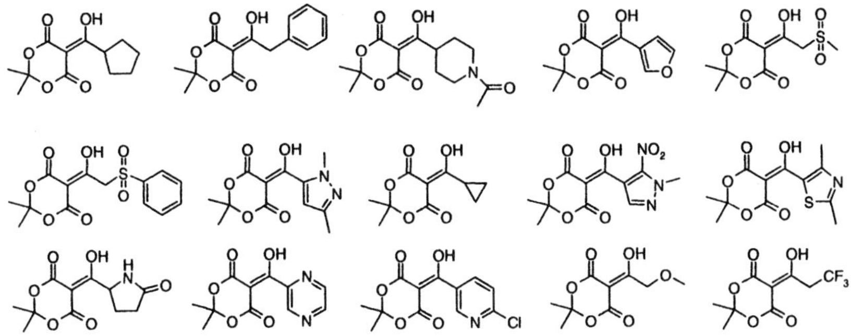

Figure 2. Library Meldrum's Acid Adduct Inputs.

We also found the manual reviews by scientists were important in defining a quality library. During the design, a significant number of molecules had MW and log P greater than the desired limits. The inputs were examined to determine which ones were responsible for the high values. Two inputs, a substituted thiophene MAA and a methyl phenyl sulfone MAA, accounted for most of the undesirable molecules (Figure 1). In particular, the thiophene MAA created both the highest MW and log P molecules in the candidate set. Since no specific biological hypothesis was to be tested with this input, it was removed from the library. In the case of the methyl phenyl sulfone MAA, it was found in other molecules with favorable properties, so it was kept in the library.

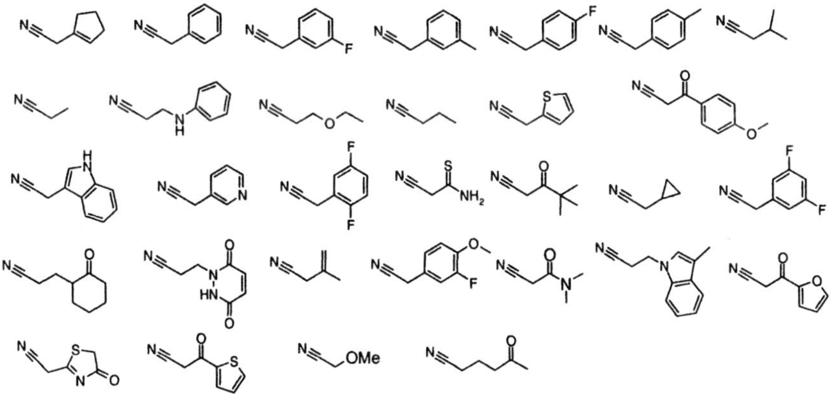

Figure 3. Library Nitrile Inputs.

The 15 Meldrum's acid derivatives and 31 nitriles selected for the final library are given in Figures 2 and 3. The majority of the molecules were predicted to have MW between 325 and 425 g/mol and log P values between 0.5 and 3.5, consistent with good lead-like and ag-like properties.

The library of 465 compounds was screened for activity in an HTS Arabidopsis assay. Thirteen active molecules were found in the library along with a clear SAR to follow. Eleven of these molecules had a cyclopropyl group at R2, and their activities were confirmed by retesting with purified samples. Based on these results, five molecules (see below) were advanced and tested in a post-emergent screen against multiple weeds.

Molecules with fluorophenyl groups at the R1 position, **1** and **5**, had activity across broad leaf and grass weed species (see Table II). The other analogs had no activity. These results were sufficient to warrant further synthesis in order to increase activity. Several molecules were made based on various hypotheses to enhance weed injury, but unfortunately, none of the attempts were successful in significantly improving performance. As a result, further work on these analogs ceased.

Although none of the attempts at optimization generated an herbicidal lead, the goal of the library as a source of new and valuable chemistry was achieved. The library had a 2.8% HTS pass rate, which is twice the historical pass rate for molecules that entered the HTS screen over a defined period (1.2%). More importantly, whole plant active molecules and trends were generated from this small library of 465 molecules that were sufficient to warrant further synthesis and exploration. This process also demonstrated the value of collaborating with the product goal teams in the design and analysis stages. Lastly, the library was designed and synthesized (including protocol development) in a short time (4 months). This demonstrated the ability of combinatorial chemistry to impact quickly the needs of the product teams.

Table II. Broad Leaf and Grass Herbicidal Activity

Molecule	Herbicidal Activity (% Visual Injury)[a]		
	Pigweed	Sunflower	Giant Foxtail
1	60	10	10
2	0	0	0
3	0	0	0
4	0	0	0
5	25	10	15

[a] data for post-emergent application at 4kg/ha

Conclusions

Lead generation combinatorial libraries are valuable sources of new chemistry for the product pipeline. Consistent with the approach used by the pharmaceutical industry, the DAS combinatorial chemistry group is focusing on creating quality libraries using both computer methods and collaboration with the product teams. Using CLASS and lead-like criteria, libraries are designed to produce molecules with desirable physical properties and structures. These features, coupled with a short time for synthesis, increase the success of a lead generation effort.

A ring-fused 2-pyridinone library was designed and synthesized for herbicide discovery. As a lead generation effort, this library achieved its goal of producing actionable chemistry that warranted synthetic follow up. Substituents for the library were chosen based on herbicide lead-like properties and input from product team scientists. The library produced molecules with herbicidal activity in broad leaf and grass weeds.

Acknowledgements

The authors would like to thank Marshall Parker, Paul Schmitzer and Fernando Valle for their assistance in this project.

References

1. Privileged structure is a substructure feature found in biologically active molecules. See for example, Lewell, X.Q.; Judd, D.B.; Watson, S.P.; Hann, M.M. *J. Chem. Inf. Comput. Sci.* **1998**, *38*, 511.
2. Beavers, M.P.; Chen, X. *J. Mol. Graphics Mod.* **2002**, *20*, 463.
3. Beno, B.R.; Mason, J.S. *Drug Disc. Today* **2001**, *6*, 251.
4. Poulain, R.; Horvath, D.; Bonnet, B.; Eckhoff, C.; Chapelain, B.; Bodinier, M.-C.; Déprez, B. *J. Med. Chem.* **2001**, *44*, 3378.
5. Cox, R.J.; O'Hagan, D. *J. Chem. Soc., Perkin Trans. 1* **1991**, 2537.
6. Casinova, C.G.; Grandolini, G.; Mercantini, R.; Oddo, N.; Olivieri, R.; Tonolo, A. *Tetrahedron Lett.* **1968**, 3175.
7. Dolle, R.E.; Nicolaou, K.C. *J. Am. Chem. Soc.* **1985**, *107*, 1691.
8. Rigby, J.; Balasubramanian, N. *J. Org. Chem.* **1989**, *54*, 224.
9. Wall, M.E.; Wani, M.C.; Cook, C.E.; Palmer, K.H.; McPhail, A.T.; Sim, G.A. *J. Am. Chem. Soc.* **1966**, *88*, 3888.
10. Comins, D.L.; Nolan, J.M. *Org. Lett.* **2001**, *3*, 4255.
11. Emtenäs, H.; Alderin, L; Almqvist, F. *J. Org. Chem.* **2001**, *66*, 6756.
12. Emtenäs, H.; Åhlin, K.; Pinkner, J.S.; Hultgren, S.J.; Almqvist, F. *J. Combin. Chem.* **2002**, *4*, 630.
13. Bajorath, J. *Nature Rev. Drug Disc.* **2002**, *1*, 882.
14. Bleicher, K.H.; Böhm, H.-J.; Müller, K.; Alanine, A.I. *Nature Rev. Drug Disc.* **2003**, *2*, 369.
15. Bajorath, J. *Drug Disc. Today* **2001**, *6*, 989.
16. Lipinski, C.A.; Lombardo, F.; Dominy, B.W.; Feeney, P.J. *Adv. Drug Del. Rev.* **1997**, *23*, 3.
17. Tice, C.M. *Pest. Manag. Sci.* **2001**, *57*, 3.
18. Teague, S.J.; Davis, A.M.; Leeson, P.D.; Oprea, T. *Angew. Chem. Int. Ed.* **1999**, *38*, 3743.
19. Cerius2 available from Accelrys, Inc., San Diego, CA.
20. Ghose, A.K.; Viswanadhan, V.N.; Wendoloski, J.J. *J. Phys. Chem. A* **1998**, *102*, 3762.
21. Pernich, D.J.; Dow AgroSciences, unpublished results.

Chapter 10

Parallel Synthesis Technologies in Lead Discovery and Optimization: Strategies and Applications

Robert J. Pasteris

Stine-Haskell Research Center, DuPont Crop Protection, P.O. Box 30, Newark, DE 19714

Combinatorial and parallel synthesis technologies have become an integral part of many discovery research organizations. This paper will discuss DuPont's entry into this field and how our strategies to capture the best value from these technologies have evolved. Applications to both lead discovery and lead optimization programs will be presented with examples taken from DuPont's fungicide, herbicide and insecticide programs. Technologies illustrated include parallel solid phase and solution phase synthesis, mixture synthesis, multi-component reactions and the design and use of combinatorial libraries. The use of contract research organizations as a way to leverage internal resources will also be discussed.

In the early 1990's, combinatorial chemistry and high throughput screening technologies were evolving rapidly and prompted new ways of thinking about discovery processes. DuPont Crop Protection was interested in exploiting these new technologies and partnered with our Central Research and Development organization to develop a core competency in these areas. Our approach was to explore and develop high throughput synthesis methods via this partnership and integrate selected technologies into our crop protection discovery programs via a combinatorial chemistry core team. The core team's focus centered on applications to hit and lead optimization and targeted discovery programs generating small, focused libraries of well characterized discrete compounds for *in vivo* testing. The team leveraged their capabilities by establishing and managing external collaborations to design and produce larger, diverse discovery libraries directed towards hit and lead generation.

By taking this approach, DuPont Crop Protection was able to assess a wide range of new chemistry technologies and their value to lead generation and optimization programs while avoiding large internal capital investment. This paper reviews a few selected examples of applications of these technologies highlighting solution phase, solid phase and mixture techniques, which helped speed the discovery process.

Scytalone Dehydratase Inhibitors via a focused mixture strategy

In order for *Magnaporthe grisea* to infect rice plants, it must melanize an infection structure to penetrate the leaf surface (1), thus making inhibition of fungal melanin biosynthesis an attractive approach for preventing rice blast disease. Scytalone dehydratase (SD) catalyzes the dehydration of scytalone and vermelone in this pathway (2). Figure 1 shows the structures of carpropamide (3) and diclocymet (4), two recently commercialized blasticides which inhibit this enzyme and compound **1**, a proprietary class of potent SD inhibitors discovered by DuPont (5).

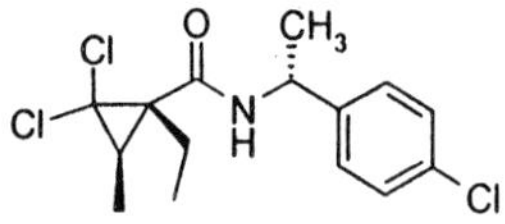

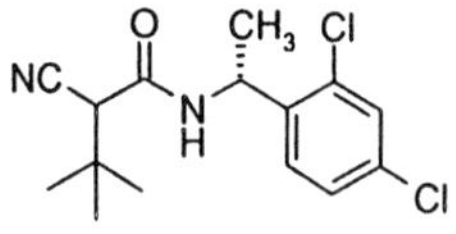

Figure 1. Selected scytalone dehydratase inhibitors

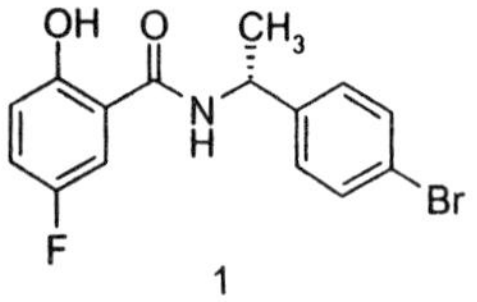

Inspection of these inhibitors show that they are all amides with the amine portion being highly conserved while the acid portion contains greater structural diversity. As a complement to DuPont's SD structure-based design program, a target-focused mixture library strategy was implemented where a diverse set of commercially available acid chlorides were reacted with an equimolar mixture of five optimized amine moieties known to be tight binders at the SD binding site.

The reactions were carried out in parallel using a filtration block and a basic amine resin to bind the HCl formed in the reaction. Filtration and concentration of each reaction well gave the desired equimolar mixture of five amides. The mixtures were assayed for their level of SD inhibition and the most active mixtures were deconvoluted by synthesis of the five individual amides and assayed to identify the inhibitor structures. Thus, 3,4-dichlorobenzoyl chloride was the acid component which gave the most active mixture in this study. Deconvolution showed that almost all the activity was derived from the 3,3-diphenylpropylamide component of that mixture. This compound **2** (Figure 2) was a 22 nM inhibitor of SD, but showed poor greenhouse level activity, most likely due to its high logP value. Replacement of the dichlorophenyl ring with the more hydrophilic dichloropyridine ring gave compound **3**, a 1 nM enzyme inhibitor with good greenhouse activity against rice blast disease.

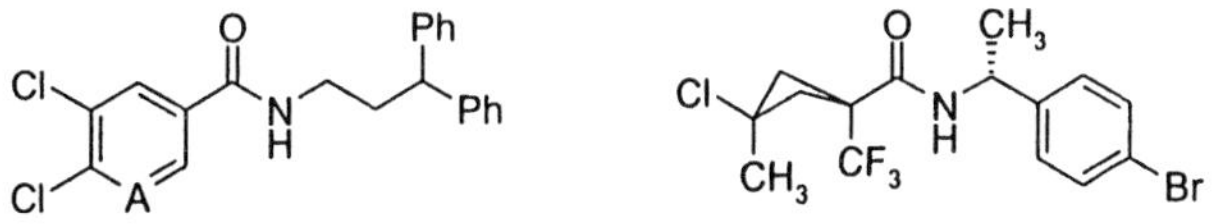

Figure 2. Selected SD inhibitors identified from a mixture strategy

The success of this mixture strategy with 131 commercially available acid chlorides prompted us to repeat this approach using DuPont's proprietary in-house acid collection from which a subset was selected based on knowledge of the nature of the binding site. Again, mixtures of five amides were prepared, assayed and deconvoluted to identify novel potent scytalone dehydratase inhibitors such as cyclobutane carboxamide **4**. Compound **4** exhibited a K_i of 0.026 nM against the enzyme (6) and had excellent greenhouse and field activity against rice blast disease.

As shown by these examples, the use of small focused mixture libraries was found to be a highly effective parallel synthesis approach by providing new, potent enzyme inhibitors while eliminating over 75% of the synthesis and testing effort that would have been required if each compound was prepared and tested individually.

Dihydropyridine Miticide Lead Optimization

The dihydropyridine **5** depicted in Figure 3 was purchased from a compound broker and found to be highly miticidal in greenhouse testing. This lead compound is prepared by a novel three-component reaction process involving an aniline, a ketone and the electron deficient olefin 1,1-dicyano-2,2-bis(trifluoromethyl)ethylene (BTF) (7). The chemistry is run most conveniently with two equivalents of BTF at room temperature in a water miscible solvent where addition of water precipitates the product. The first equivalent of BTF acts as a dehydrating agent to drive formation of an enamine intermediate, which adds across the double bond of the second BTF molecule, and ring closes onto one of the cyano groups to give the observed product.

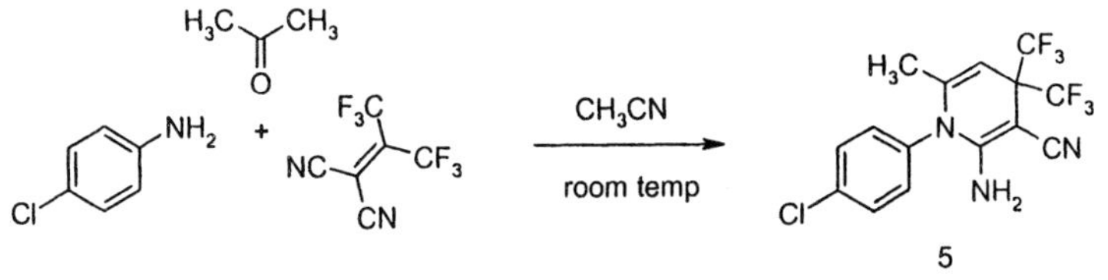

Figure 3. Synthesis of lead compound 5

The high level of diversity possible in each of the three components coupled with our desire to rapidly develop a structure activity relationship and identify potential field candidates led us to choose a positional scanning approach to narrow each reactant set. We would then do combinatorial crosses of the best amines with the best carbonyl components with the best olefins to identify the optimally active analogs. The synthesis was carried out in parallel using solution phase liquid handling methods and filtration blocks to collect the precipitated products.

Conceptually, the acetone component could be replaced with any carbonyl compound containing an adjacent methylene group which will undergo enamine formation. Carbonyl variations, which we found to readily undergo this reaction with 4-chloroaniline and BTF in dry acetonitrile, are aliphatic linear, branched and cyclic ketones, pyruvates, β-ketoesters, β-ketosulfones and 1,3-diones. Biacetyl did not provide product under these conditions. Aldehydes and acetophenones required pre-formation of the enamine intermediate prior to BTF addition and tended to proceed in lower yields. Eighty examples were chosen to test steric, electronic and lipophilic properties. In cases where the carbonyl component was unsymmetrical and could produce two isomeric enamines, multiple dihydropyridine products were formed.

4-Chloroaniline was replaced by one hundred and fifty ortho, meta and/or para substituted anilines using acetone as both the carbonyl component and the solvent. Again, substituents were chosen to test steric, electronic and lipophilic properties and all gave smooth product formation. Five- and six-membered heterocyclic amines were also explored and tended to give lower yields of less pure products. In some cases, the only product isolated derived from direct addition of the aminoheterocycle to BTF. Aliphatic amines failed in this reaction even using pre-formed enamines.

Replacement of the BTF component was less productive. Dicyanoethylenes derived from methyl or ethyl pyruvate do not form stable adducts with 4-chloroaniline in the presence of acetone (7). Dicyanoethylenes derived from trifluoroacetophenones or from chloral reluctantly gave addition products. Many other BTF replacements failed.

The structure activity relationship developed for these dihydropyridines showed that optimal activity was derived from BTF, using anilines containing a halogen, cyano or CF_3 group in the para position and small aliphatic and cyclic ketones and acetophenones.

Structures isoelectronic with the presumed enamine intermediate in this chemistry (7) were found to react readily with one equivalent of BTF in dry acetonitrile. As shown if Figure 4, amidines, guanidines, isoureas and isothioureas all gave the corresponding dihydropyrimidines at room temperature and showed good to excellent miticidal activity in greenhouse tests (8). Amide oximes required heating to form the BTF adduct and were not miticidal. Cyanoamidines would not condense with BTF.

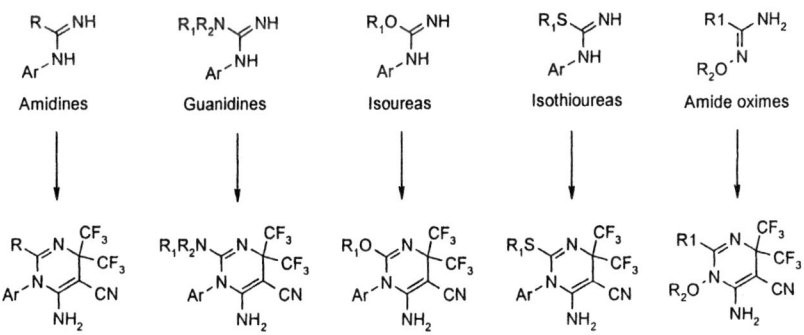

Figure 4. Dihydropyrimidines obtained via BTF condensations

The positional scanning approach and parallel synthesis methods used in this optimization program were found to be highly effective, allowing a wide range of structure variations to be explored and the structure activity relationships to be quickly defined. The best set of aniline components were reacted with the best set of carbonyl components using BTF as the olefin to give a combinatorial array of products from which the field candidates in Figure 5 were chosen and tested against a range of economically important mite species.

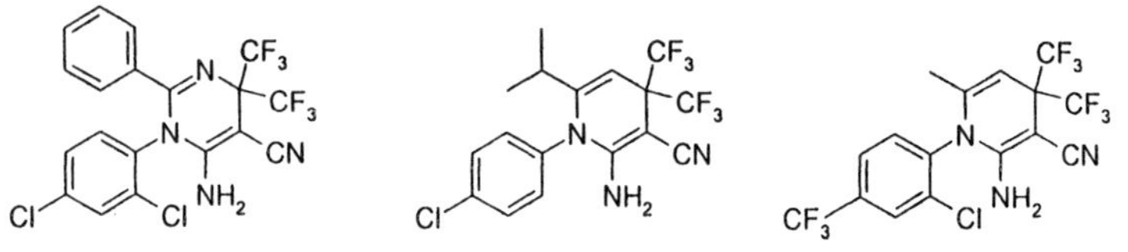

Figure 5. Insecticide field candidates.

N-Azoyl Phenoxypyrimidine Herbicide Lead Optimization

Carotenoid biosynthesis has long been a target for herbicide discovery with many commercial herbicides acting at various enzyme targets along this pathway. DuPont discovered a class of phytoene desaturase inhibitors where an azole was linked to a phenoxypyrimidine via a carbon-nitrogen bond (9). Compounds such as **6** (Figure 6) showed excellent preemergent and early postemergent herbicidal activity against broadleaf and grassy weeds with wheat safety (10). Both parallel solution phase and solid phase synthesis methods were used to help define the structure activity relationships in this active area.

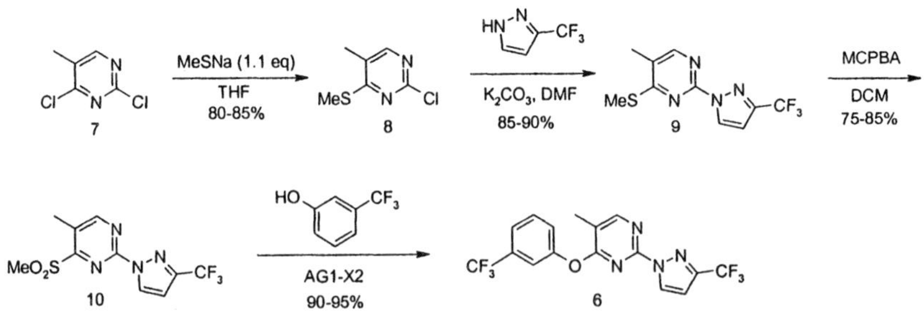

Figure 6. Solution phase synthesis of compound 6

As shown in Figure 6, a regioselective synthesis route was developed where the more reactive 4-chloro group of compound 7 was first replaced by a methylthio blocking group. Introduction of the trifluoromethyl pyrazole heterocycle in the 2 position proceeds smoothly. Oxidation of the methylthio group converts it into a methylsulfonyl group, which is readily displaced by nucleophiles, such as phenols, to give the desired compounds. The last step has conveniently been carried out in parallel by using a strongly basic ion exchange resin to give clean products by simple filtration. Eighty examples were prepared by this method.

This synthetic approach was modified to the traceless linker method outlined in Figure 7. A simple, inexpensive, high load thiol resin was easily prepared by heating a high load aminomethyl polystyrene resin 11 with γ–thiobutyrolactone for a few hours in toluene. The resulting thiobutyramide resin (TBA resin) (11) is a free-flowing, shelf-stable material which does not require protection of the thiol function and maintains its active thiol titer after a year of storage at room temperature in a sealed container.

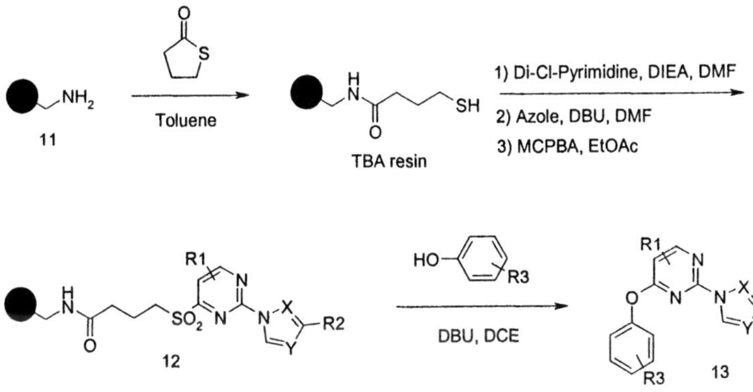

Figure 7. Solid phase synthesis method

Sequential reaction of the TBA resin with a 2,4-dichloropyrimidine and an azole followed by MCPBA oxidation gives the highly functionalized resin bound reactive intermediate 12. Reaction of 12 with phenols not only introduces the third diversity component, but also releases the final product 13 from the resin.

By using a combination of these solution and solid phase methods, the structure activity relationships were quickly developed and field candidate selection and patent exemplification was facilitated.

Oxazolidine Scaffold-Based Discovery Library

A proven method for discovery library synthesis is via a scaffold approach. The scaffold is a core structure or template upon which diverse functionality is appended. The scaffold should be small relative to the appended functionality and should be constructed by reliable chemistry from readily available inputs. Oxazolidines fit this profile -- they are compact, heterocyclic rings easily constructed by treatment of an aldehyde or ketone with an amino alcohol. The ring nitrogen can be further substituted by reaction with acid chlorides, isocyanates, etc., providing a highly functionalized molecule. An efficient, robust solution phase synthesis protocol was developed to prepare many compounds based on this scaffold in high yield and purity. Figure 8 outlines the synthesis used to prepare a 2500 compound oxazaspirodecane microtiter plate based sublibrary.

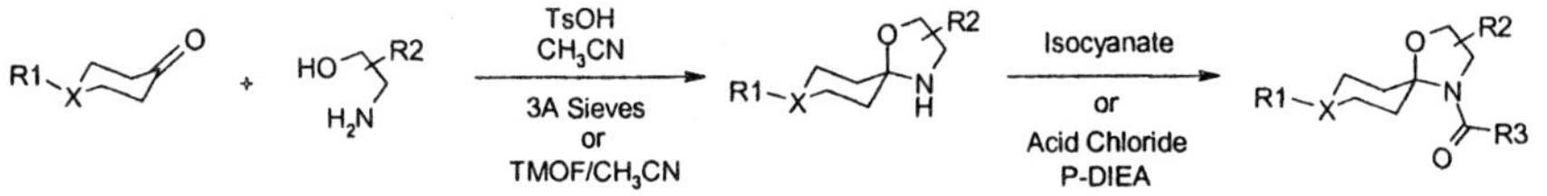

Figure 8. Oxazaspirodecane library synthesis scheme

Commercially available and DuPont proprietary inputs were assembled and subsets selected based on size and functionality considerations. A 262,000 member virtual library was enumerated and mapped into a multi-dimensional chemical property space. Final input selections were based on how well the corresponding library compounds fit DuPont's "ag like" property criteria, spanned the defined chemical property space of the virtual library and how well the inputs performed in synthesis validation studies.

The library was prepared using automated liquid handling methods and active compounds were discovered in all three disciplines -- herbicide, fungicide and insecticide. One of the library members, compound **14** depicted in Figure 9, was advanced to lead status based on its excellent preventative, curative and residual activity against wheat powdery mildew seen in our fungicide screens.

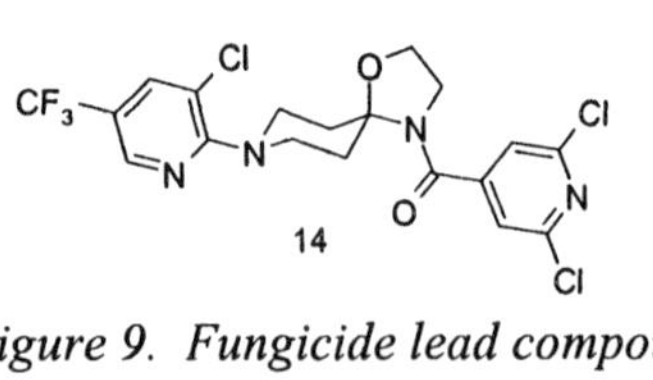

Figure 9. Fungicide lead compound.

While some discovery library design and synthesis is carried out internally, we leverage our capabilities by purchasing additional libraries externally. We acquire both off-the-shelf and DuPont-designed libraries, with vendor selection based on the science they offer and the experience they bring. Compound selection/deselection is highly computationally driven and includes diversity analysis, "ag like" property assessments, virtual screening and pattern recognition methods.

Conclusions

Parallel synthesis technologies are effective tools in the discovery of new crop protection chemistries. Use of solution phase, solid phase, supported reagents and scavengers in combination with positional scanning and mixture approaches were found to be effective in reducing synthesis and screening efforts and providing faster hit and lead assessments, more confident field candidate selections and broader support for patent filings. Leveraging our capabilities through external partners allowed better focus of internal resources and provided opportunity to assess a broader range of discovery technologies and approaches.

Acknowledgements

I wish to thank my colleagues at DuPont Crop Protection and DuPont Central Research who have contributed to and supported the work reviewed in this paper, especially Denis Amorose, Frank Coppo, Kevin Cottrell, Caleb Holyoke, Lee Jennings, John Kinney, Mike Kline, Boris Klyashchitsky, Kevin Lee, Pat Mauvais, Reza Nassirpour, Rand Schwartz, Tom Selby, Sejal Shah and Kirk Simmons.

References

1. Howard, R. J.; Ferrari, M. A. *Exp. Mycol.* **1989**, *13*, 403.
2. (a) Bell, A. A.; Wheeler, M. H. *Annu. Rev. Phytopath.* **1986**, *24*, 411. (b) Chumley, F. G.; Valent, B. *Mol. Plant-Microbe Interact.* **1990**, *3*, 135.
3. (a) Kurahashi, Y.; Sakawa, S; Kinbara, T.; Tanaka, K.;Kaguba, S. *J. Pesticide Sci.* **1997**, *22*, 108. (b) Tsuji, G.; Takeda, T.; Furusawa, I.; Horino, O.; Kubo, Y. *Pesticide Biochem. Physiol.* **1997**, *57*, 211.

4. (a) Manabe, A.; Mizutani, M.; Maeda, K.; Ooishi, T.; Takamo, H.; Kirime, O. US Patent 4,946,867 (**1990**). (b) Agrow, PJP Publications Ltd, UK, **1997**; Vol 287, pp 21-22.
5. Lundqvist, T.; Rice, J.; Hodge, C. N.; Basarab, G. S.; Pierce, J.; Lindqvist, Y. *Structure (London)* **1994**, *2*, 937.
6. Jennings, L. D.; Wawrzak, Z.; Amorose, D.; Schwartz, R. S.; Jordan, D. B. *Bioorg. Med. Chem. Lett.* **1999**, *9*, 2509.
7. Tyutin, V. Y.; Chkanikov, N. D.; Nesterov, V. N.; Antipin, M. Y.; Struchkov, Y. T.; Kolomiets, A. F.; Fokin, A. V. *Isv. Akad. Nauk, Ser. Khim.* **1993**, 552.
8. Frasier, D. A.; Holyoke, C. W.; Howard, M. H.; Lepone, G. E.; Powell, J. E.; Pasteris, R. J. (DuPont) PCT Int. Appl. WO 97 11,057 A1, 1997.
9. Selby, T. P. (DuPont) PCT Int. Pat. Appl. WO 98 40,379, 1998.
10. Selby, T. P., Drumm, J. E., Coats, R. A., Coppo, F. T., Gee, S. K., Hay, J. V., Pasteris, R. J., Stevenson, T. M. In *Synthesis and Chemistry of Agrichemicals VI*, Baker, D. R., Lahm, G. P., Selby, T. P., Stevensen, T. M. Eds.; ACS Symposium series, American Chemical Society, Washington, DC, 2001.
11. Coppo, F. T. (DuPont) US Patent 6,306,977, **2001**.

Chapter 11

A Combinatorial Synthesis Approach for Agrochemical Lead Discovery

James A. Turner, Michael R. Dick, Thomas M. Bargar, Gail M. Garvin, and Thomas L. Siddall

Dow AgroSciences Discovery Research, 9330 Zionsville Road, Indianapolis, IN 46268

Combinatorial chemistry was used as a tool for delivering 10 diverse libraries as one approach for increasing the number of compounds with *desirable* properties entering *in vivo* HTS screens. The libraries consisted of polar, weak acids – materials whose physical properties are known to confer phloem mobility, a desirable attribute for reliable herbicidal activity. Part of this effort was focused towards the preparation of a series of libraries based on pyridine carboxylic acids. Details of preparation of one such library of 5,6-disubstituted nicotinic acids are described. The screening inputs prepared using this targeted, attribute-driven combinatorial chemistry approach provided a nearly 5-fold higher pass rate than randomly selected inputs. Hits from three of these libraries showed sufficient levels of activity in higher level testing to warrant further synthesis.

INTRODUCTION

The role of combinatorial chemistry in biologically driven (agrochemical, medicinal) discovery research has undergone considerable evolution since the technology was first popularized in the early 1990s. The initial emphasis was directed towards construction of massive "diversity" libraries (thousands or even millions of compounds) for lead generation purposes, an approach that has now largely been abandoned. Instead, the current trend is towards preparation of small, highly focused libraries (dozens to hundreds of compounds) whose primary role is for use in lead optimization. *(1)* In the pharmaceutical industry, this change in strategy has been driven in large part by the realization that the rate-limiting step in the discovery process is not lead generation, but rather optimization of ADME (absorption, distribution, metabolism, excretion) properties to deliver candidates with the requisite pharmacokinetic and toxicological properties to survive the rigors of clinical testing. *(2)*

Yet agrochemical, unlike pharmaceutical, discovery still relies in large part upon *in vivo* assays to find chemical "leads", a situation that necessitates the continued availability of a large pool of novel chemical entities for screening. Whole organism activity, and at least to a certain extent, favorable pharmacokinetic profiles, are implicit in identified "leads". In addition, changes in the agrochemical marketplace have placed ever-increasing demands for better performance and profitability on each new product candidate. These demands can be met by either significantly increasing the total number of compounds available for random screening or by enriching the screens with compounds that have a higher potential for success.

This report describes the results of an attempt to use combinatorial chemistry to increase the number of compounds with *desirable* properties entering *in vivo* high throughput herbicide screens. In this "attribute-driven" approach to herbicide discovery, only those materials providing favorable ("optimized") physical properties, or other attributes, are selected as screening inputs (Figure 1). This differs from the traditional approach in which physical properties are optimized after detection of bioactivity. The expectations were two-fold: (1) that screening inputs with improved physical properties would stand a better chance of being detected in the screens, and (2) that actives detected from this new screening paradigm would serve as significantly better leads. The approach is illustrated in this report with an example of a novel pyridine carboxylic acid library that was selected and prepared because of the favorable properties of such materials. Screening results and conclusions are presented for this library as well as for a series of carboxylic and phosphonic acid libraries designed around a diverse group of templates.

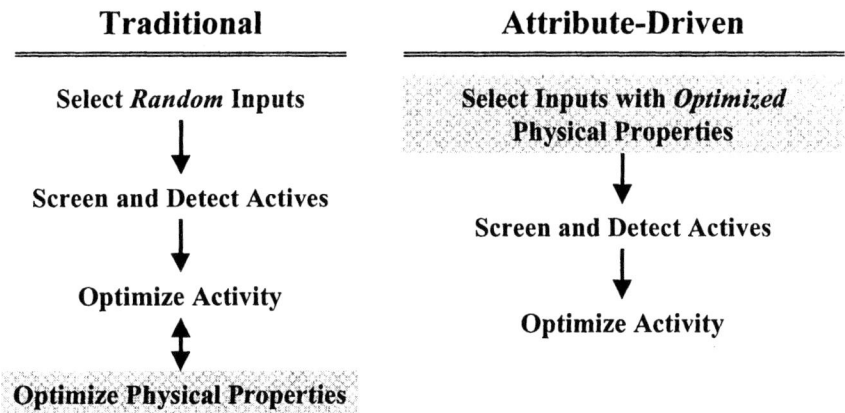

Figure 1. Traditional vs. Attribute-Driven Approach to Discovery

Designing Combinatorial Libraries for Herbicide Lead Discovery

In the "attribute-driven" approach to discovery, only those materials that have desired physical properties for a particular product goal are selected for screening. As an example, low use rates, pre- / postemergent flexibility, and reliable, robust weed control are desirable attributes in modern herbicides. Delivery of such attributes requires that the herbicide be available to the plant, through either (or both) leaf penetration or soil uptake, and that the compound then be mobile within the plant, readily moving through the phloem to the growing point. Physical properties that confer such phloem mobility attributes include weak acidity (pKa~3-6) combined with moderate lipophilicity (log P <3). *(3)* Finding phloem mobile leads by random screening is challenging for two reasons. First, the availability of screening inputs with physical properties suitable for phloem mobility (pKa~3-6) is low, representing less than 7% of available screening samples based on a recent in-house analysis. Second, experience has shown that adding acidic moieties to neutral lipophilic lead molecules almost always results in a significant reduction in bioactivity. This is hardly surprising, as random addition of such an extremely polar moiety to a neutral molecule would likely perturb the stereoelectronic features that allowed the neutral material to interact with the target protein.

For these reasons an effort was initiated to increase both the quantity and quality of herbicide screening inputs. Combinatorial chemistry was used as the tool for preparation of large numbers of compounds (11,528 compounds in 10 separate libraries) with desirable attributes (pKa <5; clog P <4.5). The largest library contained 2,112 compounds and the smallest 176. Members of the libraries were prepared as single entities using both solid phase as well as solution formats. An example of the selection, design, and preparation of one of these libraries is described below.

Preparation of 5,6-Disubstituted Nicotinic Acid Libraries

The physical properties of pyridine-containing acids make them particularly attractive starting points for combinatorial libraries for herbicide screens. Pyridines are relatively polar (log P = 0.60 for pyridine vs. 2.13 for benzene) *(4)* and the simple unsubstituted acids are weakly acidic (nicotinic acid pKa = 4.8). *(5)* Perhaps for these reasons, pyridine-containing acids have found considerable utility as herbicides. Representative examples of commercially successful, phloem mobile, pyridine acids (or their derivatives) include: picloram, clopyralid, and triclopyr (synthetic auxins); imazethapyr and nicosulfuron (ALS inhibitors); clodinafop and haloxyfop (ACCase inhibitors). Each has a direct counterpart in the corresponding benzene series but, in these instances, the pyridine analog is invariably the superior herbicide.

As a result of the asymmetry imparted by the nitrogen in the heterocycle, there are three simple pyridine carboxylic acids (CO_2H attached at C-2, C-3, and C-4 of the pyridine) as opposed to the single benzoic acid. Combining this asymmetry with additional pyridine ring substituents results in 10 isomeric mono-substituted and 30 disubstituted pyridine carboxylic acids (**1**, Figure 2). Consequently, there is the potential to generate numerous unique pyridine acid libraries through minor modifications of a single synthetic protocol resulting in considerable efficiency and savings in the time-consuming protocol development phase of library synthesis.

For these reasons part of this effort was focused towards preparation of a series of combinatorial libraries based on pyridine carboxylic acids such as **2** (Figure 2). Such libraries can nominally be prepared in two chemical steps from 6-fluoro-5-iodonicotinic acid (**3**). The fluorine serves as a facile leaving group for the installation of a variety of nucleophiles (Nu, amines in this library) while the iodine participates in one of a variety of Pd-catalyzed C-C or C-N (X) bond formation reactions. This design takes advantage of the inherent enhanced reactivity of pyridines (relative to benzenes) in each of these reactions *(6)* while strategic choice and location of the two halogens on the starting pyridine ring maximizes the rate and ensures the regioselectivity of each of the individual reactions.

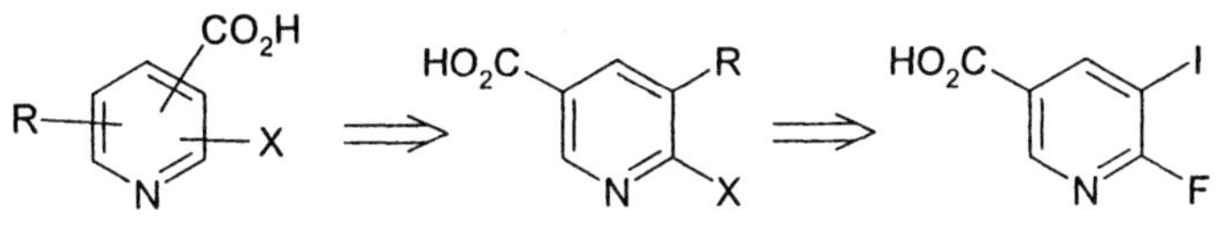

Figure 2. Pyridine Carboxylic Acids

The library was prepared on Wang resin, in polypropylene mesh containers (IRORI MiniKans[TM]), using the "mix-and-split" methodology. *(7)* Samples were tracked using encoded radio-frequency tags contained within the MiniKans. The reactions were run on the solid phase to take advantage of purification efficiencies as well as to allow the use of large excesses of reagent to drive each step to or near complete conversion. Optimized conditions to effect each of these transformations are described below (Figure 3).

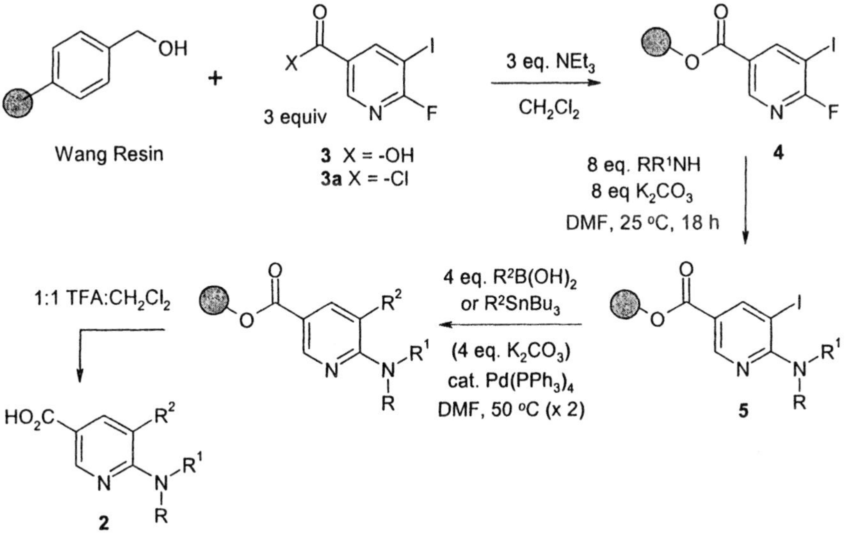

Figure 3. Nicotinic Acid Library Construction

Nicotinic acid **3** was attached to Wang resin through an ester linkage under typical acylation conditions. Thus, the resin was treated with 3 equivalents of the acid chloride prepared from **3** and excess triethylamine in dichloromethane at ambient temperature for 20 hours. Cleavage of an aliquot of this resin by treatment with a 1:1 mixture of trifluoroacetic acid and methylene chloride for 2 hours gave a quantitative recovery (based on the theoretical resin loading) of starting acid **3**; this material was essentially 100% pure as judged by HPLC and ^{1}H NMR analyses.

Resin bound fluoronicotinoate **4** was reacted with a variety of amines under mild conditions (8 eq. amine, 8 eq. K_2CO_3, DMF, 25 °C). Simple alkyl amines functioned as good nucleophiles in this reaction *(8)*; even morpholine, typically a relatively weak nucleophile *(9)*, gave excellent results under these conditions. Of the aliphatic amines surveyed, only sterically demanding diisopropylamine failed to react with the fluoropyridine substrate. On the other hand, aromatic amines, which do not commonly participate in S$_N$AR reactions *(11)* did not react under these conditions.

Palladium-catalyzed cross coupling of resin bound iodopyridine with a boronic acid (Suzuki reaction) *(12)* or with an alkenyl-, alkynyl-, or arylstannane (Stille reaction) *(13)* was effected by treatment of resin **5** with 4 equivalents of the boronic acid or stannane, 8 equivalents of K_2CO_3 (Suzuki reaction only), and catalytic Pd(PPh$_3$)$_4$ in DMF at 50 °C for 20 hours. *(14)* Best results were achieved by running each reaction twice, with an intermediate wash of the resin, in order to drive the reaction to completion. The corresponding Pd-catalyzed amination (Buchwald reaction) *(15)* worked well in protocol development but the reaction failed with the 3 amines attempted during actual library preparation.

The final library consisted of 1320 nicotinic acids of general structure **2** derived from, in part, 36 amines as nucleophiles (2 of which failed), 24 boronic acids, and 5 stannanes. Each MiniKan contained 80 mg of resin, which, with a measured loading of 1.36 mmol/g, was capable of yielding 0.1088 mmol of disubstituted nicotinic acid product. The final product acids were cleaved from the resin with trifluoroacetic acid and isolated as discrete samples. Each sample was analyzed by LCMS and the expected molecular ion was found in >80% of the samples (92% when failed inputs were eliminated). In selected instances, ^{1}H NMR analyses were also used to ascertain the success of individual reactions. Members of a 25-membered rehearsal library were fully characterized; the yield and purity of each member of this library are shown in Table 1.

Table 1. Yield and Purity of a 25-Membered Rehearsal Library of 5,6-Disubstituted Nicotinic Acids

X \ R	2-Cl-C₆H₄	4-OMe-C₆H₄	5-Cl-thienyl	thienyl	phenylacetylene (stannanes)
NH-iPr	95%[a] (99%)[b]	83% (15%)	82% (15%)	85% (99%)	100% (68%)
NH-CH₂-(4-Cl-C₆H₄)	100% (99%)	100% (97%)	84% (32%)	100% (98%)	94% (92%)
NH-C(O)-O-iPr	98% (87%)	84% (82%)	75% (20%)	97% (91%)	99% (82%)
NH-CH₂CH₂-OMe	100% (99%)	79% (99%)	92% (14%)	89% (99%)	100% (90%)
morpholino	100% (99%)	100% (92%)	91% (24%)	88% (99%)	100% (83%)

[a] mass recovery
[b] purity – LCMS with ELS detection

Screening Results

Using a strategy similar to that described for the nicotinic acids above, a series of 10 structurally diverse libraries, consisting of 11,528 carboxylic or phosphonic acids (pKa <5) and derived from a variety of templates, were designed and prepared (Table 2). All members of the libraries were prepared and tested as single entities. The results from screening the 10 combinatorial libraries in the high throughput herbicide screens (Weed Management HTS) are also shown in Table 2. Inspection of the number of passes from each library is enlightening. One library showed a remarkable 14.9% pass rate and four had a pass rate greater than 9%. In contrast, there was not a single hit from library 10, which consisted of 1408 individual molecules!

Table 2. Results of Screening the 10 Diverse Acid Libraries in Weed Management HTS

Library #	Compounds	HTS Passes[a]	%
1	1056	157	14.9
2[b]	1320	153	11.6
3	1056	100	9.5
4	176	16	9.1
5	1056	57	5.4
6	1408	53	3.8
7	1232	41	3.3
8	704	18	2.6
9	2112	48	2.3
10	1408	0	0.0
Total	**11528**	**643**	**5.6**

[a] >50% Growth inhibition of either AGSST at 25 ppm; LEMMI at 25 ppm; or ARBTH at 10 ppm. [b] 6-Amino nicotinic acids.

Pass rates for these ten acid libraries are compared in graphical form in Figure 4 to the corresponding rates for two sets of randomly selected compounds from our general screening efforts. The first collection, a set of 29,729 diverse compounds (labeled as "Random" in Figure 4), represented all inputs, regardless of source (purchased, internally prepared, etc.), which entered the WM HTS over a defined period. The second collection consisted of 10,021 diverse acids ("Acids" in Figure 4) randomly chosen from the WM HTS screening inputs. As anticipated, the average HTS pass rate for the targeted libraries was substantially higher than that for diverse, non-optimized screening inputs (1.2%) but only slightly greater than that for randomly selected acids (5.6% vs. 4.8%). Certain libraries provided much higher pass rates than random acids but the converse was also true. One possible explanation for this difference is that the degree of diversity within an individual library may be lower than that of an equal number of truly randomly selected acids. Thus if a library motif "fits" a set of criteria which delivers bioactivity then many members of the library (14.9%) may hit in the screens. Alternatively, missing bioactivity space with the library template can result in a disastrously low hit rate (0.0%).

HTS hit rates are of limited value as a potential metric for the success of these libraries for herbicide lead generation – a single hit from any one library could result in an important new lead. Further, the HTS does not sort actives on the basis of symptomology, mode-of-action, or other important attributes of the "lead" criteria. A better indicator of herbicidal activity is to consider compounds that have activity in higher level assays. A total of 58 compounds (0.5%) met this whole-plant activity criteria, again not significantly different from that of randomly acquired acids (0.6%) but at least 5-10 fold greater than the pass rate for random chemistry.

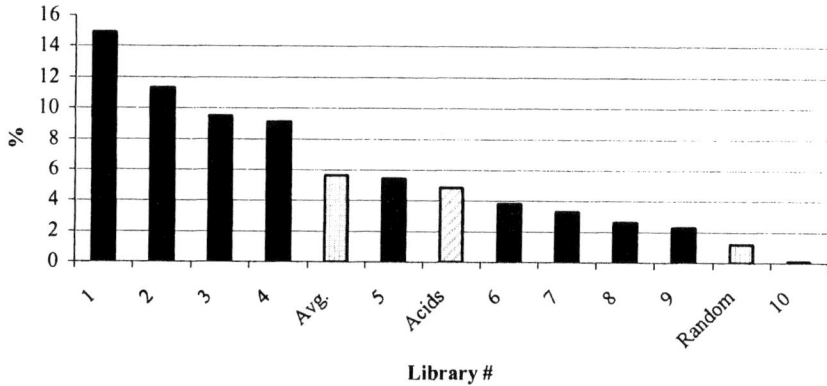

Figure 4. HTS Pass Rate of Acid Libraries vs. Random Acids and Random Chemistry

The most important measure of success of this effort is the number of leads generated. Compounds from three of the ten libraries, (Libraries 3, 6, and 7) were of sufficient interest in higher level assays to warrant some level of synthetic follow-up.

The results from screening the combinatorial libraries in the Crop Disease (CDM) and Insect Management (IM) HTS are shown in Table 3. A comparison of the screening results for the combinatorial libraries in each therapeutic area shows little overlap in bioactivity among the three areas. This is not surprising, as "optimal" physical properties for bioactivity are clearly different in each instance. (16) For example, while movement to the growing point of the plant is a requisite for a reliable herbicide, retention of the compound on or within the leaf surface is a prerequisite for a contact, foliar-applied insecticide.

An important conclusion from this comparison of hit rates of the combinatorial libraries across therapeutic areas is that missing bioactivity space for a particular area does not necessarily deem a library a failure - the library may still deliver valuable leads against another product goal. As an example, Library 10, with remarkably low 0% and 0.6% hit rates in the WM and CDM HTS respectively, showed a very robust 16.8% pass rate in the Insect Management HTS.

Table 3. HTS Screening Results for 3 Targeted Therapeutic Areas

Library #	WM (%)	Therapeutic Area CDM (%)	IM (%)
1	14.9	10.9	2.1
2	11.6	18.8	5.0
3	9.5	6.7	22.7
4	9.1	7.4	0.6
5	5.4	5.9	19.2
6	3.8	4.9	1.2
7	3.3	28.5	6.5
8	2.6	2.0	0.0
9	2.3	7.4	6.3
10	0.0	0.6	16.8
Pass Rate:	5.6%	9.6%	8.7%

Conclusions

In conclusion, a series of 10 combinatorial libraries consisting of 11,528 total compounds were designed, prepared, and screened as potential herbicide leads. The libraries were chosen to deliver compounds with desirable physical properties, in this case polar weak acids. The approach was illustrated with a series of 5,6-disubstituted nicotinic acids, prepared using solid phase, combinatorial techniques.

Results of screening these 10 libraries as potential herbicide leads demonstrated that the screening inputs prepared using this targeted, attribute-driven combinatorial chemistry approach provided a nearly 5-fold higher pass rate than randomly selected inputs. More importantly, hits from three of these libraries showed sufficient levels of activity in higher level testing to warrant further synthetic follow-up. This clearly illustrates that the attribute-driven approach can be successfully used to provide novel inputs with desirable properties for agrochemical screens.

References

1. For an analysis of the current role of combinatorial chemistry in the drug discovery process see: Dolle, R. E. *J. Comb. Chem.* **2002**, *4*, 370-418.
2. (a) Lipinski, C. A.; Lombardo, F.; Dominy, B. W.; Feeney; P. J. *Adv. Drug Deliv. Rev.* **1997**, *23*, 3-25; (b) Eddershaw, P. E.; Beresford, A. P.; Bayliss, M. K. *Drug Discovery Today* **2000**, *5*, 409-414.
3. Kleier, D. A. *Pestic. Sci.* **1994**, *42*, 1. Brudenell, A. J. P.; Baker, D. A.; Grayson, B. T. *J. Plant Growth Regulation* **1995**, *16*, 215.
4. Hansch, C.; Leo, A.; Hoekman, D. Exploring QSAR. *Hydrophobic, Electronic, and Steric Constants*; American Chemical Society Professional Reference Book, **1995**; p 12 and 18.
5. Green, R. W.; Tong, H. K. *J. Am. Chem. Soc.* **1956**, *78*, 4896.
6. Newkome, G. R.; Paudler, W. W. *Contemporary Heterocyclic Chemistry*; John Wiley & Sons, **1982**; 250-275
7. (a) Moran, E. J.; Sarshar, S.; Cargill, J. F.; Shahbaz, M. M.; Lio, A.; Mjalli, A. M. M.; Armstrong, R. W. *J. Am. Chem. Soc.* **1995**, *117*, 10787-10788. (b) Shi, S.; Xiao, X.; Czarnik, A. W. *Biotechnol. Bioeng. (Comb. Chem.)* **1998**, 61, 7-12.
8. March, J. *Advanced Organic Chemistry, 3rd Ed.*; John Wiley and Sons: New York, 1985; p. 587.
9. The nucleophilicity within a series, such as amines, is generally correlated with the basicity of the nucleophile. *(8)* The pKa of protonated morpholine is 8.3 vs. 11.1 for piperidine. *(10)*
10. Gordon, A. J.; Ford, R. A. *The Chemist's Companion: A Handbook of Practical Data, Techniques, and References*; John Wiley and Sons: New York, 1972; p. 59-60.
11. The pKa of protonated aniline is 4.6 while the corresponding pKa for protonated ethylamine is 10.8.*(10)*
12. Miyaura, N.; Suzuki, A. *Chem. Rev.* **1995**, *95*, 2457.
13. Chamoin, S.; Houldsworth, S.; Snieckus, V. *Tetrahedron Lett.* **1998**, *39*, 4175-4178.
14. Guiles, J. W.; Johnson, S. G.; Murray, W. V. *J. Org. Chem.* **1996**, *61*, 5169-5171.
15. Wolfe, J. P.; Buchwald, S. L. *Tetrahedron Lett.* **1997**, *38*, 6359-6362.
16. Tice, C. M. *Pest. Manag. Sci.* **2001**, *57*, 3-16.

Mode of Action

Chapter 12

Fungal Site of Action Determination: Integration with High-Volume Screening and Lead Progression

Steven Gutteridge*, Steve O. Pember, LiHong Wu, Yong Tao, and Mike Walker[‡]

Stine-Haskell Research Center, DuPont Crop Protection, P.O. Box 30, Newark, DE 19714

Many bioactive compounds discovered in Crop Protection are selected from hundreds of thousands of samples using automated high throughput procedures and miniaturized whole organism screens. The process is designed to rapidly reduce the number of chemical entities to a manageable number that have confirmed biological activity and likely to be amenable to synthetic embellishment. This lead identification process is applicable to all areas of crop protection interest, herbicide, fungicide and insecticide. This presentation describes some of the biochemical and genetic tools the Site of Action group is developing to provide fungal target site information earlier in discovery. Depending on their amenability, targets identified in this way are used to build assays to support optimization, screen for other small molecules or anticipate impending regulatory hurdles during candidate development. Where possible, new small molecule structural starting points are adorned further using targeted design approaches.

[‡]Deceased. See the acknowledgments.

Introduction

Three major components are required to deal with the sourcing and scheduling of the large numbers of compounds entering the screening process in Discovery. A chemical sourcing group acquires the compounds, assigns each individual sample a unique identifier and passes them to a Sample Management Facility (SMF). SMF oversees their storage, handling and dispensing from then on. Hit Generation is that part of the process involved in screening the compounds through a series of whole organism screens in all indication areas, i.e. assessing their herbicidal, fungicidal or insecticidal activity. The level 1 screens (L1) have been designed to handle the large initial volume of compounds that are acquired and have enough sensitivity to identify weak bioactivity. Subsequent screens are designed to help identify those compounds with the best potential to reach lead status. Prior to a lead being declared the compound will not only show significant biological activity but will have been assessed for whether it is active on known sites of action, its purity and whether there are any relevant literature references. Once a lead is declared, optimization chemists work on adjusting the biological activity and spectrum to conform to particular predefined product niches and market opportunities, using advanced biological screens to provide the appropriate response. At this stage of discovery the Site of Action group begins to better define and understand the mode of action of the compounds. If it is clear a lead will progress beyond the second stage of development, it is the aim of the group to define in some detail the biological process which is implicated.

Target Identification

The nature of the process associated with identifying the target of lead chemistry involves a series of chemical genetic approaches broadly termed chemistry to gene (C2G). The first series of studies require rescreening the lead and analogs in an expanded panel of assays, often against model organisms, that provide some insight into the mode of action. Further diagnostic approaches narrow the list of potential targets and ideally result in an unique identification. A search is then initiated for a suitable gene that can be used to generate recombinant material for further analyses. Confirmation that the chemistry acts at a particular target is the point where the C2G process has been considered a success. At this stage there are a number of options available, e.g. simply identifying the target can spark new ideas about chemistry, especially if it is an enzyme, or if amenable, an assay is developed to search for other structural classes different from the lead, to pursue in a fresh analoging program.

The following examples are from three different lead areas that are active as fungicides to illustrate the process and the types of approaches being employed. In all cases the structures of the leads cannot be revealed since they are still in active areas of research and synthesis, however this will not detract from how the diagnostic studies are organized. The process has been so designed to acquire the most useful information using the least intensive approaches first and build more involved techniques into the analyses only when warranted. As much of the process is performed in a parallel fashion although certain basic pieces of information need to be collected first since downstream studies key off these data, e.g. LD_{50}, growth progression, time courses etc.

Lead 1

Lead 1 is a broadly active compound with commercial levels of pathogen control that also inhibits growth of yeast. Figure 1 shows a growth curve for *Saccharomyces cerevisiae* in the presence and absence of Lead 1.

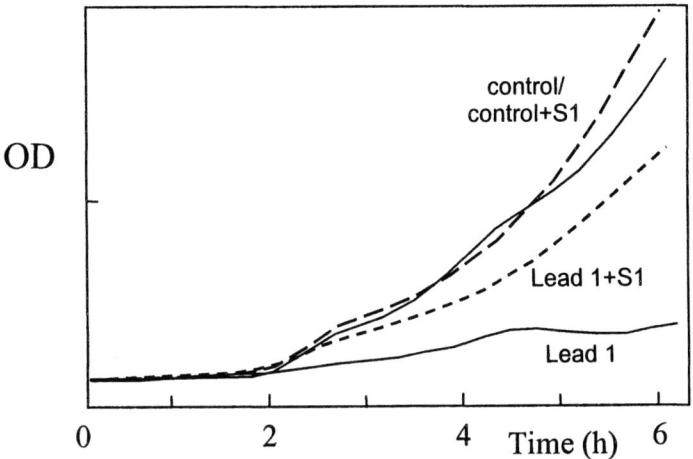

Figure 1 Growth curve of Saccharomyces cerevisiae in various conditions

Some observations from the progress curve are characteristic of this lead. The LD_{50} is ~ 2ppm in minimal medium, but the initial phase almost tracks like the untreated sample where onset of inhibition seems to be somewhat delayed. The degree of inhibition is also sensitive to the composition of the medium. In the presence of particular supplements, e.g. S1 significant reversal of growth inhibition is observed.

Supplements are pooled together in specific mixture combinations such that each one is represented in more than one pool. Reversal of inhibition by any one supplement should thus show up as growth restoration in all those pools that contain that particular supplement. For Lead 1 this was the case for all pools containing the S1 supplement. This is a plate-based system allowing fast analysis of multiple pools.

Figure 2 shows the growth response of yeast in the various pools when in the presence of Lead 1. The lead shows less inhibitory action when pools 2 and 9 are added to the medium.

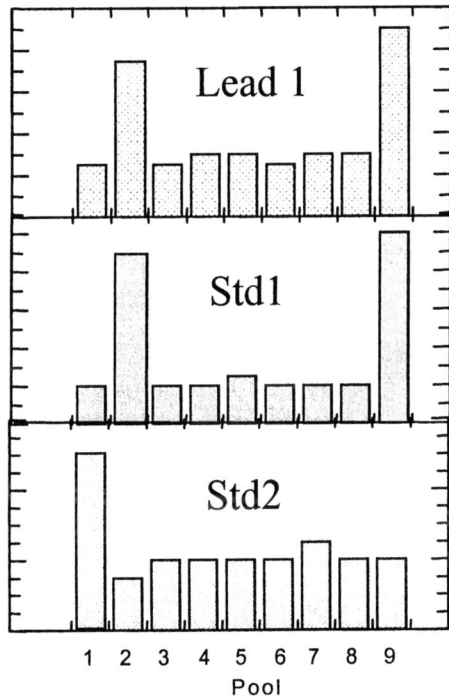

Figure 2 Effects of various supplement pools on yeast growth

This produces a diagnostic fingerprint of the mode of action of the compound even though at this stage the action cannot be defined in very precise biochemical terms. However, by comparison with standards with better defined

MoA some further insight is possible. In this case the response fingerprint to Std1 is identical to that of Lead 1, but quite distinct from Std2.

Some additional details were also evident after the supplements were analysed individually using a zone test approach where filter discs soaked with the various supplements are added to yeast plated on solid medium. With Lead 1 it was found that 4 other supplements produced some level of resistance and a further 4 supplements caused hypersensitivity to the lead.

Although the nutritional supplement work provided a basis to build hypotheses concerning the mode of action, two further types of analyses were initiated to acquire broader information that might result in the target of the chemistry being identified. Using the growth inhibition data of Figure 1 conditions were chosen, such as time, amount of treatment, to provide cells for microarray analysis. From comparative changes in mRNA expression between untreated and treated cells, analyzed using a yeast array, complementary information to that of nutritional profiling was obtained.

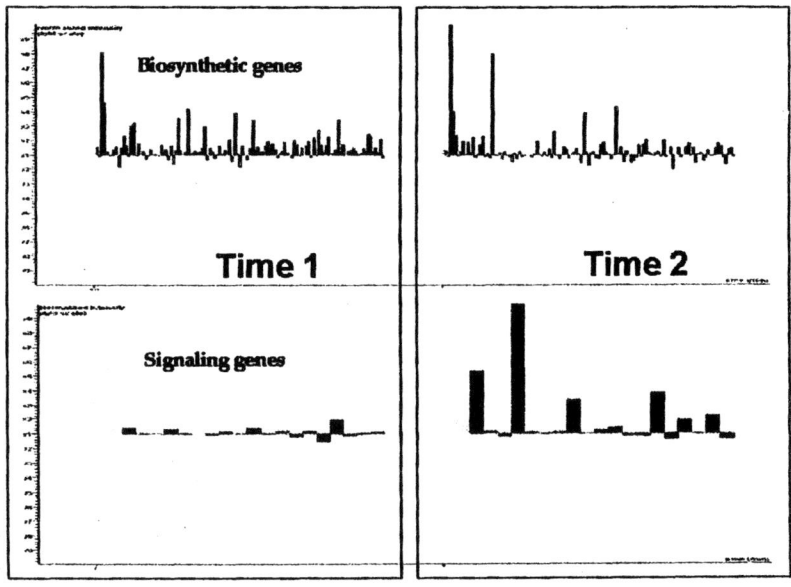

Figure 3 Effects of Lead 1 on transcript levels of specific gene clusters

Together the information was used to select a subset of genes which were cloned in suitable vectors to generate overexpression strains of yeast. The lead compound, with and without supplements was tested against these strains to find which showed the most resistance.

Figure 3 shows some of the genes that are upregulated and repressed when treated with Lead 1. There are clear clusters of related genes in two categories that are indicative of which biosynthetic processes are disturbed by action of the compound. There also seemed to be an effect of time on the response of some of the gene clusters. The choice of target of this chemistry has been narrowed to a small number of potential sites and further work is in progress to isolate which is the actual target and the basis for the delayed growth inhibition.

Lead 2

A similar series of analyses were performed with Lead 2. Nutritional reversals nor array results provided a clear indication of mode of action of the lead, but a second analog generated similar patterns amongst certain gene clusters.

In this case most insight was gained from using overexpression lines. Four genes with known function imparted some level of resistance in complemented yeast strains. Only three of the four produced some degree of resistance consistent with the potency of the compound and these were closely related in function. Unfortunately, Lead 2 was discontinued due to adverse biological properties.

As more fungal active compounds are identified that affect yeast growth we are compiling the array results into hierarchical families as a means of defining the compound mode of action and then diagnosing which particular step in the cellular function is impaired (*1*). These data are assessed along with growth response to nutritional supplements for consistency and further detailed definition of functional disruption. Typically with this volume of information, much of it helps to rule out which processes are not involved, but in combination with forward and reverse genetics the numbers of likely target candidates can be reduced to a manageable number.

Lead 3

In the case of Lead 3 a more biochemical approach was required because the pathogen controlled is not amenable to a detailed molecular genetic approach. In this case we were fortunate to have an analog of the chemistry that could be employed to specifically isolate the target using affinity techniques.

The approach was typical for affinity based methods namely immobilisation of the ligand to a column matrix through a suitable inert spacer. The column was then used to fractionate the cell contents of the pathogen and wash conditions optimized to remove all except the most tightly bound proteins. Under denaturing conditions the tightly bound fraction was isolated and purified further using gel electrophoresis. The individual protein bands were sequenced using in-gel digestion and mass spectroscopy. From these sequences the identity of the proteins was established and the genes isolated. The genes were then cloned into suitable expression vectors, the particular protein isolated and its function analysed. Generally, this is the point where an attempt is made to acquire a clear indication that the purified native protein has affinity for the class of chemistry.

The C2G approach is thus a definable process using a combination of technologies based either on profiling or biochemical approaches to more rapidly define the mode of action of an active lead or hit and ideally point to a minimum number of potential targets. Having reduced the options to a manageable number of candidates, confirmation of the identity of a site of action can employ genomic techniques, particularly where the molecule is effective on a model organism.

Where there is a need to develop an assay based on a newly identified target some are likely to be more challenging than others and screen development needs to be creative, e.g. with membrane bound proteins. In such cases, the SoA group is collaborating with CompleGen Inc. to use their cell-based technology to construct assays involving such targets.

Site Topography

Once the binding site of active compounds has been identified, there is often further interest in better defining the topography of that site. For example, famoxate and other commercial inhibitors target the Qo site of the *bc1* complex. There is thus much interest in defining any differentiating features of the interaction between compound and specific residues of the protein.

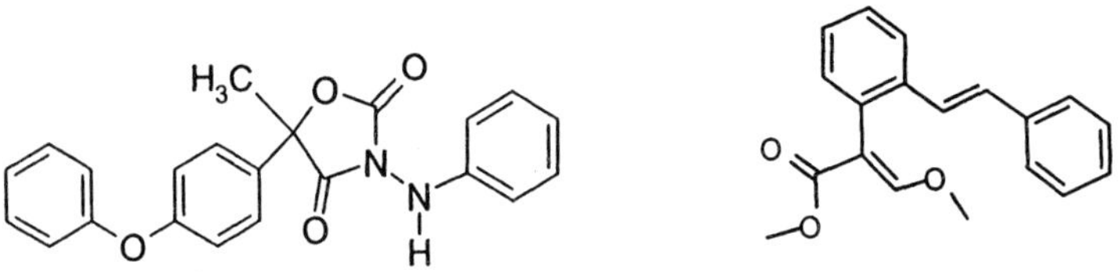

Figure 4 Famoxate and methoxyacrylate stilbene, respectively; two inhibitors of the bc1 complex.

Structurally the small molecules shown in Figure 4 are quite distinct and so intuitively one might conclude they occupy different regions of the binding pocket. Indeed, there are three sub-classes of the Qo site defined by the effects of inhibitors on the interactions of the site with the various redox active centers and the response of the kinetic parameters to certain conditions (*2,3*). For example, removal of the Rieske iron-sulfur (Fe-S) protein causes loss of stigmatellin binding but strobilurin type analogues still retain some affinity (*4*).

When the effect of methoxyacrylate stilbene (MOAS) on quinol oxidation was compared to that of famoxate using steady state conditions, it was clear that the two molecules were affecting turnover differently. Although the reciprocal analyses indicated fairly classical non-competitive inhibition of quinol binding, a study of the response of the progress curves relative to changes in complex concentration, suggested the underlying interactions for both compounds might be more complicated. In the case of famoxate the increase in oxidation rate with increased amounts of the complex was linear, whereas with MOAS, the rate changes were curved upward compared to the uninhibited rate. Additional analyses confirmed the non-competitive nature of famoxate inhibition, but MOAS acted predominantly as a competitive inhibitor.

Further insight into the nature of the binding interactions was forthcoming using red shift titrations of the heme center (*5*). With this approach, MOAS gave a clear indication of competitive binding between quinol and the inhibitor, whereas with famoxate, not only was non-competitive binding confirmed but with increasing concentrations of the small molecule some facilitative interactions were detected (*6*).

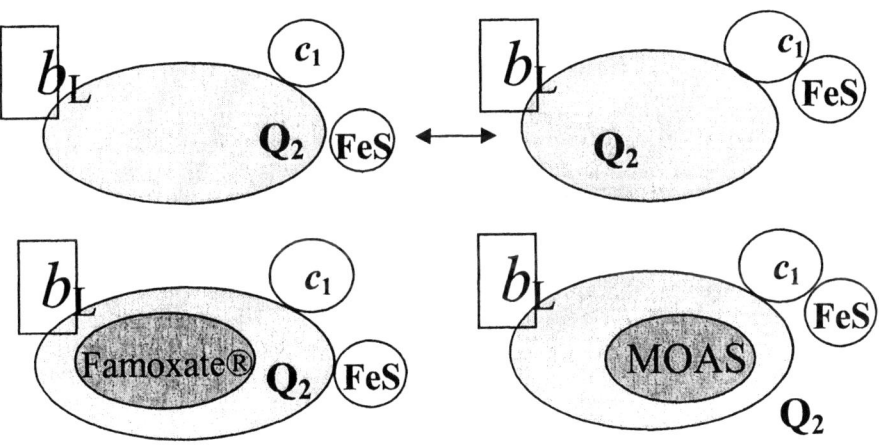

Figure 5 Famoxate and methoxyacrylate stilbene occupy slightly different regions of the Qo binding site.

The model in Figure 5 attempts to explain the differences in the action of famoxate and MOAS. The upper scheme is the process of quinol oxidation in terms of the relative locations of the redox centers to the substrate Qo binding site. Based on the nature of the inhibition of famoxate as non-competitive, presumably there is little interference with quinol substrate binding. As shown in the lower scheme, famoxate induces a conformational change that prevents the FeS protein from interacting with the $c1$ center (*7*) but allows close association with cytochrome b.

In the case of MOAS the interactions may be more complicated. Binding of MOAS promotes a 'loose' conformation of the FeS protein allowing a closer association with $c1$ (*8,9*). With this disposition it can be inferred that the substrate binding site is not preferentially formed since determinants on both cytochrome b and the FeS protein contribute to the quinol binding pocket (*10*). The quinol substrate cannot bind and ultimately the association with MOAS becomes tight enough to cause mutual depletion by effectively removing some significant proportion of the functional enzyme from the total active component.

Based on a comparison of the X-ray structures of methoxyacrylate analogs and famoxate in the $bc1$ complex, it is clear that there is only partial overlap in terms of the occupancy of the Qo binding site with the two inhibitors making different contacts with the sidechains that compose the site.

Future Directions

At present the targeted activities that are integral to the high throughput discovery process involve, in the first instance, assisting in deselecting compounds that have little activity or adverse attributes. However, once a compound has achieved lead status, the focus of the group shifts to defining in extensive detail how the compound acts and interferes in biological processes. This information is used initially to identify and generate new ideas about any other small molecules that might have similar activity, where possible support optimization and anticipate potential toxicological or future regulatory hurdles. Ultimately, the desired direction is to isolate the actual target of the most active compounds. With a gene in the bottle, a number of approaches to assay development and screening are explored with the intent of introducing a design component into compound selection and synthesis. If there is more than one structural class of small molecule that binds to a valid target, then in vitro based assays provide a fast means of finding them.

Acknowledgements

The presentation given at the Pan Pacific Conference in Honolulu and this paper are dedicated to the memory of Mike Walker, the talented synthetic chemist involved in many aspects of this work. Mike was recognized externally for his technical and leadership capabilities with a recent ACS award for Excellence in Chemistry and internally with the prestigious DuPont Scientific Leadership award. Mike died way too young and is sadly missed.

References

1. Jia, M.H.; LaRossa, R.A.; Lee, J-M.; Rafalski, A.; DeRose, E.; Gonye, G.; Xue, Z.: *Physiol. Genomics* **2000**, *3*, 83-92.
2. von Jagow, G.; Link, T.A.: *Methods Enzymol.* **1986**, *126*, 253-271.
3. Link, T.A.; Haase, U.; Brandt, U.; von Jagow, G.: *J. Bioenerg. Biomembr.* **1993**, *25*, 221-232.
4. Brandt, U.; Haase, U.; Schagger, H.; von Jagow, G.: *J. Biol. Chem.* **1991**, *266*, 19958-19964.
5. Brandt, U.; Schagger, H.; von Jagow, G.: *Eur. J. Biochem.* **1988**, *173*, 499-506.
6. Pember, S.O.; Moberg, W.K.; Walker, M.P.; Fleck, L.C.; Benner, E.A.: *Proceedings, Resistance 2001, Meeting the Challenge*, **2001**, IACR Rothamsted, Harpenden Herts, UK.
7. Gao, X.; Wen, X.; Yu, C.; Esser, L.; Tsao, S.; Quinn, B.; Zhang, L.; Yu, L.; Xia, D.: *Biochemistry* **2002**, *41*, 11692-11702.
8. Kim, H.; Xia, D.; Yu, C.A.; Xia, J.Z.; Kachurin, A.M.; Zhang, L.; Yu, L.; Deisenhofer, J.: *Proc. Natl. Acad. Sci. USA*, **1998**, *95*, 8026-8033.
9. Crofts, A.R.; Guergova-Kuras, M.; Huang, L.; Kuras, R.; Zhang, Z.; Berry, E.A.: *Biochemistry* **1999**, *38*, 15791-15806.
10. Samoilova, R.; Kolling, D.; Uzawa, T.; Iwasaki, T.; Crofts, A.; Dikanov, S.: *J. Biol. Chem.* **2002**, *277*, 4605-4608.

Chapter 13

Three-Dimensional Modeling of Cytochrome P450 14α-Demethylase (*CYP51*) and Interaction of Azole Fungicide Metconazole with *CYP51*

Atsushi Ito[1], Keiichi Sudo[1], Satoru Kumazawa[1], Mami Kikuchi[2], and Hiroshi Chuman[3]

[1]Nishiki Research Laboratories, Kureha Chemical Industry Company, Ltd., Fukushima, Japan
[2]Bio-Medical Research Laboratories, Kureha Chemical Industry Company, Ltd., Tokyo, Japan
[3]Faculty of Pharmaceutical Sciences, The University of Tokushima, Tokushima, Japan

> We have reported the quantitative structure-activity relationships (QSAR) of metconazole, a triazole fungicide and its related compounds, and proposed the interaction mode between CYP51 and metconazole. To confirm our model, we constructed three-dimensional model of phytopathogen's CYP51 by homology modeling on the basis of the recently reported crystal structure of CYP51 of *Mycobacterium tuberculosis* (MTCYP51). Plant pathogen's CYP51s exhibit 25-28% amino acid sequence identities with MTCYP51. The residues located within 8 Å from fluconazole in the crystallized complex of MTCYP51 were identified. The selected amino acids were replaced with those of CYP51 of *Botrytis cinerea* (BCCYP51), and a model of interaction between metconazole and BCCYP51 was constructed. The modeled three-dimensional structure corresponded to our putative interaction model proposed with the QSAR analyses.

1 Introduction

During the past two decades, many azole fungicides have been investigated all over the world (*1*). Azoles are a large and an important group of fungicides used for the control of fungal diseases in agriculture and medicine. The mode of action of azole fungicide is the inhibition of biosynthesis of ergosterol, which is an important component of fungal membranes (*2-4*). In the ergosterol biosynthesis, lanosterol 14α-demethylase (P450$_{14DM}$, CYP51) catalyzes the transformation of lanosterol to ergosterol by elimination of 14α-methyl group of lanosterol to give the $\Delta^{14,15}$ unsaturated sterol. Azole fungicide binds to this CYP51 and inhibits this reaction.

Metconazole and ipconazole shown in Figure 1 are Kureha's azole fungicides (*5*). Metconazole is a foliar fungicide for controlling a wide range of cereal diseases. It is particularly effective against *Fusarium*, *Septoria* and rust diseases on cereals. Ipconazole is a seed treatment fungicide for controlling seed borne rice diseases such as 'bakanae' disease, Helminthosporium leaf spot and blast.

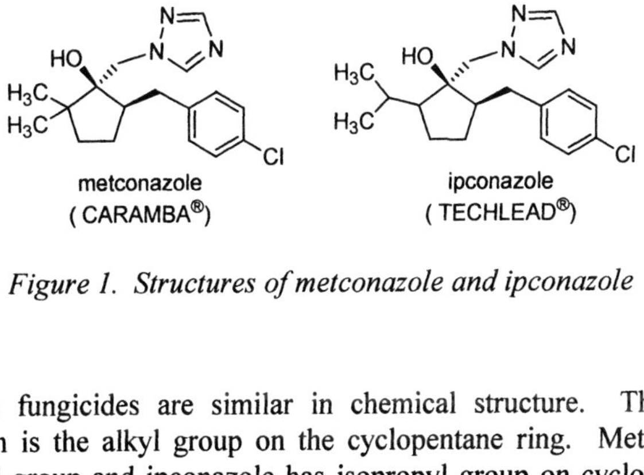

Figure 1. Structures of metconazole and ipconazole

The two fungicides are similar in chemical structure. The difference between them is the alkyl group on the cyclopentane ring. Metconazole has *gem*-dimethyl group and ipconazole has isopropyl group on cyclopentane ring, respectively. We have studied the structure-activity relationship of these fungicides and proposed the interaction mode between our azole fungicides and the target receptor, CYP51 (*6*). The interaction mode is schematically illustrated in Figure 2. The following interactions at the molecular level are proposed:

A. The nitrogen atom at the 4-position of the triazole ring coordinates to the heme iron of CYP51.

B. A hydrophobic pocket may be existed in the active site. The pocket accommodates the *para*-chlorophenyl group.

C. The increase of the activity with the introduction of *gem*-dimethyl and isopropyl groups to the cyclopentane ring indicates that these alkyl groups play an essential role in binding to CYP51. These structures are possibly assigned as another hydrophobic interaction site.

D. The presence of the hydroxyl group at 1-position of the cyclopentane ring was experimentally confirmed to be important for the activity. This group is considered to be involved in either hydrogen bond or electrostatic interaction with an amino acid residue in the active site of CYP51.

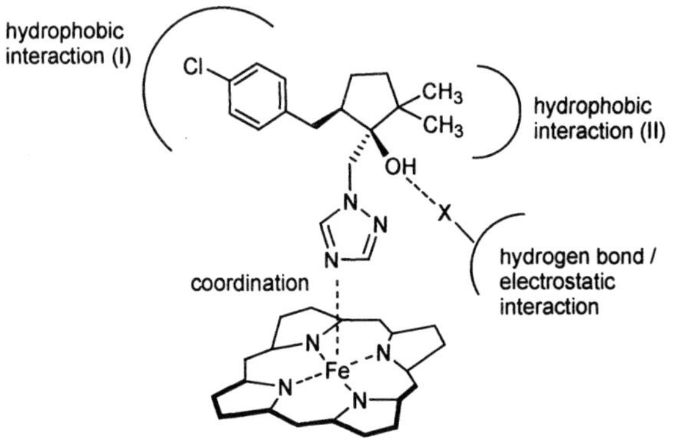

Figure 2. A proposed interaction mode for metconazole to the target enzyme, CYP51

2 Molecular Modeling of CYP51

To confirm our model, we attempted to construct three-dimensional model of plant pathogen's CYP51 by homology modeling. Some homology models of CYP51 have been reported. Most of them are models of *Candida* species studied mainly in pharmaceutical research. Because the target enzyme *Candida* CYP51 is a membrane-bound enzyme, it is difficult to crystallize for X-ray analysis. So these models have been constructed based on the crystal structure of prokaryotic cytochrome P450 family, such as P450cam, P450terp, P450eryF and P450BM3. Amino acid sequence identities between these four P450s and *Candida* CYP51 are low (18-20%). But their topologies have been reported to be quite similar, so their crystal structure data have been used in homology modeling (*7-12*). And recently crystal structure of CYP51 from *Mycobacterium tuberculosis* (MTCYP51) complexed with azole fungicide fluconazole was determined and interaction mode was reported in 2001 (*13*). So we tried to construct the model of plant pathogen's CYP51 on the basis of the crystal structure of MTCYP51.

2.1 Amino Acid Sequence Alignment

Amino acid sequences of plant pathogen's CYP51s were taken from GenBank / EMBL / DDBJ databases. There are twelve plant pathogens' CYP51s: *Botrytis cinerea* (AAF85983), *Erysiphe graminis* (ACC97606), *Septoria tritici* (AAF74756), *Tapesia yallundae* (AAG44832), *Tapesia acuformis* (AAF18468), *Uncinula necator* (O14442), *Venturia inaequalis* (AAF71293), *Venturia nashicola* (CAC85409), *Monilinia fructicola* (AAL79180), *Ustilago maydis* (CAA88176), *Penicillium digitatum* (BAB03658) and *Penicillium italicum* (CAA89824). Table I lists the fungicidal activities of metconazole against these plant pathogens. Although there are not all fungicidal data to those pathogens, metconazole shows high fungicidal activity.

Table I. Fungicidal activity of metconazole against plant pathogen

Plant pathogen [a]	MIC (mg/l) in vitro	ED_{50} (ppm) in vivo
Botrytis cinerea	0.8	15.4
Erysiphe graminis	- [b]	5.2
Septoria triticii	0.1	-
Tapesia yallundae	0.8	-
Tapesia acufomis	-	-
Uncinula necator	-	-
Venturia inaequalis	-	-
Venturia nashicola	-	-
Monilinia fructicola	< 0.2	-
Ustilago maydis	6.3	-
Penicillium digitatum	0.2	-
Penicillium italicum	3.1	-

a) CYP51s of these plant pathogens were registered in database
b) no data

Sequences alignment was performed using the program ClustalX (*14*) and results are shown in Table II. Plant pathogen's CYP51s exhibit 25-28% amino acid sequence identities with MTCYP51. High sequence identities (44 to 92%) were found among plant pathogen's CYP51s.

Table II. Amino acid sequence homology of CYP51 between plant pathogens and *Mycobacterium tuberculosis*

(%)

	B.c	E.g	S.t	T.y	U.n	V.i	M.f	U.m	P.d	P.i
B.c	/	72	61	75	70	61	92	49	60	59
E.g		/	57	71	72	60	72	46	59	57
S.t			/	60	59	60	60	45	56	56
T.y				/	69	63	75	46	57	58
U.n					/	59	69	47	60	57
V.i						/	62	46	58	57
M.f							/	48	59	58
U.m								/	45	44
P.d									/	87
P.i										/
M.t	27	26	26	28	25	27	27	27	27	27

2.2 Determination of Binding Site with Fluconazole

The crystallographic coordinates of MTCYP51 (2.2 Å resolution, Rcryst=0.20) were obtained from the Protein Databank, entry 1EA1. Heme is located in the middle of this structure, and fluconazole exists above the heme, as shown in Figure 3. Binding site of plant pathogen's CYP51 was defined by the heme group and by amino acid residues located within 8 Å from fluconazole in the crystallized complex of MTCYP51.

Figure 3 shows six domains which are considered to interact with fluconazole. Domain 5 is on opposite site of fluconazole across the heme. Domain 5 is called cyctein pocket and is assumed to interact with not azole fungicide but with heme. Domain 5 is assumed to play an important role in interaction with heme.

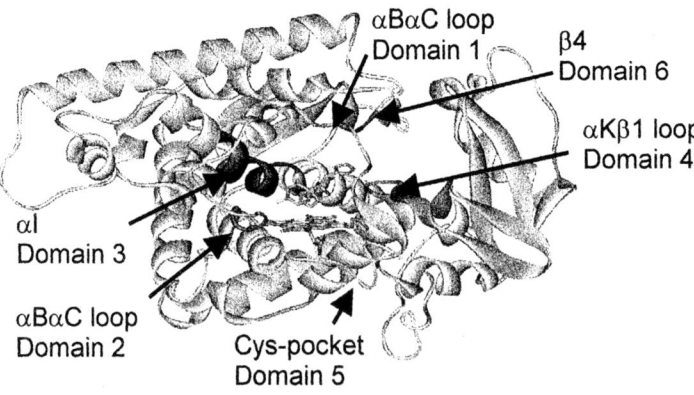

Figure 3. Binding site of MTCYP51 within 8 Å of fluconazole

2.3 Amino Acid Sequence Alignment of the Binding Site

Amino acid sequence alignment of the binding site is shown in Figure 4. Average sequence identity in the binding sites between plant pathogen's CYP51s and MTCYP51 is about 35%. Domain 5 (cyctein pocket) showed high homology between *M. tuberculosis* and plant pathogens. Among plant pathogens high sequence identities were shown. The broad fungicidal spectrum by azole compound is considered to be related to the high degree of sequence identity in the binding site. From among these plant pathogen's CYP51s, we selected the sequence of CYP51 from *Botrytis cinerea* (BCCYP51) for this homology modeling.

```
Domain 1  αBαC loop                 Domain 2  αBαC loop            Domain 3  αl
M.t  :  72  QAKAYPFMT-PIF  83       M.t  :  95  RR—KEMLH  101      M.t  :  252  SMMFAGHHT  260
              *  ***                                                              ** *
B.c  : 118  AEEIYTVLTTPVF  130      B.c  : 142  KLMEQKKFM  150     B.c  : 304  ALLMAGQHS  312
E.g  :      AEEIYTVLTTPVF            E.g  :     KLMEQKKFM           E.g  :      ALLMAGQHS
U.n  :      AEEIYTNLTTPVF            U.n  :     KLMEQKKFM           U.n  :      ALLMAGQHS
S.t  :      AEEIYSPLTTPVF            S.t  :     KLMEQKKFV           S.t  :      ALLMAGQHS
M.f  :      AEEIYTVLTTPVF            M.f  :     KLMEQKKFM           M.f  :      ALLMAGQHS
T.y  :      AEEIYTVLTTPVF            T.y  :     KLMEQKKFM           T.y  :      ALLMAGQHS
V.i  :      AEEIYSPLTTPVF            V.i  :     KLMEQKKFV           V.i  :      ALLMAGQHS
U.m  :      AEDAYTHLTTPVF            U.m  :     VFMEQKKFV           U.m  :      ALLMAGQHT
P.d  :      AEEIYGKLTTPVF            P.d  :     KLMEQKKFI           P.d  :      TLLMAGQHS

Domain 4  αKβ1 loop                 [ Domain 5  Cys pocket ]       Domain 6  β4
M.t  : 320  PLIILM  325             M.t  : 387  FGAGRHRCVGAA  398  M.t  : 433  MVV  435
              *  *                              ********
B.c  : 373  PIHSIM  378             B.c  : 460  FGAGRHRCIGEQ  471  B.c  : 507  LFT  509
E.g  :      PIHSIL                   E.g  :     FGAGRHRCIGEQ        E.g  :      MFS
U.n  :      PIHSIM                   U.n  :     FGAGRHRCIGEQ        U.n  :      LFS
S.t  :      PIHSIL                   S.t  :     FGAGRHRCIGEQ        S.t  :      LFS
M.f  :      PIHSIM                   M.f  :     FGAGSHRCIGEQ        M.f  :      LFT
T.y  :      PIHSIM                   T.y  :     FGAGRHRCIGEQ        T.y  :      LFS
V.i  :      PIHSIL                   V.i  :     FGAGRHRCIGEQ        V.i  :      LFS
U.m  :      PLHSIM                   U.m  :     FGAGRHRCIGEQ        U.m  :      LFS
P.d  :      SIHTLM                   P.d  :     FGAGRHRCIGEK        P.d  :      LFS
```

Figure 4. Multiple sequence alignment of the binding site

2.4 Conformation Analysis of Metconazole

Before replacing fluconazole with metconazole in the binding site, we carried out conformational analysis for metconazole and fluconazole. Initial conformations were generated by a grid search method using MMFF molecular force field (*15*), and the obtained conformations were optimized by Hartree Fock MO calculations with the 3-21G* basis set in the Gaussian 98 package (*16*). The conformations of metconazole and fluconazole were superimposed using flexible fitting method (*6*). Fitting points of them were triazole ring, benzene ring and hydroxyl group, as shown in Figure 5. So we found that there are three pairs having similar conformation each other and one of them was most similar to the conformation of fluconazole in crystal structure of MTCYP51. And then we used this active conformation of metconazole for modeling.

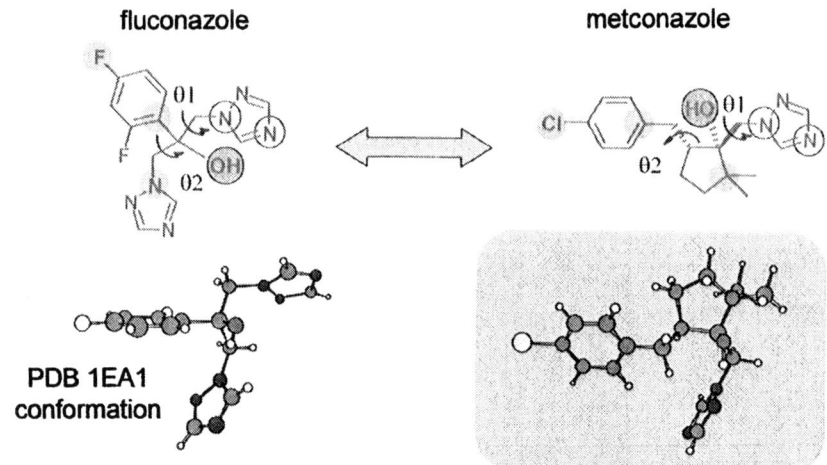

Figure 5. Conformation analysis of metconazole

2.5 Substitution of Amino Acid Residues in the Binding Site and Docking Model of Metconazole in Complex with Plant Pathogen's CYP51

Amino acid residues of MTCYP51 were substituted for those of BCCYP51. The structure was optimized using Tripos force field. The above-mentioned active conformer of metconazole was docked into the binding site of BCCYP51 and a docking model of metconazole in the binding site of BCCYP51 was obtained (Figure 6).

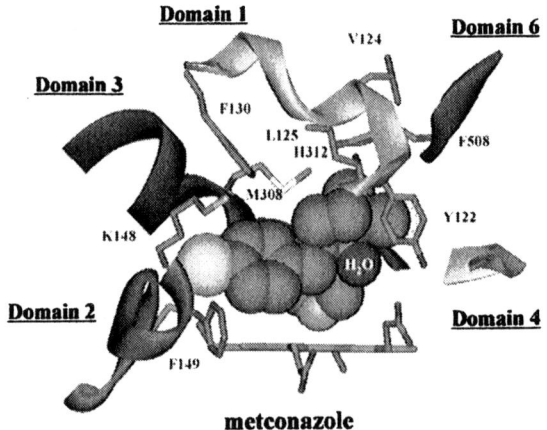

Figure 6. Docking of metconazole into the binding site of BCCYP51 model. Heme and residues within a distance of 8 Å from metconazole are shown.

3 Conclusion

The three-dimensional model of BCCYP51 was built on the basis of the sequence homology with the recently reported crystal structure of the MTCYP51. The model of BCCYP51 corresponded to our putative interaction mode proposed with the QSAR analyses. The interaction mode for metconazole to BCCYP51 is illustrated in Figure 7. The *para*-chlorophenyl group in metconazole is surrounded by the hydrophobic amino acid residues, two phenylalanines and leucine (Phe130, Leu143, and Phe149). The *gem*-dimethyl group linked to cyclopentane ring interacts with hydrophobic amino acid residues, leucine, isoleucine and phenylalanine (Leu125, Ile374 and Phe508). Leu125 closely contacts cyclopentane moiety. Hydroxyl group of metconazole forms a hydrogen bond with Tyr122 through a water molecule. In the complex with MTCYP51, hydroxyl group of fluconazole also forms a hydrogen bond with Tyr76 through a water molecule. Although the binding mode predicted through molecular modeling studies needs to be validated by X-ray crystallography, these results will be useful for the rational design of more potent inhibitors of cyctochrome P450.

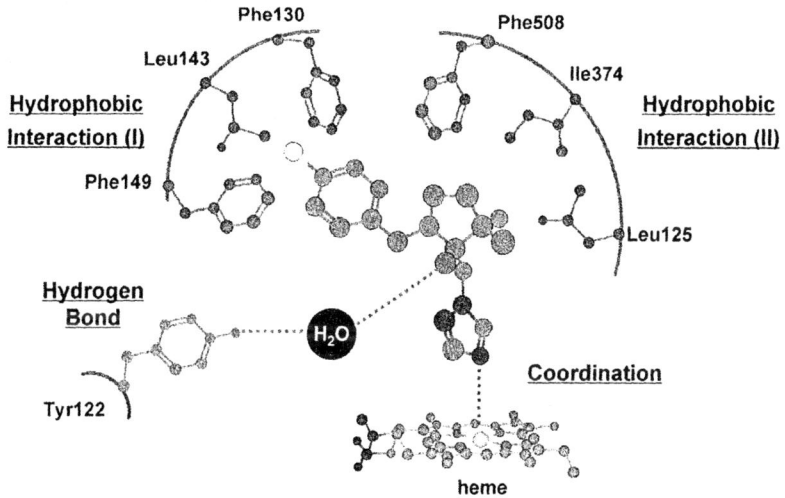

Figure 7. Interaction mode for metconazole to BCCYP51

4 References

1. Köller, W. *Pestic. Sci.*, **1987**, *18*, 129.
2. Köller, W. In *Target Sites of Fungicide Action*; Köller, W., Ed.; CRC Press, Boca Raton, **1992**; pp 119-206.

3. Kwok, Y.; Loeffler, R. T. *Pestic. Sci.*, **1993**, *39*, 1.
4. Kuck, K. in *Modern Selective Fungicides*; Lyr, H., Ed.; Gustav Fischer Verlag, Jena; Stuttgart; New York, **1995**; pp 171-185.
5. Kumazawa, S.; ITO, A.; Saishoji, T.; Chuman, H. *J. Pesticide Sci.*, **2000**, *25*, 321.
6. Chuman, H.; ITO, A.; Saishoji, T.; Kumazawa, S. In *Classical and 3D QASR in Agrochemistry*; Hansch, C.; Fujita, T., Eds.; ACS Symp. Ser. 606; American Chemical Society: Washington, D.C., **1995**; pp 171-185.
7. Boscott, P.E.; Grant, G.H. *J. Mol. Graphics*, **1994**, *12*, 185.
8. Tsukuda, T.; Shiratori, Y.; Watanabe, M.; Ontsuka, H.; Hattori, K.; Shirai, M.; Shimma, N. *Bioorganic & Medicinal Chemistry Letters*, **1998**, *8*, 1819.
9. Höltje, H. D.; Fattorusso, C. *Pharmaceutica Acta Helvetiae*, **1998**, *72*, 271.
10. Talele, T. T.; Hariprasad, V.; Kulkarni, V. M. *Drug Design and Discovery*, **1998**, *15*, 181.
11. Lewis, D. F.; Wiseman, A.; Tarbit, M. H. *J Enzyme Inhib.*, **1999**, *14*, 175.
12. Ji, H.; Zhang, W.; Zhou, Y.; Zhang, M.; Zhu, J.; Song, Y.; Lu, J. *J Med Chem.* **2000**, *43*, 2493.
13. Podust, L. M.; Poulos, T. L.; Waterman, M. R. *Proc. Natl. Acad. Sci.*, **2001**, *98*, 3068.
14. Thompson, J.D.; Higgins, D. G.; Gibson, T.J. *Nucleic Acids Res.*, **1994**. *22*, 4673.
15. Halgren, T. A., *J. Computational Chem.*, **1996**, *17*, 490.
16. Gaussian Inc., Pittsburgh, USA.

Chapter 14

New Herbicide Target Sites from Natural Compounds

Stephen O. Duke, Franck E. Dayan, Isabelle A. Kagan, and Scott R. Baerson

NPURU, Agricultural Research Service, U.S. Department of Agriculture, P.O. Box 8048, University, MS 38677

Introduction

The relatively weak effort to use natural compounds as pesticide leads has led to several commercial herbicides, including glufosinate, bialaphos, the triketones, and pelargonic acid. Furthermore, natural phytotoxins have often been found to have molecular target sites that were previously unknown as viable sites for herbicides. Commercially available herbicides target only about 20 enzymes or energy transfer processes. Even if the natural compound is not a competitive herbicide, the knowledge of a new target site can be used to develop new synthetic herbicides with novel modes of action (*1*). Discovery of new herbicide target sites is even more important since the U.S. Food Quality Protection Act combined food tolerance levels of pesticides with the same molecular target sites.

Natural products have not been an area of focus for discovery efforts because of the relatively high cost of bioassay-directed isolation, the probability of rediscovery, the often prohibitively complex structure of the active molecule, and the often more uncertain legal aspects of a patent. Furthermore, the approaches for discovery of natural product-based pesticide discovery are different from those for synthetic compounds, requiring a different laboratory infrastructure, including emphasis on natural product chemistry, rather than synthetic chemistry. The time and effort required for natural product-based herbicide discovery programs have been reduced by new methods, including semi-automated bioassays, automated chemical extraction and analysis, and molecular biology methods to probe biosynthesis and mode of action. We will focus on our own work in this area in this short review.

New Target Sites

Inhibition of the fructose-1,6-bisphosphate aldolase (FBPase; EC 4.1.2.13) step of glycolysis is known to have long-ranging repercussions on the physiology of plants, leading to reduced photosynthetic efficiency, impaired sugar and starch metabolism, and ultimately to reduced growth (2). Efforts have been made to identify inhibitors of animal FBPase to develop novel pharmaceutical compounds. Although this enzyme will probably not be developed as a potential herbicide target site because it is too ubiquitous, our studies have shown that some natural phytotoxins may target this enzyme.

The fructose analogue 2,5-anhydro-D-glucitol (AhG) produced by the plant pathogen *Fusarium solani* (Mart.) Sacc. NRRL 18883 is phytotoxic (3), inhibiting lettuce root growth with a I_{50} of 1.6 mM. The mechanism of action of this sugar analog is relatively complex, involving its sequential bisphosphorylation by hexokinase and phosphofructokinase to yield AhG-1,6-bisphosphate (Figure 1). This phosphorylated sugar analogue is a competitive inhibitor of FBPase (I_{50} of 570 μM and K_i value of 103 μM) (4). The catalytic mechanism of FBPase involves the formation of a covalent bond between the anomeric hydroxyl group of FPB and the gamma amino functionality of a lysine residue of the enzyme. AhG (and AhG-1,6-bisphosphate) do not have this anomeric hydroxyl group, preventing the normal catalytic function of aldolase.

Ceramide synthase biosynthesis is another potential target site. Ceramides are essential lipids in plants and animals. They are particularly important to the integrity of nervous system cells and tissues of animals. Their function in plants is not well understood, although they are significant constituents of the plasma membrane, and inhibition of their synthesis causes rapid cell death. The fumonisins, potent inhibitors of ceramide synthase, are highly phytotoxic (5,6), causing rapid cellular leakage and cell death at low (submicromolar) concentrations. AAL-toxin, a plant pathogen-produced phytotoxin and structural analogue of the fumonisins, produces physiological effects on plants similar to those of the fumonisins (7). Fumonisins and AAL-toxin, both structural analogues of sphinganine, cause rapid and dramatic increases in phytosphingosine and sphinganine in treated plants (8). These ceramide precursors accumulate when ceramide synthase is inhibited (Figure 2).

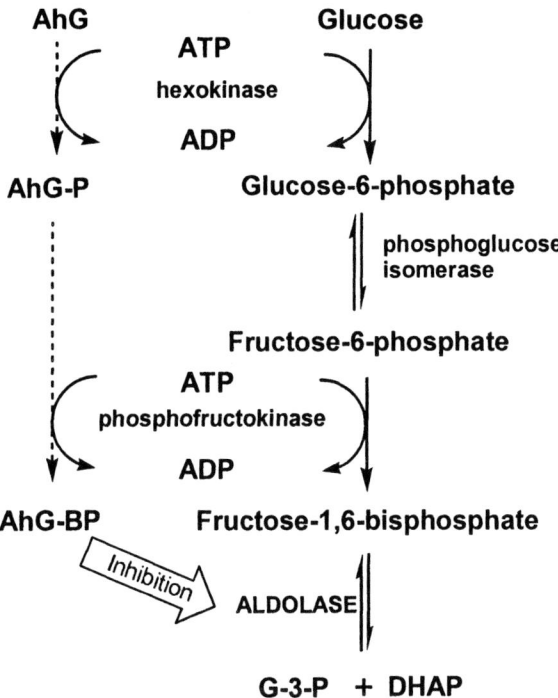

Figure 1. Schematic of the mode of action of AhG with the normal glycolytic pathway on the right (solid arrows) and the consecutive bioactivation of AhG into AhG-BP and the ultimate inhibition of the target site on the left (dashed arrows).

Symptoms of ceramide synthase inhibition could be caused by depletion of ceramide and ceramide derivatives or by toxic increases in sphinganine and its derivatives. The effects of these phytotoxins are so rapid, that it is unlikely that depletion of ceramides is responsible for cellular death. Others have invoked induction of apoptosis in the mode of action of these compounds (e.g., 9); however, treatment of plant tissues with phytosphingosine and sphinganine causes symptoms very similar to those caused by inhibition of ceramide synthase (10). The very rapid plasma membrane damage at one micromolar and higher concentrations makes it unlikely that apotosis is involved in the main phytotoxic effect.

The ceramide synthase pathway is clearly a viable site for herbicides, provided an inhibitor can be found that is plant specific. We have been unsuccessful in finding a ceramide synthase inhibitor with high phytotoxicity, but low mammalian toxicity (*e.g.*, *11*). Even inhibitors with little structural similarity to the fumonisins, such as australifungin (*12*), have relatively little difference between mammalian and plant toxicity. This topic is considered in more detail in a recent review (*13*).

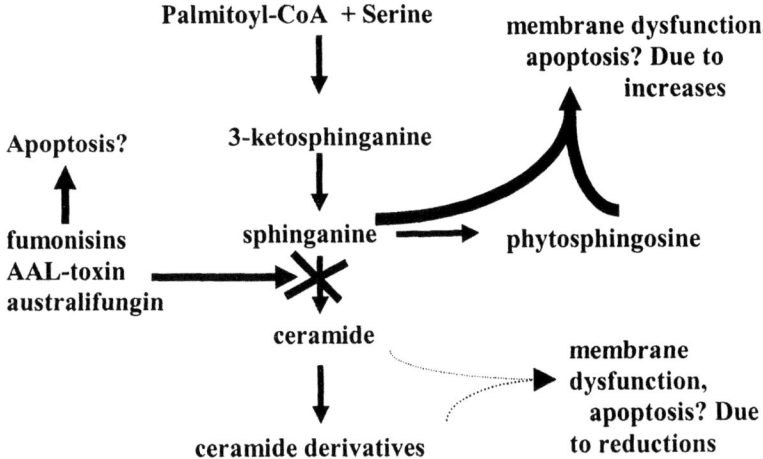

Figure 2. Mechanism of action of ceramide synthase inhibitors in plants. Thick lines indicate increased level effects, whereas thin lines represent decreased level effects.

We have found that the key enzyme in asparagine synthesis, asparagine synthetase (AS; EC6.3.5.4), is yet another potential herbicide target site. AS was hypothesized as a potential herbicide target site when it was discovered that the inhibition of growth caused by natural phytotoxin 1,4-cineole (or its synthetic analogue cinmethylin) could be reversed with exogenous supply of asparagine (*14*). Root uptake of asparagine increased in the presence of the inhibitors, suggesting that the supply of asparagine was low in the treated seedlings. Uptake of aspartate was not affected by the treatments. The *in vitro* AS activity was inhibited by 1,4-cineole (Figure 3A). The synthetic analogue cinmethylin did not inhibit AS, suggesting that it must be metabolized into a 1,4-cineole analogue, ostensibly 2-hydroxy-1,4-cineole, prior to inhibiting AS activity (Figure 3B).

This hypothesis was tested by measuring the inhibitory activity of *cis*- and *trans*-2-hydroxy-1,4-cineole. The *cis*-form was more effective against AS than 1,4-cineole by more than an order of magnitde (Figure 3). The addition of the hydroxyl group to the molecule renders it less volatile, therefore probably allowing more of the compound to react with the site of inhibition. The trans-form of 2-hydroxy-1,4-cineole was less active than either the cis-diastereomer or 1,4-cineole (not shown). Greenhouse studies had demonstrated that the cis-form of cinmethylin was more active than its trans-diastereomer, further supporting the current hypothesis that the orientation of the hydroxyl group may play a key role in the potency of this metabolic intermediate.

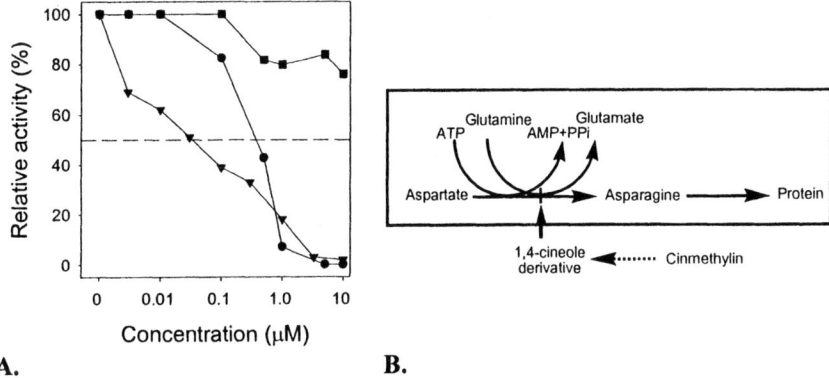

Figure 3. A. Effect of 1,4-cineole (circles), cis-2-hydroxy-1,4-cineole (triangles) and the commercial herbicide cinmethylin (squares) on the activity of asparagine synthetase from lupin. The dotted line represents 50% inhibition of enzyme activity. B. Scheme for proposed mode of action of cinmethylin.

This hypothesis was tested by measuring the inhibitory activity of *cis*- and *trans*-2-hydroxy-1,4-cineole. The *cis*-form was more effective against AS than 1,4-cineole by more than an order of magnitde (Figure 3). The addition of the hydroxyl group to the molecule renders it less volatile, therefore probably allowing more of the compound to react with the site of inhibition. The trans-form of 2-hydroxy-1,4-cineole was less active than either the cis-diastereomer or 1,4-cineole (not shown). Greenhouse studies had demonstrated that the cis-form of cinmethylin was more active than its trans-diastereomer, further supporting the current hypothesis that the orientation of the hydroxyl group may play a key role in the potency of this metabolic intermediate.

Old Target Sites

Several natural phytotoxins have been found by us and others to target the same molecular target site as a class of commercial, synthetic herbicides. In one case below, the compound targets two herbicide target sites at low doses, something that no commercial herbicide does. Thus, even when natural products target old sites, they may have utility as new herbicides or herbicide classes for multiple target sites. They could also be used as allelochemicals that could be genetically engineered into crops.

The allelochemical sorgoleone (2-hydroxy-5-methoxy-3-[(8'Z,-11'Z)- 8'11'14'-pentadecatriene]-*p*-benzoquinone; Figure 4) is exuded in oily droplets from the root hairs of *Sorghum* species. These droplets are up to 90% sorgoleone and its analogues. It is relatively stable in soil and accumulates in the rhizosphere. This molecule is highly phytotoxic to broadleaf and grass weeds at concentrations as low as 10 µM (*15*), and its presence in the soil surrounding sorghum plants inhibits the growth of weeds.

Sorgoleone was initially found to inhibit mitochondrial respiration, but it was later found to be a more potent inhibitor of photosynthetic electron transport of photosystem II (PSII) (*15, 16*). Sorgoleone is structurally similar to plastoquinone (PQ), a benzoquinone involved in photosynthetic electron transport. Sorgoleone competes for the PQ binding site of the D-1 protein in a manner similar to most commercial photosynthetic inhibitors (*17*). The *in vitro* PSII inhibiting activity of sorgoleone is similar to some of the commercial herbicides targeting this site (*e.g.*, atrazine and diuron).

Sorgoleone and other natural quinones can inhibit the enzyme *p*-hydroxyphenylpyruvate dioxygenase (HPPD) (Figure 4), the key enzyme in plastoquinone synthesis (*18*). In addition to acting as a redox reagent in PSII, plastoquinone is a cofactor for phytoene desaturase, the target of many carotenoid synthesis inhibiting herbicides (*19*). Without plastoquinone, phytoene desaturase does not function, resulting in cessation of carotenoid synthesis. Most commercial HPPD inhibitors (*e.g.*, sulcotrione (Figure 4) and isoxaflutole) are competitive, time-dependent (tight-binding) inhibitors. As such, these herbicides bind to the enzyme very tightly with $t_{\frac{1}{2}}$ of dissociation ranging from a few hours to several days, as opposed to milliseconds for traditional reversible inhibitors. Sorgoleone does not behave as these herbicides and appears to be a reversible inhibitor of HPPD. This quinone is structurally more planar than the traditional HPPD inhibitors, so it may not form a stable tightly-binding reaction intermediate. Instead, its backbone may resemble the conformation of one of the later intermediate step in the reaction mechanism of HPPD.

Usnic acid [2,6-diacetyl-7,9-dihydroxy-8,9b-dimethyl-1,3(2H9βH)-dibenzofurandione] (Figure 4) is one of the most common secondary metabolites found in the lichen genus *Usnea*. Seedlings grown in the presence of usnic acid develop chlorosis in the cotyledonary tissues. Loss of chlorophylls in response to phytotoxins can be associated with light-dependent destabilization of cellular and sub-cellular membranes, but usnic acid apparently acts differently since it causes membrane leakage in the absence of light, though the loss of membrane integrity is increased in light. There is a strong decrease in β-carotene in plants treated with usnic acid, and total carotenoids decrease with increased usnic acid concentration. Inhibition of carotenoid synthesis is known to be accompanied with the destruction of chlorophyll (bleaching) due to the destabilization of the photosynthetic apparatus.

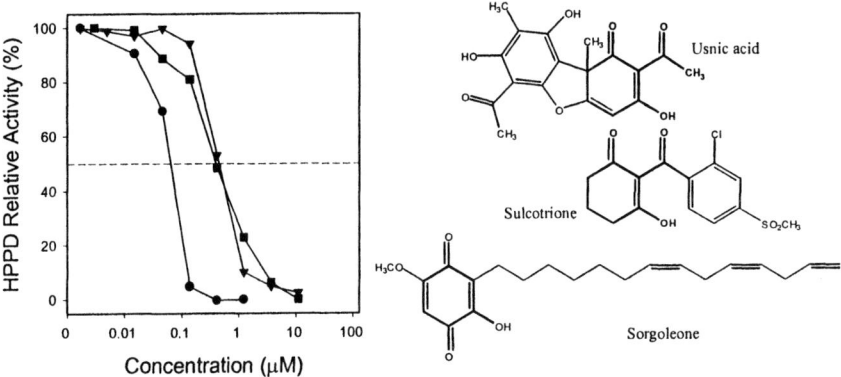

Figure 4. Effect of the β-triketone (-)-usnic acid (circles), the benzoquinone sorgoleone (triangles) and the commercial herbicide sulcotrione (squares) on the activity of *p*-hydroxyphenylpyruvate dioxygenase. The dotted line represents 50% inhibition of enzyme activity.

Although most herbicides that cause normally green tissues to be white target the enzyme phytoene desaturase that converts phytoene to carotenes, this symptom is also associated with inhibition of HPPD, the enzyme responsible for plastoquinone biosynthesis (*20*). Usnic acid possesses a 2-keto-cyclohexane-1,3-dione substructure common to many triketone HPPD inhibitors (*e.g.*, sulcotrione and mesotrione) (*21*). When tested on this enzyme, usnic acid is strongly inhibitory, with an I_{50} of 70 nM, surpassing the activity obtained with the commercial herbicide sulcotrione (Figure 4) (*22*).
Despite high *in vitro* activity, usnic acid was poorly active on 3-week-old morningglory, barnyardgrass, velvetleaf, sicklepod, and yellow nutsedge seedlings in a preliminary greenhouse study. The discrepancy between the high *in vitro* activity on HPPD and the low *in vivo* activity is most likely due to poor

foliar uptake associated with inadequate physicochemical properties of usnic acid.

New Methods for Target Site Determination

Molecular biology is having profound effects on the pesticide industry. In addition to its use in producing herbicide-resistant crops, improved biocontrol agents, and more allelopathic or competitive crops (*23*), it can be used in both the search for new molecular target sites and the elucidation of modes of action of natural phytotoxins.

Using DNA microarray technology for transcriptional profiling is a potentially effective means for elucidating inhibitor modes of action. Theoretically, specific patterns of up- and down-regulated genes might be associated with particular molecular target sites. This could allow the researcher to quickly determine if a new phytotoxin fits a particular transcriptional pattern associated with a target site from a gene transcription library of profiles of compounds with known modes of action. If the results do not fit one of these profiles, the phytotoxin may target a new molecular site, and its effects on gene transcription could provide clues as to what that target is. In addition to genes specifically associated with the molecular target site, stress-related and detoxification-associated genes are likely to be found associated non-specifically in response to exposure to toxic xenobiotics. Separation of target site-specific effects on transcription from effects of these less specific, general toxicant-related genes will simplify this potentially very complicated tool.

Another issue has been determination of proper doses and time points for sampling. The later the data are taken, the more genes unrelated to the primary target site are affected. With plants, an issue is whether to use whole plants or cell cultures. In whole plants, herbicides sometime target only cells of certain tissues, so that whole plant extaction of RNA dilutes the RNA of interest. Uniform and rapid uptake of the phytotoxin is also problematic with whole plants.

We are in the early stages of building gene expression profile libraries for both fungicides and herbicides, using whole genome DNA microarrays for *Saccharomyces cerevisiae* and *Arabidopsis thaliana*, respectively. When generating data on tens of thousands of genes, ordinary statistics will provide hundreds of false positives and negatives (*24*). Experiments must be well designed and executed to provide robust transcriptional profile fingerprints. In many cases, microarray results for genes of interest must be verified with quantitative real-time RT-PCR (*e.g., 25*)

With *Saccharomyces cerevisiae* microarrays, we have found good gene transcription fingerprints for certain molecular target sites. Results with *Arabidopsis thaliana* as a model for studying herbicidal modes of action have thus far been encouraging, although our studies with this organism are less developed than with *S. cerevisiae*. Current experiments should allow us to distinguish between gene expression changes that occur in direct response to the presence of a specific inhibitor, versus those that occur indirectly or due to stress.

Conclusions

Powerful, new tools and approaches to natural product discovery and utilization are rapidly expanding our knowledge of chemical diversity, the biological activity, and the molecular genetics of naturally occurring compounds. Many of these compounds are phytotoxic and have the potential for use in weed management, either as a herbicide or as a genetically engineered allelochemical. But, still, compared to the synthetic herbicide discovery and herbicide-resistant crop development efforts, little has been done with natural products. This knowledge gap and the availability of new capabilities should attract future researchers to focus on discovery and development of natural compounds for weed management.

References

1. Duke, S.O.; Dayan, F.E.; Rimando, A.M. In *Herbicides and Their Mechanisms of Action*; Cobb, A.H.; Kirkwood, R.C. Eds.; Sheffield Academic Press, Ltd., Sheffield, UK, 2000, pp. 105-133.
2. Haake, V.; Zrenner, R.; Sonnewald, U.; Stitt, M. *Plant J.* **1998**, 14, 147-157.
3. Tanaka, T.; Hanato, K.; Watanabe, M; Abbas, H.K. *J.Nat. Toxins* **1996**, 5, 317-329.
4. Dayan, F.E.; Rimando, A.M.; Tellez, M.R.; Scheffler, B.E.; Roy, T.; Abbas, H.K.; Duke, S.O. *Z. Naturforsch.* **2002**, 57c: 645-653,
5. Abbas, H.K.; Boyette, C.D. *Weed Technol.* **1992**, 6, 548-543.
6. Abbas, H.K.; Paul, R.N.; Boyette, C.D.; Duke, S.O.; Vesonder, R.F. *Can. J. Bot.* **1992**, 70, 1824-1833.
7. Abbas, H.K.; Vesonder, R.F.; Boyette, C.D.; Peterson, S.W. *Can. J. Bot.* **1993**, 71, 155-160.

8. Abbas, H.K.; Tanaka, T.; Duke, S.O.; Porter, J.K.; Wray, E.M.; Hodges, L.; Sessions, A.E.; Merrill, A.H.; Riley, R.T. *Plant Physiol.* **1994**, 106, 1085-1093.
9. Gilchrist, D.G. *Cell Death Differ.* *1997*, 4, 689-698.
10. Tanaka, T.; Abbas, H.K.; Duke, S.O. *Phytochemistry* **1993**, 33, 779-785.
11. Abbas, H.K.; Tanaka, T.; Shier, W.T. *Phytochemistry* **1995**, 35, 1681-1689.
12. Abbas, H.K.; Duke, S.O.; Merrill, A.H.; Want, W.; Shier, W.T. *Phytochemistry* **1998**, 47, 1509-1514.
13. Abbas, H.K.; Duke, S.O.; Shier, W.T.; Duke, M.V. In *Advances in Microbial Toxin Research and its Biotechnological Exploitation*; Upadhyay, R.K., Ed.; Kluwer, Amsterdam, 2002, pp. 211-229.
14. Romagni, J.G.; Duke, S.O.; Dayan, F.E. *Plant Physiol.*, **2000**, 123, 725-732.
15. Nimbal, C.I.; Yerkes, C.N.; Weston, L.A.; Weller, S.C. *Pestic. Biochem. Physiol.* **1996**, 54, 73-83.
16. Einhellig, F.A.; Rasmussen, J.A.; Hejl, A.M.; Souza, I.F. *J. Chem. Ecol.* **1993**, 19, 369-375.
17. Gonzalez, V.M.; Kazimir, J.; Nimbal, C.I.; Weston, L.A.; Cheniae, G.M. *J. Agric. Food Chem.* **1997**, 45, 1415-1421.
18. Meazza, G.; Scheffler, B.E.; Tellez, M.R.; Rimando, A.M.; Nanayakkara, N.P.D.; Khan, I.A.; Abourashed, E.A.; Romagni, J.G.; Duke, S.O.; Dayan, F.E. *Phytochemistry* **2002**, 59, 281-288.
19. Devine, M.D.; Duke, S.O.; Fedtke, C. *Physiology of Herbicide Action*. Prentice Hall, Englewood Cliffs, NJ, **1993**, pp. 395-424.
20. Pallett, K.E.; Little, J.P.; Sheekey, M.; Veerasekaran, P. *Pestic. Biochem. Physiol.* **1998**, 62, 113-124.
21. Lee, D.L.; Knudsen, C.G.; Michaely, W.J.; Chin, H.-L.; Nguyen, N.H.; Carter, C.G.; Cromartie, T.H.; Lake, B.H.; Shribbs, J.M.; Fraser, *Pestic. Sci.* **1998**, 54, 377-384.
22. Romagni, J.G.; Meazza, G.; Nanayakkara, N.P.D.; Dayan, F.E. *FEBS Lett.* **2000**, 480, 301-305.
23. Duke, S.O. *Trends Biotechnol.* **2003**, 21, 192-195.
24. Beneš, V.; Muckenthaler, M. *Trends Biotechnol.* **2003**, 28: 244-249.
25. Agarwal, A.K.; Rogers, P.D.; Baerson, S.R.; Jacob, M.R.; Barker, K.S.; Cleary, J.D.; Walker, L.A.; Nagle, D.G.; Clark, A.M. *J. Biol. Chem.* **2003**, In press.

Chapter 15

Mode of Action of Pyrazole Herbicides Pyrazolate and Pyrazoxyfen: HPPD Inhibition by the Common Metabolite

Hiroshi Matsumoto

Institute of Applied Biochemistry, University of Tsukuba, Tsukuba, Ibaraki 305–8572, Japan

Pyrazole herbicides pyrazolate [4-(2,4-dichlorobenzoyl)-1,3-dimethyl-5-pyrazolyl *p*-toluenesulfonate] and pyrazoxyfen [2-(4-(2,4-dichlorobenzoyl)-1,3-dimethyl-pyrazoyl-5-yloxy) acetophenone] have been widely used for annual and perennial weed control in paddy rice. In plants, both herbicides were metabolized to the same compound, 4-(2,4-dichlorobenzoyl)-1,3-dimethyl-5-hydroxypyrazole, and this metabolite has been suggested as the actual active principle of the herbicides. Pyrazolate is only slightly soluble in water, but once dissolved, rapidly hydrolyzed to the herbicidally active metabolite. In contrast, pyrazoxyfen is considerably stable in aqueous solution. When roots of early watergrass (*Echinochloa oryzicola* Vasing.) were treated with pyrazolate at the two-leaf stage, greater accumulation of phytoene was observed in the third leaves 3 days after treatment and the high level of phytoene in the leaves was kept until 9 days. Homogentisate treatment reduced this accumulation in the leaves suggesting the site of action of the herbicide located in the pathway of homogentisate synthesis. The 4-hydroxyphenylpyruvate dioxygenase (HPPD) assay revealed that the metabolite inhibited the enzyme activity with the IC_{50} value of 13 nM. The values of pyrazolate and pyrazoxyfen were 52 nM and 7.5 µM, respectively. In pyrazolate solution used in the assay, a part of the herbicide was possibly hydrolyzed to the metabolite. Analysis of herbicides solution with HPLC showed that the 52 nM of pyrazolate and 7.5 µM pyrazoxyfen solutions contained the metabolite at the similar concentration with its IC_{50} value. These data strongly suggest that these pyrazole herbicides inhibit HPPD after conversion to the herbicidally active metabolite in aqueous solution and/or in plants.

Herbicides interfering carotenoid biosynthesis induce bleaching of growing leaves that develop after the herbicide treatment. The best-studied site of herbicide action in the carotenoid biosynthesis pathway is inhibition of phytoene desaturase. Inhibition of this enzyme causes a huge accumulation of phytoene in treated plants. The enzyme is involved in two desaturation reactions (two double bonds formed), with the end product being ζ-carotene. Therefore, phytofluene also accumulates to some extent if phytoene desaturase is inhibited. Although not infallible, the accumulation of phytoene is the most commonly reported proof for the herbicide action at phytoene desaturase site. The other target site of herbicidal compounds in carotenogenesis-related pathways is 4-hydroxyphenylpyruvate dioxygenase (HPPD). HPPD catalyzes conversion of 4-hydroxyphenylpyruvate to homogentisate in the biosynthetic pathway of plastoquinone and α-tocopherol. This inhibition leads to decreased plastoquinone content that impairs phytoene desaturation since plastoquinone is needed as a key cofactor for accepting hydrogen from phytoene desaturase. Herbicides sulcotrirone and isoxaflutole that belong to triketone and benzoylisoxazole classes, respectively, inhibit HPPD as the primary target (1-3).

Pyrazolate [4-(2,4-dichlorobenzoyl)-1,3-dimethyl-5-pyrazolyl p-toluenesulfonate] and pyrazoxyfen [2-(4-(2,4-dichlorobenzoyl)-1,3-dimethyl-pyrazoyl-5-yloxy)acetophenone] have has been used for annual and perennial weeds control in paddy rice. Both herbicides were metabolized in plants to the same compound, 4-(2,4-dichlorobenzoyl)-1,3-dimethyl-5-hydroxypyrazole, and this metabolite has been suggested as the actual active principle of the herbicidal activity (Fig. 1). Pyrazolate is only slightly soluble in water but once dissolved, it is rapidly hydrolyzed to the herbicidally active metabolite (4). In contrast, pyrazoxyfen is considerably stable in aqueous solution. Since the metabolite has similar substructure of 2-benzoylethen-1-ol with sulcotrirone, Lee *et al.* (5) suggested the possibility that the metabolite acted as HPPD inhibitor. To obtain more information on the mode of action of the pyrazole herbicides, pigment contents in each leaf blade were separately analyzed in susceptible weed, early watergrass (*Echinochloa oryzicola* Vasing.). Furthermore, their *in vitro* effect on HPPD was determined by extracting the enzyme activity from suspension-cultured carrot cells.

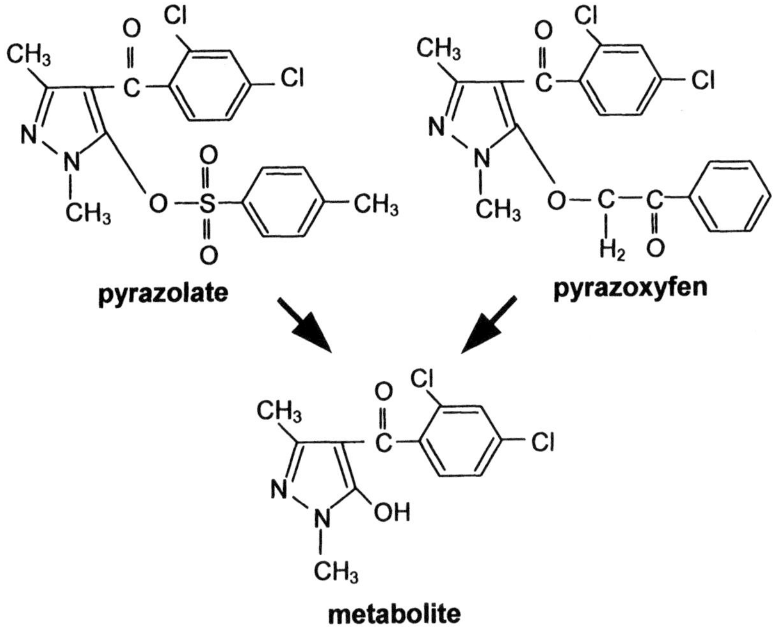

Fig.1. Chemical structures of pyrazolate, pyrazoxyfen and their metabolite.

MATERIALS AND METHODS

Plant Materials

Early watergrass (*Echinochloa oryzicola* Vasing.) was germinated in darkness at 30°C and grown hydroponically with modified Kasugai nutrient solution (6) to the 2-leaf stage in growth chambers under a 12hr, 25 °C day (white light of 360 μ mol m^{-2} s^{-1} PAR) / 20 °C night cycle.

Herbicides Treatment

Roots of early watergrass were soaked for 24 hr into pyrazolate or pyrazoxyfen solution (2.5×10^{-5}M, 5×10^{-5}M) containing 0.5% volume of acetone. The solutions were prepared just before the herbicide treatments. Treated plants were then transferred to an herbicide-free nutrient solution and grown until 9 DAT. The nutrient solution renewed every 2 days.

Pigments Extraction and determination

The second or third leaf-blades (500 mg fresh weight) of early watergrass were separately sampled from 7 plants at immediately, 3, 6, and 9 days after herbicides treatment. The leaf tissues were homogenized in 4 ml of chloroform: methanol (1:1, v/v) using a chilled mortar and pestle under dim light. The homogenates were then transferred to test tubes, to which were added 5 ml distilled water. After vigorous shaking, they were centrifuged at 5500 rpm for 7 min to separate chloroform and water phases. The chloroform phases (1 ml) were evaporated to dryness under a stream of nitrogen gas. The remaining residues were dissolved in 1 ml of a solution of acetonitrile: methanol: chloroform (67.5:22.5:10, v/v/v) and its 20 µl aliquot was supplied for pigment analysis. Separation, identification, and quantification of pigments were performed by HPLC system (Waters 600E) equipped with a photodiode array detector (Waters 996). Pigments were separated by a Waters 150x3.6 mm Nova-Pak C18 column preceded by a Nova-Pak C18 pre-column with a solution of acetonitrile: methanol: chloroform (67.5:22.5:10, v/v/v) at a flow rate of 1.0 ml min^{-1}. Each pigment was identified by its retention time and absorption spectra monitored simultaneously at 287 (phytoene), 430 (chlorophyll a), and 455 nm (chlorophyll b and β-carotene). β-Carotene standard (Wako Pure Chemical, Japan) was used for its identification and quantification. Phytoene was identified by comparison of absorption spectra with that in the literature (7). All quantitative comparisons of pigment levels were made using areas under selected peaks determined by integration at each wavelength.

Cell Growth Condition

The suspension cultures of carrot (*Daucus carota* L. cv. Harumakigosun) cells were grown in Murashige and Skoog medium, pH 5.8, containing 30 g liter^{-1} sucrose and 2 mg liter^{-1} of 2,4-dichlorophenoxyacetic acid. The cultures were maintained on a gyratory shaker agitating 110 rpm under dim light at 25 °C and sub-cultured 3 weeks interval.

HPPD Preparation and Assay

HPPD activity was isolated form the carrot cells according to the methods reported by Viviani *et al.* (8) with minor modification. One liter of carrot cells grown for 7 days after sub-culturing were harvested through one layer of Miracloth and gently squeezed to remove excess liquid medium. Cells were then homogenized for 30 s twice at 4 °C with a Hitachi HG-30 homogenizer at a maximum speed in 1 ml g^{-1} fresh weight of an extraction buffer. The extraction buffer was 0.4 M Tris-HCl, pH8.0, containing dithiothreitol (1 mM), EDTA (1 mM), EGTA (1 mM), aminocapronate (5 mM), benzamide (1 mM), and phenylmethylsulfonyl fluoride (PMSF; 0.1 mM). The cell homogenate was centrifuged for 5 min at 10,000 g at 4 °C and the resulting supernatant was again centrifuged for 60 min at 100,000 g at 4 °C. The clear supernatant was then subjected to a 35-60 % ammonium sulfate precipitation and centrifuged or 5 min at 100,000 g. The final pellet was gently re-suspended in a minimum volume of storage buffer 0.1 M Tris-HCl, pH 7.0, to obtain a homogeneous protein solution. This was centrifuged for 5 min at 20,000 g at 4 °C and then 0.6 volume of glycerol was added to the clear supernatant obtained. This solution was stored at -80 °C until supplied for HPPD assay. The protein concentration was determined with the Bradford method (9). An assay for HPPD activity was carried out in a final volume of 60 µl in a micro test tube containing 0.1 mM Tris-HCl, pH 7.2, ascorbate (10 mM), glutathione (10 mM), 20 µg of protein, and 4-hydroxyphenylpyruvate (HPPA; 0.1 mM). The enzyme was pre-incubated in the buffer containing ascorbate and glutathione for 5 min at room temperature before starting the reaction by addition of HPPA. The herbicide was added to the solution as 0.5 % dimethyl sulfoxide (DMSO) solution. Incubation was continued up to 7.5 min at 30 °C with gentle agitation and stopped by addition of 10 µl of a 25% perchloric acid solution. The tube was then centrifuged at 3,000 g for 10 min and the supernatants subjected to HPLC analysis using the system as described above. Assay solution (10 µl) was injected onto a Waters 150x3.6 mm Symmetry C18 column equilibrated in mobile phase of acetonitrile : trifluoroacetic acid (TFA) : water (10 : 0.1 : 89.9, v/v/v) and eluted at a flow rate of 1.0 ml min^{-1}. The enzymatic production of HGA was monitored at 292 nm with a comparison of authentic sample of HGA for retention time and absorbance spectrum.

Determination of metabolite in aqueous solution.

The concentration of the metabolite in pyrazolate and pyrazoxyfen solutions used in the HPPD assay were determined by the HPLC system with a YMC J'sphere ODS-M80 column and mobile phase of acetonitrile : water (6:4, v/v).

RESULTS AND DISCUSSION

In the whole plant studies, pyrazolate or pyrazoxyfen was treated to roots of rice and early watergrass when their second leaves were fully expanded. Both herbicides caused severe growth retardation and bleaching on the third leaves of early watergrass while no effect on rice (data not shown).

The effects of pyrazolate and norflurazon on pigment contents in the second and third leaves of early watergrass are shown in Table 1. The third leaves could not be harvested until 3 DAT since they developed after the herbicides treatment. At the first sampling time (3 DAT), norflurazon caused significant decrease of chlorophyll in the third leaves. In pyrazolate-treated plants, however, chlorophyll content in the third leaves was not different with the untreated control 3 DAT and it was decreased between 3 and 6 DAT. There was no noteworthy difference in chlorophyll contents in the second leaves between the herbicides throughout 9 days although slight decrease was observed during 6 and 9 DAT. The declining pattern of β−carotene in the third leaf of early watergrass was very similar to that of chlorophyll (Table 1). In the second leaves, although some of β−carotene decreased during the 24 hr herbicides treatment (0 day), there was no remarkable decrease throughout 9 days. These results indicated that, at used concentrations, pyrazolate as well as norflurazon caused severe bleaching in the third leaves but the bleaching of the developing leaves appeared later in pyrazolate-treated plants. Norflurazon interferes with carotenoids synthesis by inhibiting phytoene desaturase and accumulate phytoene (10). Therefore a pattern of phytoene accumulation by pyrazolate was compared with that by norflurazon. Both herbicides induced greater accumulation of phytoene in the third leaves of early watergrass 3 DAT and the high levels were kept until 9 DAT (Table 1). In the second leaves, however, phytoene accumulation was observed only in norflurazon-treated plants. Although the final symptom in early watergrass was very similar between herbicides and the symptom was considered to be due to the inhibition of carotenoid biosynthesis, these data suggest that the primary site of pyrazolate differed from that of norflurazon (phytoene desaturase).

The earlier study on mode of action of pyrazolate by Kawakubo and Shindo (11) had suggested that the herbicidally active metabolite of pyrazolate blocked protochlorophyllide synthesis by removing Mg^{2+} from Mg-protoporphyrins. They concluded that the metabolite worked as an acid (donating H^+ to chlorophyll) to form pheophytin. We also tried to reconfirm the formation of pheophytin by the metabolite *in vivo* and *in vitro*. When it was added to the pigment solution at 5×10^{-4} M and kept in a dark cold room for 24 hr, a dramatic accumulation of pheophytin (43-fold) with a concomitant depletion of

Table 1. Change in chlorophyll, β-carotene and phytoene contents in the second or third leaves of early watergrass treated with pyrazolate (5×10^{-5}M) or norflurazon (10^{-5}M) at the second leaf-stage.

	0 day	3 day	6 day	9 day
Chlorophyll Content (μg/mg fresh tissue ± SE)				
Second leaf				
Control	0.217±0.0060	0.225±0.0072	0.232±0.0026	0.150±0.0023
Pyrazolate	0.157±0.0026	0.177±0.0035	0.188±0.0116	0.165±0.0017
Norflurazon	0.186±0.0027	0.182±0.0096	0.176±0.0006	0.156±0.0075
Third leaf				
Control	-	0.128±0.0010	0.210±0.0021	0.182±0.0116
Pyrazolate	-	0.117±0.0010	0.067±0.0022	0.063±0.0060
Norflurazon	-	0.051±0.0003	0.030±0.0010	0.022±0.0021
β-Carotene Content (ng/mg fresh tissue ± SE)				
Second leaf				
Control	0.070±0.0035	0.070±0.0027	0.073±0.0023	0.051±0.0015
Pyrazolate	0.047±0.0005	0.049±0.0013	0.055±0.0031	0.048±0.0077
Norflurazon	0.049±0.0010	0.034±0.0020	0.041±0.0035	0.039±0.0020
Third leaf				
Control	-	0.038±0.0010	0.059±0.0024	0.050±0.0042
Pyrazolate	-	0.031±0.0001	0.020±0.0007	0.017±0.0015
Norflurazon	-	0.009±0.0001	0.008±0.0003	0.006±0.0007
Phytoene Content (peak area x 10^3/ mg fresh tissue ± SE)				
Second leaf				
Control	0.6± 0.15	3.7± 0.10	2.7± 0.64	3.6± 0.16
Pyrazolate	0.6± 0.08	4.3± 0.07	1.8± 0.10	11.4± 2.30
Norflurazon	49.8± 2.86	955.2±58.02	1243.3±42.75	1316.2±203.2
Third leaf				
Control	-	1.8± 0.10	0.7± 0.05	3.5± 0.92
Pyrazolate	-	91.3± 0.26	109.7± 8.21	87.6± 3.25
Norflurazon	-	455.0± 9.22	275.2±18.35	315.2±12.72

chlorophyll was detected(data not shown). In contrast, norflurazon did not cause any effect on the pigment content. However, there was no accumulation of pheophytin in the second and third leaves of early watergrass *in vivo* when the metabolite or norflurazon was treated to intact early watergrass as same as the growth response study. This indicates that the removal of Mg^{2+} is hardly occurred in the intact plants. Therefore, decrease of chlorophyll content in early watergrass shown in Table1 is not considered to be due to conversion of chlorophyll to pheophytin.

The phytoene content in the third leaves of early watergrass treated with the mixture of homogentisate and the herbicides was determined at 6 DAT. Adding homogentisate to pyrazolate reduced the phytoene accumulation in early watergrass (Fig. 2). However, the reduction was not observed in the plants treated with norflurazon-homogentisate mixture. This characteristic effect of homogentisate on phytoene content in pyrazolate-treated early watergrass together with the recent identification that HPPD as a new target for herbicides (1,2), and structural similarity of the metabolite with HPPD inhibitor sulcotrirone (2) led to determination of effect of pyrazolate, pyrazoxyfen and the metabolite on this enzyme. For HPPD assay, the enzyme activity extracted from carotenogenic carrot cell culture was used. The cell is highly susceptible to the herbicide metabolite and the growth was remarkably suppressed at 10 µM (Fig.3). In vitro assay of the enzyme revealed that the metabolite inhibited the enzyme with an IC_{50} value of 13 nM (Fig. 4). To our knowledge, this is the first data showing direct inhibition of HPPD by the metabolite of pyrazole herbicides. The assay also showed that the IC_{50} values for pyrazolate and pyrazoxyfen were 52 nM and 7.5 µM, respectively (Fig. 4). In the pyrazoxyfen solution used in the study, a part of the compound might be hydrolyzed to the metabolite due to its physicochemical property (4). Analysis of herbicides solution with HPLC showed that the 52 nM of pyrazolate and 7.5 µM pyrazoxyfen solutions contained the metabolite at the similar concentration with its IC_{50} value (Fig. 5) From all of the data obtained, it is strongly suggested that that these pyrazole herbicides inhibit HPPD after conversion to the herbicidally active metabolite in aqueous solution and/or in plants. The inhibition of HPPD may cause the reduction in plastoquinone levels and this results in the *in vivo* inhibition of phytoene desaturase and accumulation of phytoene. This indirect effect of the metabolite on phytoene desaturase may lead to slower bleaching in the developing leaves compared with the direct phytoene desaturase inhibitor norflurazon.

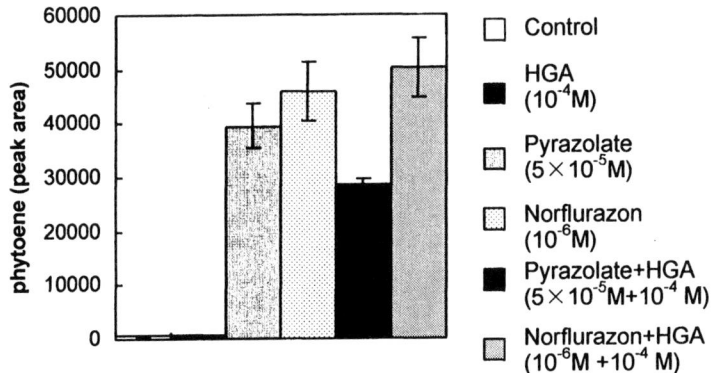

Fig. 2. Effect of homogentisate (HGA) on phytoene accumulation in the third leaves of early watergrass 6 days after treatment. Error bars are SE of the means. Reproduced with the permission from ref. 12. Copyright Blackwell Publishing Ltd.

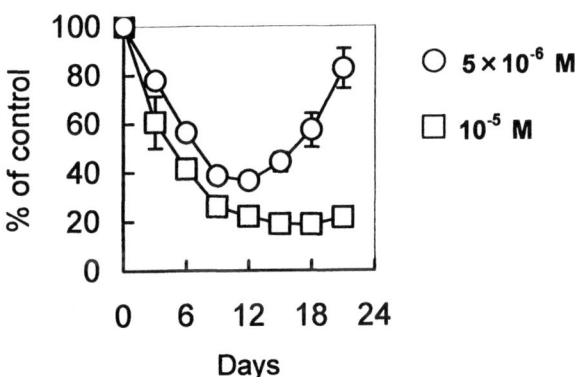

Fig. 3. Effect of the metabolite on growth of suspention-cultured carrot cells. Error bars are SE of the means.

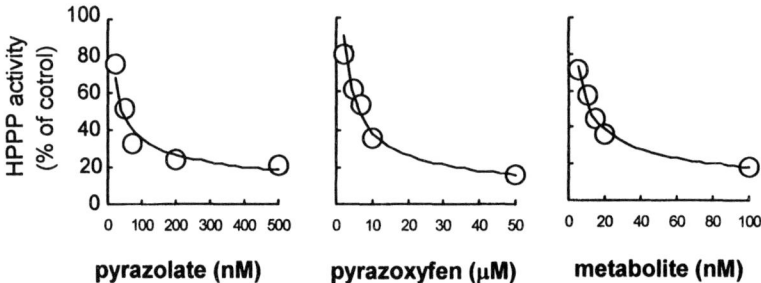

Fig.4. Effect of pyrazolate, pyrazoxyfen and their hydrolytic metabolite on 4-hydroxyphenylpyruvate dioxygenase (HPPD) activity extracted from carrot suspension-cultured cells. The enzyme activity of non-treated control was 4.9 nmol homogentisate mg protein^{-1} min^{-1}. Some error bar (SE) is obscured by the datum symbol.

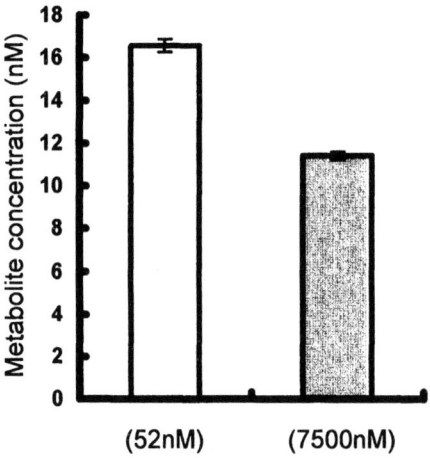

Fig. 5. Concentration of metabolite in pyrazole herbicides solution used in HPPD assay.   Error bars are SE of the means.

REFERENCES

1. Schulz, A.; Orr, O.,; Beyer, P.; Kleinig, H. *FEBS Lett.* **1993**, 318, 162.
2. Secor, J. *Plant Physiol.* **1994**, 106, 1429.
3. Pallett, K. E.; Little, J. P.; Sheekey, M.; Veerasekaran, P. *Pestic. Biochem. Physiol.* **1998**, 62, 113.
4. Yamaoka, K.; Nakagawa, M.; Ishida, M. *J. Pestic. Sci.* **1987**, 12, 209.
5. Lee, D. L.; Prisbylla, M. P.; Cromartie, T. H.; Dagarin, D. P.; Howard, S. W.; Provan, W. M.,; Ellis, M. K.; Frazer, T.; Mutter, L. C. *Weed Sci.* **1997**, 45, 601.
6. Ohta, Y.; Yamamoto, K.; Deguchi, M. *Jpn. J. Soil Sci. Plant Nutri.* **1970**, 41, 19 (in Japanese).
7. Böger, P.; Sandmann, G. *Photosynthetica* **1993**, 28, 481.
8. Viviani, F.; Little, J. P.; Pallett, K. E.. *Pestic. Biochem. Physiol.* **1998**, 62, 125.
9. Bradford, M. M. *Anal. Biochem.* ,**1976**, 72, 248.
10. Sandmann, G.,; Linden, H.; Böger, P. *Z. Naturforsch.* **1989**, 44c, 787.
11. Kawakubo, K.; Shindo, M. *Plant Physiol.* **1979**, 64, 774.
12. Matsumoto, H.; Mizutani, M.; Yamaguchi, T.; Kadotani J. *Weed Biol. Manag.* **2002**, 2, 39.

Chapter 16

Mechanism of Selective Actions of Neonicotinoids on Insect Nicotinic Acetylcholine Receptors

Kazuhiko Matsuda[1] and David B. Sattelle[2]

[1]Department of Agricultural Chemistry, Faculty of Agriculture, Kinki University, Nara, Japan
[2]MRC Functional Genetics Unit, Department of Human Anatomy and Genetics, University of Oxford, South Parks Road, Oxford OX1 3QX, United Kingdom

Neonicotinoid insecticides act selectively on insect nicotinic acetylcholine receptors (nAChRs), but little is known about the mechanism of selectivity. To elucidate the mechanism, structural features of neonicotinoids and insect nAChRs contributing to this selectivity have been examined. Using molecular-oribital calculations, electrostatic interactions and hydrogen-bond formation of neonicotinoids with insect nAChRs were postulated to contribute to the selectivity of neonicotinoid-nAChR interactions. Also, the use of voltage-clamp electrophysiology combined with molecular biology showed that replacement of the vertebrate α4 subunit in the α4β2 nAChR by *Drosophila* α subunits and mutation to basic residues of an amino acid in loop D of the α7 nAChR enhanced neonicotinoid sensitivity of the nAChRs. These findings suggest important roles for α and non-α subunits in the selective actions of neonicotinoids on insect nAChRs.

Introduction

Nicotinic acetylcholine receptors (nAChRs) are ligand-gated ion channels that play central roles in mediating fast cholinergic synaptic transmission not only in vertebrate, but also invertebrate nervous systems (*1*). The cation-selective ion channels of nAChRs open in response to the binding of ACh, inducing depolarization of the postsynaptic membranes. Unlike ACh, which is hydrolyzed by acetylcholine esterases, nicotinic agonists and antagonists lacking the ester linkage can generate, respectively, sustained activation, or block of nAChRs. Thus, their application to the nervous system results in modulation of cholinergic transmission. Among the nicotinic ligands, nicotine has been used to control pests but is only of historical interest when compared to the intensive use of other neurotoxic insecticides such as organophosphates and synthetic pyrethroids. Therefore, until recently, nAChRs have been undervalued as targets for developing new pesticides.

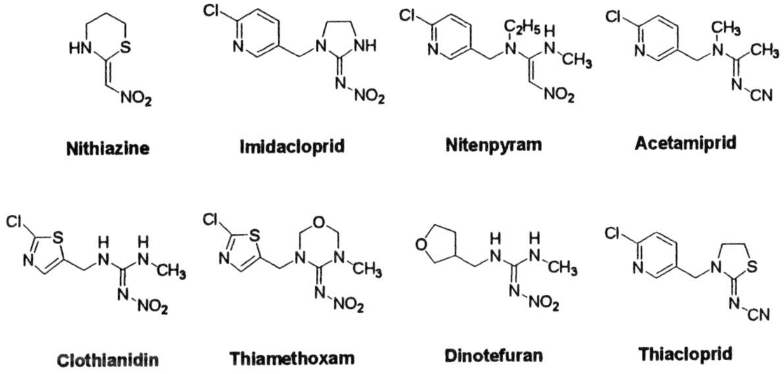

Figure 1. Chemical structures of neonicotinoids.

Soloway *et al.* reported that nithiazine (Fig. 1) exhibited good insecticidal activity but its mammalian toxicity was low (*2*). Schroeder and Flattum investigated the actions of nitromethylene compounds on the nerve cords of American cockroaches (*3*). One of the compounds tested elicited excitation at the 6th abdominal ganglion which was blocked by *d*-tubocurarine, but had no direct effect on the giant fiber axons, suggesting that the targets of this compound were the cholinergic synapses (*3*). Sattelle *et al.* investigated the mode of action of nithiazine by electrophysiology and binding experiments (*4*). Nithiazine was found to depolarize the fast coxal depressor motor neuron Df, an action blocked by mecamylamine and α-bungarotoxin (α-Bgt), and to enhance binding of [^{3}H]H$_{12}$-histrinicotoxin (*4*). Nithiazine was also tested on the locust αL1 nAChR expressed in *Xenopus* oocytes and shown to activate the expressed receptor (*5*). All these results indicate that nAChRs are the target site of nithiazine.

Although nithiazine was not a commercial product, Dr. Kagabu and coworkers synthesized a number of nitromethylene, nitroimine and cyanoimine compounds to develop novel, practical insecticides (*6*). Among these compounds, imidacloprid (Fig. 1) was most effective on the sucking insect pests tested and its toxicity to mammals was minimal. Since imidacloprid was more resistant to photo-decomposition than nithazine and showed good systemic distribution in crop plants, it was introduced commercially (*6*). Bai *et al.* found that imidacloprid and its nitromethylene analog PMNI with no chlorine atom in the pyridine ring depolarized the Df motor neuron of American cockroaches, which was blocked by mecamylamine and dihydro-β-erythroidine (*7*). In addition, these compounds also displaced [^{125}I]α-Bgt binding, suggesting that, like nithiazine, they act as agonists on nAChRs (*7*). Using a patch clamp technique, Zwart *et al.* compared agonist actions of PMNI on nAChRs in locust ganglion neurons with those on mouse BC2H1 muscle cells and mouse NIE-115 neuroblastoma cells (*8*). The compound exhibited much stronger agonist actions on locust nAChRs than on nAChRs of vertebrate cells (*8*). Liu and Casida deployed [^{3}H]imidacloprid to show that the binding affinity of the ligand for house-fly membrane preparations was much higher than that for rat brain membrane preparations (*9*). These two studies suggested that the selective toxicity of the insecticide is attributable at least in part to its more potent agonist actions on insect nAChRs compared with vertebrate nAChRs.

Since the discovery of imidacloprid, diverse imidacloprid-related insecticides referred to as neonicotinoids have been synthesized (Fig. 1). There are several lines of evidence for their nicotinic actions (*10*) but much less is known about the mechanism of selectivity to insect nAChRs. Therefore, structural features of neonicotinoids and insect nAChRs have been investigated.

Structural and Physicochemical Properties of Neonicotinoids Contributing to Their Selective Actions

Neonicotinoids possess nitroimine, nitromethylene or cyanoimine moieties (Fig. 1) as one of their key structural features (*6*). Since the two nitrogens in the imidazolidine ring of imidacloprid are conjugated with the nitro group, the

2-nitroimino-imidazolidine moiety of imidacloprid has a planar structure as demonstrated by X ray crystallography (*11*). In addition to this structural feature, electrostatic properties of neonicotinoids differ strikingly from those of acetylcholine, nicotine and epibatidine. At neutral pH, imidacloprid is substantially free from protonation because of its low pKa values (*6*), while the nitrogen in the pyrrolidine ring of nicotine is mostly protonated, having a positive charge. Yamamoto *et al.* used ^{15}N-NMR to show that the nitrogen at the 1 position in the imidazolidine ring of imidacloprid is slightly positive (*12*). However, Tomizawa *et al.* deployed semi-empirical molecular-orbital calculations to show that the atomic charge of the N-1 nitrogen in the imidazolidine ring is not positive (*13*), and therefore undervalued the electrostatic nature of the nitrogen. Nevertheless, it was found, by semi-empirical molecular-orbital calculations, that the nitrogens in the imidazolidine ring of imidacloprid are changed to positive once the nitro group oxygen forms a hydrogen bond with the positively-charged ammonium cation (Fig. 2) (*10*). Such a change does not occur when the hydrogen bond is formed with the hydroxyl group of phenol and methanol, although the hydrogen bonding can stabilize the nAChR-imidacloprid complex (*14*). These results suggest that the basic residues in insect nAChRs are likely to play an important role in determining their neonicotinoid sensitivity.

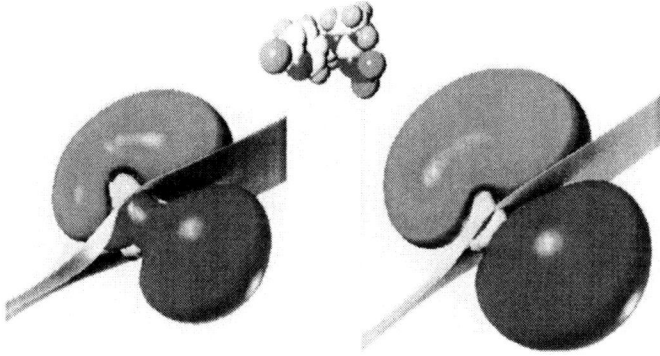

*Figure 2. Contour plots for electrostatic potential of imidacloprid calculated by a semi-empirical molecular orbital method PM3 combined with the AMSOL programme. The plot on the left shows unpolarized imidacloprid, whereas that on the right shows the compound polarized by an ammonium located in the vicinity of the nitro group. Contours are plotted at +1 (red), zero (yellow) and -1 kcal/mol (blue). A space-filling model of imidacloprid is shown between the two plots. Reproduced from ref. 10 with permission of Elsevier.
(See page 2 of color insert.)*

Interactions of α subunits with neonicotinoids

The nAChRs are glycoproteins consisting of two α and three non-α subunits (Fig. 3) (*15-19*). From vertebrates, muscle (α1, β1, δ, γ and ε) and neuronal (α2-α10 and β2-4) nAChR subunits have been isolated. Of these, α7-α9 subunits are able to form homomers when expressed alone in *Xenopus laevis* oocytes (*20-22*). Agonists bind at the interface of α and non-α subunits of the heteromeric receptors or between each subunit of the homomeric receptors. Six loops (A-F) in the long N-terminal domain of nAChR subunits

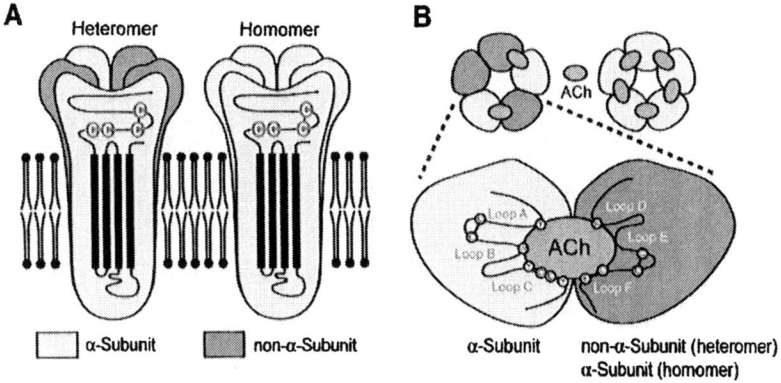

Figure 3. Schematic representation of nicotinic acetylcholine receptors. Side (A) and top (B) views of heteromeric and homomeric nicotinic receptors. Loops A-F forming the agonist binding site of heteromeric receptors and key amino acid residues that have been shown to contribute to the agonist binding are also illustrated in B. Abbreviation: ACh, acetylcholine. Reproduced from Ref. 10 with permission of Elsevier. (See page 2 of color insert.)

have been identified as regions that contribute to the formation of the agonist binding site (Fig. 3) (*16-18*). In contrast with the intensive studies on the vertebrate nAChRs, the functions and pharmacology of insect nAChRs are much less understood. Although a number of genes encoding insect nAChR subunits have been isolated (*10*) and the *Drosophila melanogaster* genome-sequencing project has revealed the complete family of nAChR genes (*23*), it is

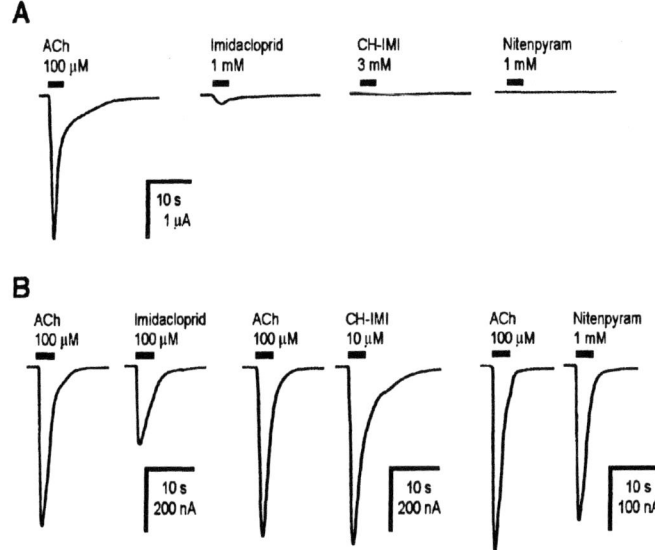

Figure 4. Agonist actions of imidacloprid and related compounds on chicken α4β2(A) and Drosophila SAD-chicken β2 (B) nicotinic acetylcholine receptors expressed in Xenopus oocytes. Abbreviations: ACh, acetylcholine; CH-IMI, a nitromethylene analog of imidacloprid. Reproduced from ref. 14 with permission of Elsevier.

still difficult to express functional insect nAChRs not only in *Xenopus laevis* oocytes, but also in several insect cell lines. Nevertheless, *Drosophila* α subunits ALS and SAD were shown to form functional nAChRs when co-expressed with a vertebrate non-α (β2) subunit in *Xenopus* oocytes (*24*).

By exploiting the *Drosophila*-vertebrate hybrid receptors, the neonicotinoid sensitivity of the chicken α4β2 nAChR was compared with that of *Drosophila* SAD-chicken β2 and *Drosophila* ALS-chicken β2 hybrid receptors expressed in *Xenopus* oocytes (*14, 25*). Two-electrode voltage-clamp (TEVC) electrophysiology was used to show that neonicotinoids activated the SADβ2 hybrid nAChR at much lower concentrations than those required to activate the chicken α4β2 receptor (Fig. 4). Imidacloprid was not effective in activating the ALSβ2 hybrid nAChR but blocked the ACh-induced response at low concentrations (*14*), suggesting that the *Drosophila* α subunits tested have structural features favorable for interactions with neonicotinoids.

Interactions of loop F with neonicotinoids

Since the α7 nAChR subunit forms a homomer when expressed in *Xenopus* oocytes (20), it has, not only loops A-C, but also loops D-F normally seen in non-α subunits. Of these, loop F has been proposed to interact electrostatically with the quaternary ammonium moiety of ACh (15). Since the electrostatic property of imidacloprid differs considerably from that of ACh, interactions of loop F of the chicken α7 nAChR with neonicotinoids were investigated by site-directed mutagenesis combined with TEVC electrophysiology (26). G189D and G189E mutations markedly reduced responses of the α7 nAChR to imidacloprid and nitenpyram, whereas G189N and G189Q mutations were scarcely influenced the responses. By contrast, agonist actions of desnitro-imidacloprid, an imidacloprid derivative lacking the nitro group (Fig. 1), were hardly affected by the G189D and G189E mutations, demonstrating that the reduction of the α7 nAChR responses to neonicotinoids can be attributed to the electrostatic repulsion between the negatively-charged oxygens of the nitro group in the neonicotinoids and the negatively-charged oxygens of the carboxylate group in the newly-added acidic residues (26). Therefore, loop F of the α7 nAChR was postulated to be located close to the nitro group of the insecticides.

Interactions of loop D with neonicotinoids

Although no high-resolution, three-dimensional structure of nAChRs has yet been reported, an acetylcholine binding protein (AChBP) from the pond snail *Lymnaea stagnalis* was recently crystallized (27). AChBP from the glia of the snail regulates the ACh concentrations in the synapse (28), is homologous to the N-terminus agonist binding domain of the α7 subunit and possesses loops A-F. In addition, AChBP is able to form a homo-pentamer like the α7 subunit (28). In the crystal structure of AChBP, Y164 in loop F, corresponding to G189 of the α7 subunit, faces the agonist binding site (27). It was found that Q55, corresponding to Q79 in loop D of the α7 nAChR, was located close to Y164 in loop F of the AChBP (29). If G189 is located close to the nitro group of imidacloprid in the imidacloprid-α7 nAChR complex, then Q79 is also likely to be located in the proximity of the nitro group.

Based on this hypothesis, the role of Q79 in loop D of the α7 nAChR in its interactions with neonicotinoids was investigated by TEVC electrophysiology combined with site-directed mutagenesis (29). The agonist responses of the α7 nAChR to imidacloprid and nitenpyram were markedly reduced by the Q79E mutation, whereas the responses were increased by mutations Q79K and Q79R. By contrast, agonist actions of desnitro-imidacloprid lacking the nitro group were increased by the Q79E mutation, whereas responses were reduced by the Q79K and Q79R mutations (Fig. 5) (29). The results indicate that the changes in neonicotinoid sensitivity resulting from the mutations of the α7 nAChR are attributable to electrostatic interactions of the nitro group of neonicotinoids with the added residues. Most of the amino acid residues in loop D of insect non-α subunits are basic (Table 1). Therefore, such basic residues are likely to play a key role in strengthening the interactions with neonicotinoids.

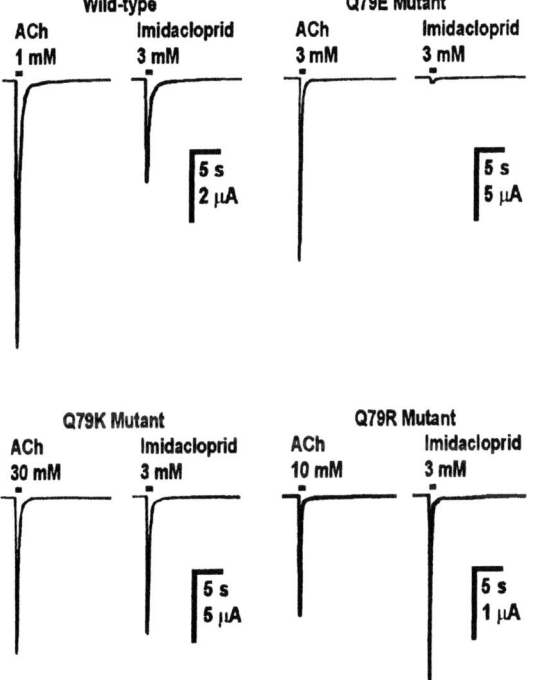

Figure 5. Agonist actions of imidacloprid on the wild-type and mutant (Q79E, Q79K and Q79R) α7 nicotinic acetylcholine receptors expressed in Xenopus oocytes. Abbreviation: ACh, acetylcholine. Reproduced from ref. 29 with permission of Nature Publishing Group.

Table 1. Amino Acid Sequences in Loop D of Vertebrate and Insect Nicotinic Acetylcholine Receptor Subunits

Subunit	Sequence
	↓*
Chicken α7	TNIWLQMYWTD
AChBP	VVFWQQTTWSD
Chicken β2	TNVWLTQEWED
Chicken β4	TNVWLNQEWID
Human β2	TNVWLTQEWED
Torpedo δ	SNVWMDHAWYD
Torpedo γ	TNVWIEIQWND
Drosophila ARD	SNVWLRLVWYD
Drosophila SBD	TNLWVKQRWFD
Drosophila β3	THCWLNLRWRD
Locusta β	SNVWLRLVWND
Myzus β1	SNVWLRLVWRD

*The arrow points Q79 in loop D of the α7 subunit and the corresponding amino acid residues of vertebrate and insect non-α subunits.
Reproduced from ref. 29 with permission of Nature Publishing Group.

Conclusion

We have found that (1) the electrostatic potentials of the nitrogens in the imidazoline ring of imidacloprid turn from negative to positive once the nitro group forms a complex with protonated basic residues, which can strengthen neonicotinoids- nAChRs interactions (Induced-fit concept), and that (2) both α and non-α subunits possess structural properties favorable for interactions with neonicotinoids. In addition, (3) we have identified basic residues in loop D of insect nAChRs (see Table 1), which have been predicted by the chemical calculations to contribute to strengthening interactions with neonicotinoids *via* electrostatic interactions and hydrogen bond formation with the nitro group of neonicotinoids. Because mutations of the basic residues would result in reduction of neonicotinoid sensitivity of insect nAChRs, it will be of interest to examine the amino acid sequence in loop D of nAChRs in insect species that develop target-site resistance to neonicotinoids.

Acknowledgements

We thank Mr. Masaru Shimomura and Mr. Makoto Ihara and other students of Kinki University for their important contributions. We also thank Professor Koichiro Komai of Kinki University for his advice and support. These studies have been supported in part by a Grant-in-Aid for Scientific Research from the Japan Society for Promotion of Science and the Program for Promotion of Basic Research Activities for Innovative Biosciences from Bio-oriented Technology Research Advancement Institution as well as the Medical Research Council of the UK.

References

1. Breer, H.; Sattelle, D. B. *J. Insect Physiol.* **1987**, *33*, 771-190.
2. Soloway, S. B.; Henry, A. C.; Kollmeyer, W. D.; Padgett, W. M.; Powell, J. E.; Roman, S. A.; Tieman, C. H.; Corey, R. A.; Horne, C. A. In *Advances in Pesticide Science* Part 2; Geissbühler, H.; Brooks, G. T.; Kearney, P. C. Eds.; Pergamon Press: Oxford, 1979, pp. 206-217.
3. Schroeder, M. E.; Flattum, R. F. *Pestic. Biochem. Physiol.* **1984**, *22*, 148-160.
4. Sattelle, D. B.; Buckingham, S. D.; Wafford, K. A.; Sherby, S. M.; Bakry, N. M.; Eldefrawi, A. T.; Eldefrawi, M. E.; May, T. E. *Proc. Royal Soc. Lond. B* **1989**, *237*, 501-514.
5. Leech, C. A.; Jewess, P.; Marshall, J.; Sattelle, D. B. *FEBS Lett.* **1991**, *290*, 90-94.
6. Kagabu, S. *Rev. Toxicol.* **1997**, *1*, 75-129.
7. Bai, D.; Lummis, S. C. R.; Leicht, W.; Breer, H.; Sattelle, D. B. *Pestic. Sci.* **1991**, *33*, 197-204.

8. Zwart, R.; Ootgiesen, M.; Vijverberg, H. P. M. *Eur. J. Pharmacol.* **1992**, *228*, 165-169.
9. Liu, M.-Y.; Casida, J. E. *Pestic. Biochem. Physiol.* **1993**, *46*, 40-46.
10. Matsuda, K.; Buckingham, S. D.; Kleier, D.; Rauh, J. J.; Grauso, M.; Sattelle, D. B. *Trends Pharmacol. Sci.* **2001**, *22*, 573-580.
11. Kagabu, S.; Matsuno, H. *J. Agric. Food Chem.* **1997**, *45*, 276-281.
12. Yamamoto, I.; Yabuta, G.; Tomizawa, M.; Sato, T.; Miyamoto, T.; Kagabu, S. *J. Pestic. Sci.* **1995**, *20*, 33-40.
13. Tomizawa, M.; Lee, D. L.; Casida, J. E. *J. Agric. Food Chem.* **2000**, *48*, 6016-6024.
14. Ihara, M.; Matsuda, K.; Otake, M.; Kuwamura, M.; Shimomura, M.; Komai, K.; Akamatsu, M.; Raymond, V.; Sattelle, D. B. *Neuropharmacology* **2003**, *45*, 133-144.
15. Karlin, A.; Akabas, M. H. *Neuron* **1995**, *15*, 1231-1244.
16. Galzi, J.-L.; Changeux, J.-P. *Neuropharmacology* **1995**, *34*, 563-582.
17. Corringer, P.-J.; Le Novère, N.; Changeux, J.-P. *Annu. Rev. Pharmacol. Toxicol.* **2000**, *40* 431-458.
18. Itier, V.; Bertrand, D. *FEBS Lett.* **2001**, *504*, 118-125.
19. Karlin, A. *Nat. Rev. Neurosci.* **2002**, *2*, 102-114.
20. Couturier, S.; Bertrand, D.; Matter, J.-M.; Hernandez, M.-C.; Bertrand, S.; Millar, N.; Valera, S.; Barkas, T.; Ballivet, M. *Neuron* **1990**, *5*, 847-856.
21. Gerzanich, V.; Anand, R.; Lindstrom, J. *Mol. Pharmacol.* **1994**, *45*, 212-220.
22. Elgoyhen, A. B. Johonson, D. S.; Boulter, J.; Vetter, D. E.; Heinemann, S. *Cell* **1994**, *79*, 705-715.
23. Littleton, J. T.; Ganetzky, B. *Neuron* **2000**, *26*, 35-43.
24. Bertrand, D.; Ballivet, M.; Gomez, M.; Bertrand, S.; Phannavong, B.; Gundelfinger, E. D. *Eur. J. Neurosci.* **1994**, *6*, 869-875.
25. Matsuda, K.; Buckingham, S. D.; Freeman, J. C.; Squire, M. D.; Baylis, H. A.; Sattelle, D. B. *Br. J. Pharmacol.* **1998**, *123*, 518-524.
26. Matsuda, K.; Shimomura, M.; Kondo, Y.; Ihara, M.; Hashigami, K.; Yoshida, N.; Raymond, V.; Mongan, N. P.; Freeman, J. C.; Komai, K.; Sattelle, D. B. *Br. J. Pharmacol.* **2000**, *130*, 981-986.
27. Brejc, K.; van Dijk, W. J.; Klaassen, R.V.; Schuurmans, M.; van der Oost, J.; Smit, A. B.; Sixma, T. K. *Nature* **2001**, *411*, 269-276.
28. Smit, A. B.; Syed, N. I.; Schaap, D.; van Minnen, J.; Klumperman, J.; Kits, K. S.; Lodder, H.; van der Schors, R. C.; van Elk, R.; Sorgedrager, B.; Brejc, K.; Sixma, T. K.; Geraerts W. P. *Nature* **2001**, *411*, 261-268.
29. Shimomura, M.; Okuda, H.; Matsuda, K.; Komai, K.; Akamatsu, M.; Sattelle, D. B. *Br. J. Pharmacol.* **2002**, *137*, 162-169.

Chapter 17

Expression of a *Bombyx mori* Tyramine Receptor in HEK-293 Cells and Action of a Formamidine Insecticide

Yoshihisa Ozoe[1], Hiroto Ohta[1], Idumi Nagai[1], and Toshihiko Utsumi[2]

[1]Department of Life Science and Biotechnology, Faculty of Life and Environmental Science, Shimane University, Matsue, Shimane 690–8504, Japan
[2]Department of Biological Chemistry, Faculty of Agriculture, Yamaguchi University, Yamaguchi 753–8515, Japan

B96Bom was cloned from the silkworm *Bombyx mori* and used for stable expression in HEK-293 cells. The expressed B96Bom receptor led to an attenuation of forskolin-stimulated intracellular cAMP production in response to tyramine. The attenuation was abolished by yohimbine. Octopamine was less effective than tyramine. Tyramine inhibited [^{3}H]tyramine binding to the B96Bom receptor more potently than octopamine and dopamine. The acaricide/insecticide chlordimeform and its metabolite demethylchlordimeform inhibited [^{3}H]tyramine binding but did not show significant agonist activity. The stable expression system might prove useful for the discovery of novel ligands and bioactive compounds.

Biogenic amines play important roles as neurotransmitters, neuromodulators, and neurohormones, and are implicated in a variety of physiological events from energy metabolism and muscle contraction to learning and memory (*1,2*). Octopamine (Figure 1) is one of the well-studied biogenic amines in invertebrates (*3,4*). In the 1970s and 1980s, the octopamine receptor was extensively studied as a target of formamidine acaricides/insecticides such as chlordimeform and amitraz (*5*). On the other hand, tyramine (Figure 1) was regarded as a mere precursor of the biosynthesis of octopamine, because the ß-hydroxylation of tyramine results in octopamine (*3,4*). Recently, however, evidence has been accumulating that tyramine also functions as a chemical messenger in insects (*6,7*). The biological functions of octopamine and tyramine are generally to increase and decrease intracellular levels of cAMP, respectively, by interacting with distinct seven-transmembrane, G protein-coupled, adenylate cyclase-linked receptors (*2*). In view of a close structural similarity of these two amines, we are interested in examining how their receptors differentiate the two amines and their analogues. In this article, we describe the cloning and expression of an insect tyramine receptor, and the actions of chlordimeform and a demethylated analogue (demethylchlordimeform) (Figure 1) on the tyramine receptor.

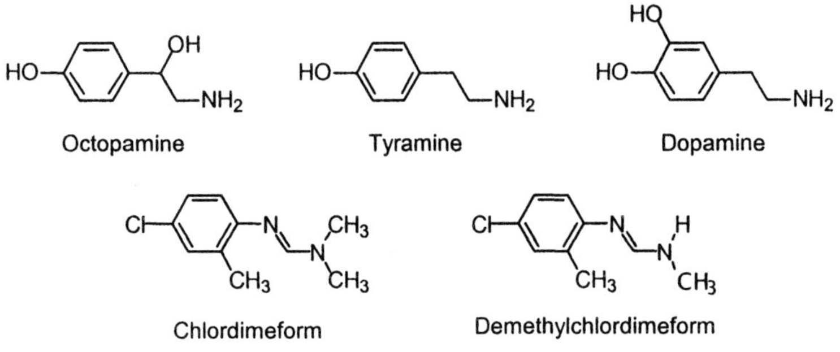

Figure 1. Structures of biogenic amines and formamidines.

Molecular Cloning of B96Bom from the Silkworm

As lepidopteran insects are particularly sensitive to octopaminergic insecticides (*5*), we decided to clone B96Bom (Accession No. X95607), the sequence of which had been deposited in the EMBL/GenBank/DDBJ database as encoding an octopamine receptor of the silkworm *Bombyx mori* but the functional properties of the encoding product of which were not reported (*8*). The PCR-amplified full-length cDNA from the heads of the fifth-instar silkworm larvae (the Kinshu-Showa strain) was ligated into a pT7Blue vector and sequenced by standard techniques. B96Bom was found to code 479 amino acids

(Figure 2). The deduced amino acid sequence is the same as that in the database, except four amino acids (see the caption) (Accession No. AB162828). The difference is probably due to the difference of the strains used. The sequence reveals seven transmembrane domains characteristic of G protein-coupled receptors, as well as conserved amino acid residues; e.g., Asp134 and Ser222, which are thought to interact with the amino group and the *p*-hydroxy group of biogenic amines, respectively, from analogy to adrenergic receptors.

MGQAATHDANNYTSINYTEIYDVIEDEKDVCAVADEPKYPSSFGISLAVP

EWEAICTAIILTMIIISTVVGNILVILSVFTYKPLRIVQNFFIVSLAVAD

LTVAILVLPLNVAYSILGQWVFGIYVCKMWLTCDIMCCTSSILNLCAIAL

DRYWAITDPINYAQKRTLERVLFMIGIVWILSLVISSPPLLGWNDWPEVF

EPDTPCRLTSQPGFVIFSSSGSFYIPLVIMTVVYFEIYLATKKRLRDRAK

ATKISTISSGRNKYETKESDPNDQDSVSSDANPNEHQGSTRLVAENEKKH

RTRKLTPKKKPKRRYWSKDDKSHNKLIIPILSNENSVTDIGENLENRNTS

SESNSKETHEDNMIEITEAAPVKIQKRPKQNQTNAVYQFIEEKQRISLTR

ERRAARTLGIIMGVFVVCWLPFFVIYLVIPFCVSCCLSNKFINFITWLGY

VNSALNPLIYTIFNMDFRRAFKKLLFIKC

Figure 2. Amino acid sequence of the B96Bom receptor. The seven putative transmembrane domains are underlined. Amino acid residues in bold are thought to function in binding tyramine. Dots on four letters refer to amino acid residues different from those reported by von Nickisch-Rosenegk et al. (8); the observed replacements are N38K, I39Y, C41S, and G289S.

Functional Properties of the B96Bom Receptor Expressed in a Cell Line

B96Bom was subcloned into the expression vector pcDNA3 and introduced into a human embryonic kidney cell line (HEK-293) using LipofectAmine (*9*). HEK-293 cells stably expressing the B96Bom receptor (B96Bom/HEK-293 cells) were selected in the presence of the antibiotic G-418. First of all, we examined the effects of biogenic amines on cAMP production in B96Bom/HEK-293 cells. However, neither octopamine nor tyramine elicited an increase in intracellular cAMP levels over basal levels (*9*). Therefore, we next examined the effects of biogenic amines on cAMP production stimulated by forskolin, a direct activator of adenylate cyclase. In response to octopamine and tyramine, the B96Bom receptor led to an attenuation of forskolin-stimulated cAMP production in a dose-dependent manner (Figure 3). Both amines apparently exhibited a maximum attenuation level of approximately 40%. However, tyramine was found to be at least two orders of magnitude more effective than octopamine, when the concentrations needed to reach the maximum attenuation level were compared. Furthermore, the tyramine receptor antagonist yohimbine abolished the tyramine-induced attenuation in the low tyramine concentration range. The effect of yohimbine was not observed in the high tyramine concentration range, indicating that yohimbine is a competitive antagonist for the B96Bom receptor.

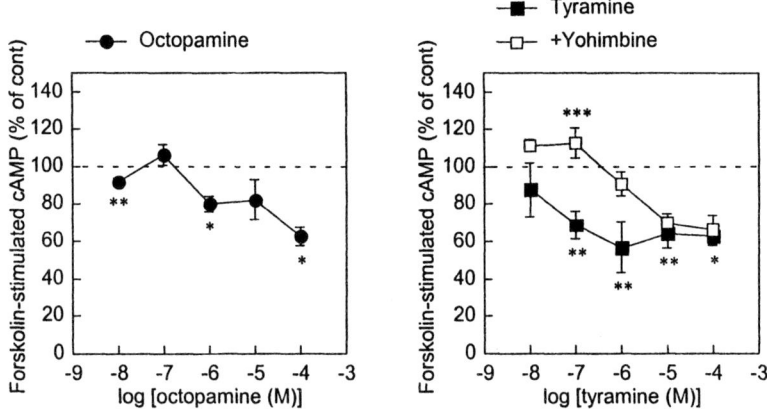

*Figure 3. Dose-response curves of octopamine and tyramine in attenuating forskolin-stimulated cAMP production in B96Bom/HEK-293 cells. Forskolin (10 µM)-stimulated cAMP levels are shown as 100%. Yohimbine (10 µM) was tested as an antagonist. Data represent means ± SE of at least four independent experiments. *p < 0.01 vs control; **p < 0.05 vs control (paired t test); ***p < 0.01 vs tyramine-attenuated levels (unpaired t test). (Reproduced with permission from reference 9. Copyright 2003 Blackwell.)*

Affinity of Biogenic Amines for the B96Bom Receptor

We investigated how the B96Bom receptor discriminates between octopamine, tyramine, and dopamine, which are different from each other by the presence or the position of only one hydroxy group. Figure 4 displays the dose-response curves for the inhibitory effects of the three amines on the binding of [^{3}H]tyramine to the B96Bom receptor. Although all three biogenic amines inhibited the binding of [³H]tyramine in a dose-dependent manner, the most potent amine was tyramine with an IC_{50} of approximately 5 nM. Both octopamine and dopamine were two orders of magnitude less potent than tyramine. Thus, the B96Bom receptor clearly discriminates tyramine from the other two amines. The hydroxy groups at the *m*-position and the ß-position of phenylethylamines are detrimental in their binding to the B96Bom receptor. No specific binding of [³H]octopamine to the membranes of HEK-293 cells expressing the B96Bom receptor was observed under the same conditions. Taken together with the results of the functional assay described above, these findings indicate that the B96Bom receptor is not an octopamine receptor but a tyramine receptor negatively coupled to the effector adenylate cyclase.

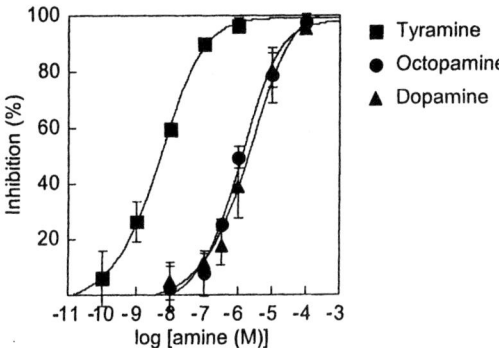

Figure 4. Dose-response curves of octopamine, tyramine, and dopamine in inhibiting specific [³H]tyramine (3 nM) binding to the membranes of B96Bom/HEK-293 cells. Data represent means ± SD of at least two or three experiments, each done in duplicate. The specific to total binding ratio was 0.94 under the conditions used. The $IC_{50}s$ of octopamine, tyramine, and dopamine were estimated to be 1.4 µM, 5.2 nM, and 1.7 µM, respectively. (Reproduced with permission from reference 9. Copyright 2003 Blackwell.)

Actions of a Formamidine Insecticide on the *B. mori* Tyramine Receptor

Chlordimeform is a formamidine acaricide/insecticide or a pestistatic chemical (*10*). Demethylchlordimeform is a biologically active form of chlordimeform and is known as a partial agonist of octopamine receptors (*11-13*). We examined if the *B. mori* tyramine receptor recognizes the formamidines as ligands. Figure 5 shows the dose-response curves of the formamidines in inhibiting the binding of [³H]tyramine to the *B. mori* tyramine receptor. The results of the binding assay indicate that the formamidines are capable of interacting with the tyramine receptor. The IC_{50} of chlordimeform estimated in this assay is one order of magnitude smaller than that reported in the inhibition of the specific binding of [^{3}H]octopamine to the homogenates of *Drosophila* heads and those of firefly light organs (*14,15*). The IC_{50} of demethylchlordimeform is one order of magnitude smaller than that reported in the inhibition of [^{3}H]yohimbine binding to a cloned *Drosophila* tyramine receptor (*16*) but two orders of magnitude larger than that in the inhibition of specific [^{3}H]octopamine binding to the homogenates of firefly light organs (*15*). The conclusion that demethylchlordimeform interacts with tyramine receptors with a lower affinity than with octopamine receptors is also the conclusion obtained with membranes from the brain of the locust (*17*).

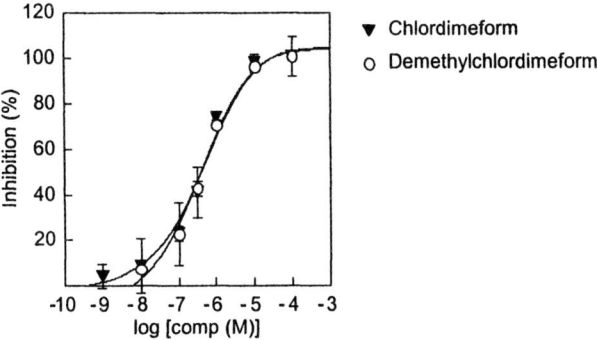

Figure 5. Dose-response curves of chlordimeform and demethylchlordimeform in inhibiting specific [^{3}H]tyramine (3 nM) binding to the membranes of B96Bom/HEK-293 cells. Data represent means ± SD of two to three experiments, each done in duplicate. The IC_{50}s of chlordimeform and demethylchlordimeform were determined to be 0.30 µM and 0.35 µM, respectively.

We next performed a functional assay to see whether the formamidines act as agonists in the tyramine receptor. Figure 6 shows the effects of demethylchlordimeform on forskolin-stimulated cAMP production in normal HEK-293 cells and HEK-293 cells stably expressing the *B. mori* tyramine receptor. Demethylchlordimeform at 100 µM attenuated forskolin-stimulated cAMP production by approximately 20 % in HEK-293 cells expressing the tyramine receptor. However, the attenuation was observed in normal HEK-293 cells as well. These findings indicate that demethylchlordimeform does not function as an agonist for the tyramine receptor, although the compound is capable of interacting with the receptor. Chlordimeform had no effects on both cell lines (data not shown). It remains to be solved whether the formamidines act as antagonists in the tyramine receptor. We are investigating the structure-activity relationships of synthetic amine analogues and the ligand recognition of mutated tyramine receptors.

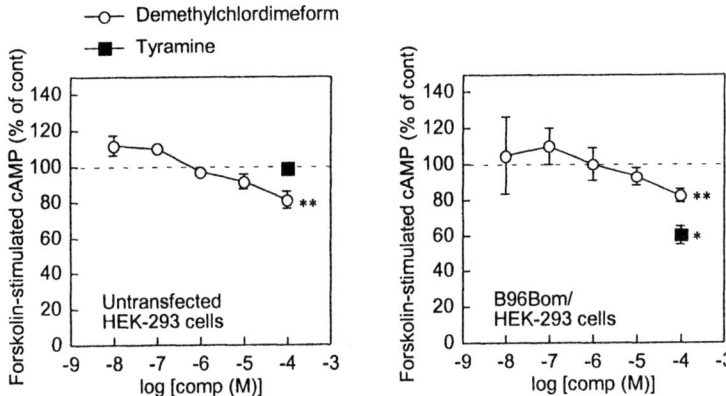

Figure 6. Effects of demethylchlordimeform on forskolin-stimulated cAMP production in normal HEK-293 cells and HEK-293 cells stably expressing a B. mori tyramine receptor. Forskolin (10 µM)-stimulated cAMP levels are shown as 100%. *$p < 0.01$ vs control; **$p < 0.05$ vs control (paired t test). Data represent means ± SE of two to five experiments, except data at 0.1 µM and 1 µM in normal HEK-293 cells (n=1).

Conclusion

Most of reported data, including those of our studies, indicate that the invertebrate octopamine and tyramine receptors convey the activation and inactivation signals for adenylate cyclase, respectively (2,9,18). It is generally difficult to discover compounds acting selectively for either the octopamine or tyramine receptor by conventional cAMP assays using membrane homogenates from nervous tissues, because membrane homogenates might contain both receptors that have the opposite effects on cAMP production (18). The use of stable expression systems of cloned biogenic amine receptors would facilitate the discovery of specific receptor ligands as well as novel insect bioregulators.

Acknowledgments

We would like to thank Ms. F. Ozoe (Shimane University) for helpful advice on molecular biological experiments. This study was supported in part by a grant (to H.O.) from Heart Co. Ltd., Japan.

References

1. Evans, P. D. *Adv. Insect Physiol.* **1980**, *15*, 317-473.
2. Blenau, W.; Baumann, A. *Arch. Insect Biochem. Physiol.* **2001**, *48*, 13-38.
3. Robertson, H. A.; Juorio, A. V. *Int. Rev. Neurobiol.* **1976**, *19*, 173-224.
4. Roeder, T. *Prog. Neurobiol.* **1999**, *59*, 533-561.
5. Hollingworth, R. M.; Johnstone, E. M.; Wright, N. In *Pesticide Synthesis Through Rational Approaches*; Magee, P. S.; Kohn, G. K.; Menn, J. J., Eds.; ACS Symposium Series No. 255; American Chemical Society: Washington, D.C., 1984; pp 103-125.
6. Nagaya, Y.; Kutsukake, M.; Chigusa, S. I.; Komatsu, A. *Neurosci. Lett.* **2002**, *329*, 324-328.
7. Blumenthal, E. M. *Am. J. Physiol. Cell Physiol.* **2003**, *284*, C718-C728.
8. von Nickisch-Rosenegk, E.; Krieger, J.; Kubick, S.; Laage, R.; Strobel, J.; Strotmann, J.; Breer, H. *Insect Biochem. Mol. Biol.* **1996**, *26*, 817-827.
9. Ohta, H.; Utsumi, T.; Ozoe, Y. *Insect Mol. Biol.* **2003**, *12*, 217-223.
10. Hollingworth, R. M. *Environ. Health Perspect.* **1976**, *14*, 57-69.
11. Hollingworth, R. M.; Murdock, L. L. *Science* **1980**, *208*, 74-76.
12. Evans, P. D.; Gee, J. D. *Nature* **1980**, *287*, 60-62.
13. Nathanson J. A.; Hunnicutt, E. J. *Mol. Pharmacol.* **1981**, *20*, 68-75.
14. Dudai, Y.; Zvi, S. *Comp. Biochem. Physiol.* **1984**, *77C*, 145-151.
15. Hashemzadeh, H.; Hollingworth, R. M.; Voliva, A. *Life Sci.* **1985**, *37*, 433-440.
16. Robb, S.; Cheek, T. R.; Hannan, F. L.; Hall, L. M.; Midgley, J. M.; Evans, P. D. *EMBO J.* **1994**, *13*, 1325-1330.
17. Hiripi, L.; Nagy, L.; Hollingworth, R. M. *Acta Biol. Hung.* **1999**, *50*, 81-87.
18. Khan, M. A. A.; Nakane, T.; Ohta, H.; Ozoe, Y. *Arch. Insect Biochem. Physiol.* **2003**, *52*, 7-16.

Chapter 18

Measurement of Receptor-Binding Activity of Non-Steroidal Ecdysone Agonists Using in vitro Expressed Receptor Proteins (EcR/USP Complex) of *Chilo suppressalis* and *Drosophila melanogaster*

Chieka Minakuchi[1], Yoshiaki Nakagawa[1], Manabu Kamimura[2], and Hisashi Miyagawa[1]

[1]Graduate School of Agriculture, Kyoto University, Kyoto 606–8502, Japan
[2]National Institute of Agrobiological Sciences, Tsukuba 305–8634, Japan

N-tert-Butyl-*N,N'*-dibenzoylhydrazine and its analogs are agonists of insect molting hormone (20-hydroxyecdysone: 20E). In order to elucidate the mode of action of these non-steroidal ecdysone agonists at the molecular level, we expressed the receptor protein for 20E – the heterodimer of ecdysone receptor (EcR) and ultraspiracle (USP) – *in vitro*, and analyzed their binding affinities to ligands. For the lepidopteran *Chilo suppressalis*, we showed quantitatively that the receptor-binding activity of non-steroidal ecdysone agonists determines the strength of their molting hormonal and insecticidal activities. The binding activity of five representative non-steroidal ecdysone agonists to *C. suppressalis* EcR/USP was higher than that to dipteran *Drosophila melanogaster* EcR/USP, which probably results in the selective toxicity between these two insects.

Introduction

Insect molting is regulated via binding of molting hormone (20-hydroxyecdysone: 20E; Figure 1, **I**: R = OH) to its receptor proteins – the heterodimer of ecdysone receptor (EcR) and ultraspiracle (USP) (1-3). The 20E-EcR/USP complex binds to the ecdysone response element (EcRE) in the promoter of target genes and regulates their transcription. cDNAs of EcR and USP were first cloned for *Drosophila melanogaster* (4-7), then for other insect species (8). Recently, the crystal structure of EcR/USP heterodimer was solved, and ligand-binding and DNA-binding models are proposed by Billas et al. (9) and Devarakonda et al. (10), respectively.

N-tert-Butyl-*N,N'*-dibenzoylhydrazine (RH-5849) (11,12) and its analogs

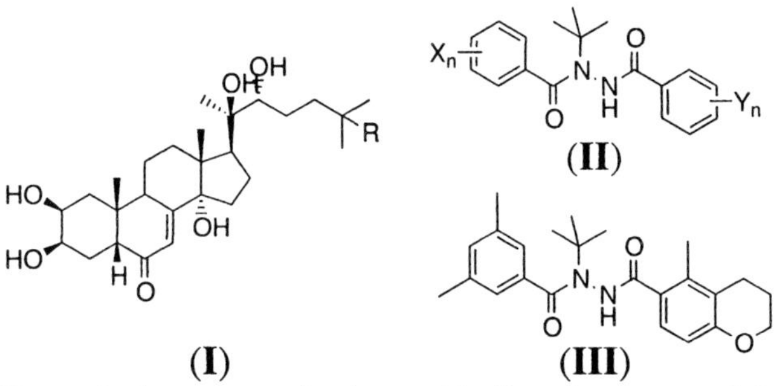

Figure 1. Structures of ecdysteroids (I) and non-steroidal ecdysone agonists (II, III).

(Figure 1, **II, III**) mimic 20E, and possess molting hormonal and insecticidal activities. Four potent analogs, tebufenozide [RH-5992; Figure 1, **II**: X_n = 3,5-$(CH_3)_2$, Y_n = 4-C_2H_5], methoxyfenozide [RH-2485; Figure 1, **II**: X_n = 3,5-$(CH_3)_2$, Y_n = 2-CH_3-3-OCH_3], halofenozide (RH-0345; Figure 1, **II**: X_n = H, Y_n = 4-Cl) and chromafenozide (ANS-118; Figure 1, **III**), are currently on the market as safer insecticides with reduced mammalian toxicity (13-17). There are several reports that these dibenzoylhydrazine-type, non-steroidal ecdysone agonists compete with labeled ponasterone A (PonA; Figure 1, **I**: R = H), a potent molting hormone analog, for binding in a cell-free preparation of insect tissue (18), in an insect cell line (11,13,19-21) or to the EcR/USP protein which was synthesized by *in vitro* transcription/translation reaction (14). It was also reported that non-steroidal ecdysone agonists regulate the ecdysone-inducible gene transcription (22-24). These results support the idea that the binding sites of these non-steroidal ecdysone agonists would be the same as that of 20E. Moreover, recent crystal structure analysis disclosed that the binding pockets in the EcR ligand-binding domain partially overlap between PonA and a non-steroidal ligand BY106830 (9).

To date, the insecticidal activity of non-steroidal ecdysone agonists are measured against various insects. Interestingly, structure-activity relationships (SARs) of these compounds are quite different among insect orders. For example, some analogs such as tebufenozide, methoxyfenozide and chromafenozide are more potent in Lepidoptera (butterflies and moths) than in other insect orders, while halofenozide acts on Coleoptera (beetles) but not on Lepidoptera. The insecticidal activity of these four analogs is not high against another insect order Diptera (flies and mosquitoes). For a series of non-steroidal ecdysone agonists, we clearly showed that the SARs are quite different between lepidopteran *Chilo suppressalis* and coleopteran *Leptinotarsa decemlineata* (25,26), while SARs are similar between the two lepidopterans, *C. suppressalis* and *Spodoptera exigua* (27,28). The selective toxicity of RH-5849 or tebufenozide between Lepidoptera and Coleoptera cannot be attributed to pharmacokinetic and metabolic differences, but presumably to other factors such as the difference in ligand-receptor binding affinity (29,30). However, it is unclear whether the insecticidal potency of a series of non-steroidal ecdysone agonists is affected significantly by the ligand-receptor binding affinity or not.

Recently we found using *in vitro* transcribed-translated EcR/USP of *C. suppressalis* that the receptor-binding activity of ecdysone agonists determines the strength of their molting hormonal activity (18). Here we compare the binding activity of ecdysone agonists to the *in vitro* transcribed-translated EcR/USP complex between a lepidopteran *C. suppressalis* and a dipteran fruit fly, *D. melanogaster*, to discuss the selective toxicity between these two insects.

Expression and functional analysis of EcR and USP proteins

To date, two isoforms of *C. suppressalis* EcR (CsEcR-A and CsEcR-B1) (31) and three isoforms of *D. melanogaster* EcR (DmEcR-A, DmEcR-B1 and DmEcR-B2) (7) have been identified, but we exclusively used EcR-B1 for protein expression in this study. The amino acid identity between CsEcR-B1 (31) and DmEcR-B1 (7) is high, especially in the DNA-binding and ligand-binding domains (95% and 71%) (Figure 2). The amino acid sequences of *C. suppressalis* USP (CsUSP) (32) and *D. melanogaster* USP (DmUSP) (4-6) are also homologous in the DNA-binding domain (97%), whereas the identity in their ligand-binding domain is relatively low (47%) (Figure 2).

The full coding regions of EcR-B1 and USP were cloned into plasmid vectors, and *in vitro* transcription/translation reaction was performed using a rabbit reticulocyte system (Promega) under T7 promoter. Early studies on the *D. melanogaster* EcR/USP heterodimer showed that PonA specifically bound to *in vitro* expressed DmEcR but not to DmUSP, and that the specific binding of PonA to DmEcR was remarkably enhanced by adding DmUSP (3,33). Recently we also reported that the binding affinity of PonA to CsEcR-B1 was remarkably

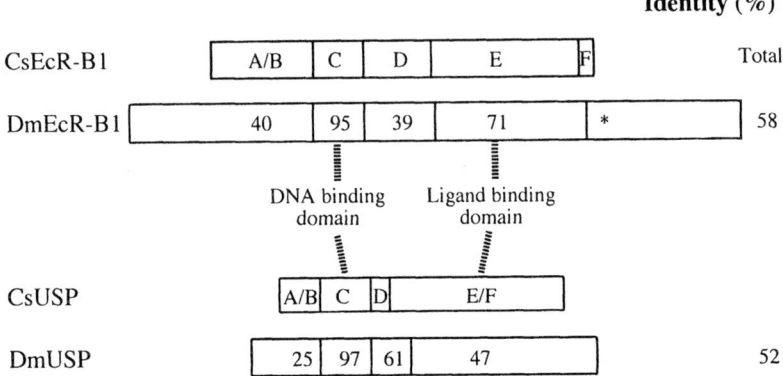

Figure 2. Comparisons of the primary sequences of EcRs and USPs. Numbers in boxes show the percentage amino acid identity in each region between Chilo suppressalis and Drosophila melanogaster.
* F region of CsEcR-B1 is too short to calculate identity.

enhanced in the presence of CsUSP (18). These results indicate that PonA is able to bind to EcR, but not to USP, and that the binding affinity between EcR and PonA is enhanced by the allosteric interaction between EcR and USP.

Relationships among receptor-binding, molting-hormonal and insecticidal activities against *C. suppressalis*

The binding activity of a series of ecdysone agonists to the *in vitro* expressed CsEcR-B1/CsUSP heterodimer was quantitatively evaluated as the reciprocal logarithm of the 50% inhibition concentration (pIC_{50}) for the competition of [^{3}H]PonA (18) (Table 1). Tebufenozide (no. 4; see Table 1) and chromafenozide (no. 5) bound to CsEcR-B1/CsUSP heterodimer with very high affinity, being about 200-fold higher than that of 20E (no. 7). Among the ecdysteroids, the order of the binding activity was PonA > 20E cyasterone > makisterone A > ecdysone. We previously reported that the binding activity of ecdysone agonists against the CsEcR-B1/CsUSP heterodimer is linearly correlated with the binding activity against a cell-free preparation of *C. suppressalis* integument, indicating that *in vitro* expressed CsEcR-B1/CsUSP heterodimer functions as the receptor for ecdysone agonists (18).

We showed that the receptor-binding activity is linearly correlated with the molting hormonal activity (34-38) which is evaluated as the reciprocal logarithm of the 50% effective concentration (pEC_{50}) for the induction of chitin synthesis in cultured integument of *C. suppressalis* (18) (Figure 3A). We also compared the receptor-binding activity of non-steroidal ecdysone agonists with the

Table 1. Binding activities (pIC_{50} values) of ecdysone agonists against *in vitro* transcribed-translated EcR-B1/USP

No.	Compounds $X_n^{a)}$	$Y_n^{a)}$	pIC_{50} (M) CsEcR-B1/CsUSP[b]	DmEcR-B1/DmUSP
1 (RH-5849)	H	H	6.50	5.16
2	H	4-C_2H_5	7.95	5.87
3 (halofenozide)	H	4-Cl	6.92	5.95
4 (tebufenozide)	3,5-$(CH_3)_2$	4-C_2H_5	8.85	6.01
5 (chromafenozide)[c]	3,5-$(CH_3)_2$	2-CH_3-3,4-[-$(CH_2)_3$O-]	9.13	6.54
6	Ponasterone A (PonA)		8.08	8.27
7	20-Hydroxyecdysone (20E)		6.66	7.03
8	Cyasterone		6.65	7.07
9	Makisterone A		6.33	6.87
10	Ecdysone		4.70	5.24

[a] See the structure in Figure 1 (**II**).
[b] From Ref (18).
[c] See the structure in Figure 1 (**III**).

insecticidal activity which was evaluated as the reciprocal logarithm of the 50% lethal dose (pLD_{50}) by topical application to *C. suppressalis* larvae (39,40) (Figure 3B). Ecdysteroids such as 20E and PonA were not included in this plot because their insecticidal activity is too low to obtain pLD_{50} (unpublished data). Excellent correlation was obtained between insecticidal and receptor-binding activities of non-steroidal ecdysone agonists (Figure 3B). These results indicate that the receptor-binding affinity of non-steroidal ecdysone agonists determines the strength of their molting hormonal and insecticidal activities against *C. suppressalis*.

In some insect species, it has been reported that these non-steroidal ecdysone agonists would be metabolized and excreted to some extent (29,30). However, we assumed that the metabolism and excretion of non-steroidal ecdysone agonists in *C. suppressalis* were suppressed to very low, insignificant levels because piperonyl butoxide, an inhibitor of oxidative metabolism, was used in the measurement of the insecticidal activity (39,40). In addition to metabolism and excretion, we assumed that the physicochemical properties of compounds (such as hydrophobicity) might affect their uptake into the target cells, which would eventually vary the insecticidal activity and molting hormonal activity. Although the hydrophobicity of tested non-steroidal ecdysone agonists varied by 1,000-fold, hydrophobicity was not a factor in the linear correlation between the binding and molting hormonal activities (Figure 3A) or between the binding and insecticidal activities (Figure 3B). We

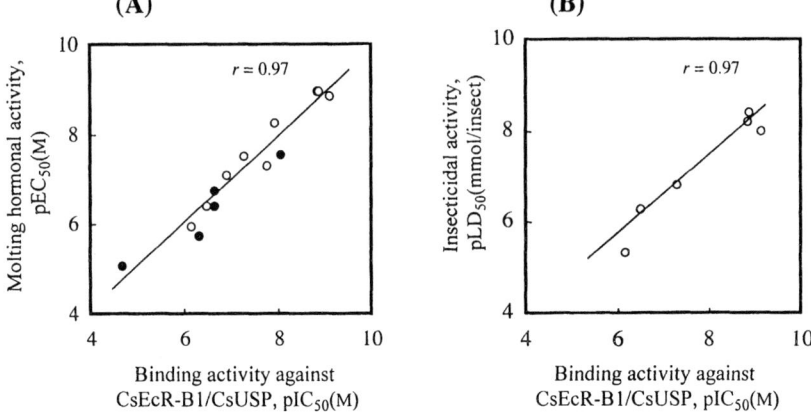

Figure 3. (A) Relationship between molting hormonal activity (pEC_{50}) and receptor-binding activity (pIC_{50}) of ecdysone agonists against C. suppressalis. Solid circles, ecdysteroids. Open circles, non-steroidal ecdysone agonists. [Reproduced with the permission from reference (18). Copyright Federation of European Biochemical Societies] (B) Relationship between insecticidal activity (pLD_{50}) and receptor-binding activity (pIC_{50}) of non-steroidal ecdysone agonists against C. suppressalis.

therefore concluded that hydrophobicity would probably be an important factor for the ligand-receptor interaction of non-steroidal ecdysone agonists, but not for their incorporation or transport process.

As stated above, the insecticidal activity of natural ecdysteroid is very low. One of the reasons for the low insecticidal activity is probably the poor penetration of hydrophilic ecdysteroids through insect cuticle. Other reasons may be related to the metabolism and excretion. According to Blackford et al. (41), tomato moth larvae are able to feed on a diet containing 400-ppm 20E without any adverse effects on growth and development. Also, when a high dose of 20E was injected into the final instar larvae of the silkworm, *Bombyx mori*, the hemolymph 20E titer decreases immediately (42). These results indicate that 20E when applied exogenously into insects by either oral administration or injection can be easily metabolized and excreted.

Relationship of receptor-binding activities between insect orders

As stated above, it is well known that the insecticidal activity of most non-steroidal ecdysone agonists is higher to Lepidoptera than to Diptera. Using *in vitro* expressed EcR/USP complex and cellular extracts, Dhadialla et al. showed that the receptor-binding activities of non-steroidal ecdysone agonists are diverse among insect orders, suggesting that the selective insecticidal toxicity of non-steroidal ecdysone agonists might be attributed to the difference in their receptor-binding affinity (14). To investigate this point further, we compared the binding activity of a series of non-steroidal ecdysone agonists between

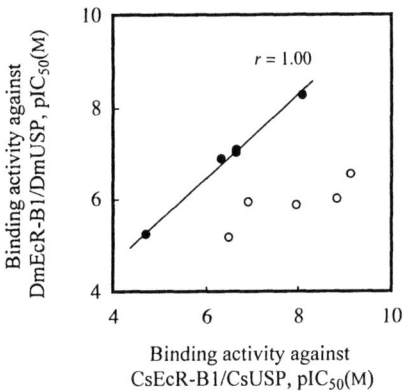

Figure 4. Relationship of receptor-binding activities (pIC_{50}) between C. suppressalis and D. melanogaster. Solid circles, ecdysteroids. Open circles, non-steroidal ecdysone agonists.

lepidopteran C. suppressalis and dipteran D. melanogaster. As listed in Table 1, the pIC_{50} of tebufenozide (no. 4) differs by 2.84 between C. suppressalis and D. melanogaster, indicating that CsEcR-B1/CsUSP binds it with 700-fold higher affinity than does DmEcR-B1/DmUSP. Importantly, the binding activity of PonA (no. 6) does not vary between C. suppressalis (pIC_{50}=8.08) and D. melanogaster (pIC_{50}=8.27). Thus, SAR for the binding activity of non-steroidal ecdysone agonists (no. 1-5) is very different between C. suppressalis and D. melanogaster, whereas SAR for ecdysteroids (no. 6-10) is very similar (Table 1 and Figure 4). We assumed that the selective insecticidal toxicity of non-steroidal ecdysone agonists would be attributed to the difference in their binding affinity to the receptor protein (EcR/USP heterodimer) between C. suppressalis and D. melanogaster.

The amino acid identity between CsEcR-B1 and DmEcR-B1 is very high in the DNA-binding domain (95%), but is lower in the ligand-binding domain (71%) (Figure 2). Generally, the amino acid sequences in the ligand-binding domains of various EcRs are highly conserved among closely related insects, but the amino acid identity decreases to some extent between different insect orders (8). Likely the small changes in the amino acid sequence in the binding site cause the variation in the receptor-binding affinity of non-steroidal ecdysone agonists, although they do not significantly affect the receptor-binding affinity of ecdysteroids.

Concluding Remarks

As discussed above, we have clearly shown that the receptor-binding affinity of non-steroidal ecdysone agonists determines their biological activities for *C. suppressalis*, and that the difference in their receptor-binding affinity results in the selective toxicity of non-steroidal ecdysone agonists between *C. suppressalis* and *D. melanogaster*. In addition to crystal structure analysis of the EcR/USP complex, three-dimensional quantitative structure-activity relationship (3D-QSAR) studies for the binding activity against wild-type or mutated EcR/USP protein would be fruitful to disclose the amino acid residues that interact with ecdysteroid and non-steroidal ecdysone agonists. These studies would provide more insight regarding the mechanism of selective toxicity of non-steroidal ecdysone agonists at the molecular level.

Acknowledgements

We are thankful to Professor Lynn M. Riddiford of the University of Washington for carefully reviewing this manuscript. We also express our sincere gratitude to Professor Margarethe Spindler-Barth, Dr. Marco Grebe (University of Ulm), Dr. Shuichiro Tomita and Dr. Atsushi Seino (National Institute of Agrobiological Sciences) for helpful comments for the binding assay, Dr. Lucy Cherbas (Indiana University) for the gift of constructs (DmEcR-B1 and DmUSP), and Sankyo Agro Co., Ltd. and Nippon Kayaku Co., Ltd. for the gift of chromafenozide.

References

1. Yao, T.-P.; Segraves, W. A.; Oro, A. E.; McKeown, M.; Evans, R. M. *Cell* **1992**, *71*, 63-72.
2. Thomas, H. E.; Stunnenberg, H. G.; Stewart, A. F. *Nature* **1993**, *362*, 471-475.
3. Yao, T.-P.; Forman, B. M.; Jiang, Z.; Cherbas, L.; Chen, J.-D.; McKeown, M.; Cherbas, P.; Evans, R. M. *Nature* **1993**, *366*, 476-479.
4. Henrich, V. C.; Sliter, T. J.; Lubahn, D. B.; MacIntyre, A.; Gilbert, L. I. *Nucleic Acids Res.* **1990**, *18*, 4143-4148.

5. Oro, A. E.; McKeown, M.; Evans, R. M. *Nature* **1990**, *347*, 298-301.
6. Shea, M. J.; King, D. L.; Conboy, M. J.; Mariani, B. D.; Kafatos, F. C. *Genes Dev.* **1990**, *4*, 1128-1140.
7. Koelle, M. R.; Talbot, W. S.; Segraves, W. A.; Bender, M. T.; Cherbas, P.; Hogness, D. S. *Cell* **1991**, *67*, 59-77.
8. Riddiford, L. M.; Cherbas, P.; Truman, J. W. *Vitam. Horm.* **2001**, *60*, 1-73.
9. Billas, I. M.; Iwema, T.; Garnier, J. M.; Mitschler, A.; Rochel, N.; Moras, D. *Nature* **2003**, *426*, 91-96.
10. Devarakonda, S.; Harp, J. M.; Kim, Y.; Ozyhar, A.; Rastinejad, F. *EMBO J.* **2003**, *22*, 5827-5840.
11. Wing, K. D. *Science* **1988**, *241*, 467-469.
12. Wing, K. D.; Slawecki, R. A.; Carlson, G. R. *Science* **1988**, *241*, 470-472.
13. Hsu, A. C.-T.; Fujimoto, T. T.; Dhadialla, T. S. In *Phytochemicals for pest control. ACS symposium series;* Hedin, P. A.; Hollingworth, R. M.; Master, E. P.; Miyamoto, J.; Thompson, D. G. Eds.; American Chemical Society: Washington DC, 1997; Vol. 658; pp 206-219.
14. Dhadialla, T. S.; Carlson, G. R.; Le, D. P. *Annu. Rev. Entomol.* **1998**, *43*, 545-569.
15. Tanaka, K.; Tsukamoto, Y.; Sawada, Y.; Kasuya, A.; Hotta, H.; Ichinose, R.; Watanabe, T.; Toya, T.; Yokoi, S.; Kawagishi, A.; Ando, M.; Sadakane, S.; Katsumi, S.; Masui, A. *Annu. Rep. Sankyo Res. Lab.* **2001**, *53*, 1-49.
16. Toya, T.; Fukasawa, H.; Masui, A.; Endo, Y. *Biochem. Biophys. Res. Commun.* **2002**, *292*, 1087-1091.
17. Sawada, Y.; Yanai, T.; Nakagawa, H.; Tsukamoto, Y.; Tamagawa, Y.; Yokoi, S.; Yanagi, M.; Toya, T.; Sugizaki, H.; Kato, Y.; Shirakura, H.; Watanabe, T.; Yajima, Y.; Kodama, S.; Masui, A. *Pest Manag. Sci.* **2003**, *59*, 49-57.
18. Minakuchi, C.; Nakagawa, Y.; Kamimura, M.; Miyagawa, H. *Eur. J. Biochem.* **2003**, *270*, 4095-4104.
19. Mikitani, K. *J. Seric. Sci. Jpn.* **1996**, *65*, 141-144.
20. Nakagawa, Y.; Minakuchi, C.; Ueno, T. *Steroids* **2000**, *65*, 537-542.
21. Nakagawa, Y.; Minakuchi, C.; Takahashi, K.; Ueno, T. *Insect Biochem. Mol. Biol.* **2002**, *32*, 175-180.
22. Mikitani, K. *J. Seric. Sci. Jpn.* **1995**, *64*, 534-539.
23. Retnakaran, A.; Hiruma, K.; Palli, S. R.; Riddiford, L. M. *Insect Biochem. Mol. Biol.* **1995**, *25*, 109-117.
24. Kumar, M. B.; Fujimoto, T.; Potter, D. W.; Deng, Q.; Palli, S. R. *Proc. Natl. Acad. Sci. USA* **2002**, *99*, 14710-14715.
25. Nakagawa, Y.; Smagghe, G.; Kugimiya, S.; Hattori, K.; Ueno, T.; Tirry, L.; Fujita, T. *Pestic. Sci.* **1999**, *55*, 909-918.
26. Nakagawa, Y.; Smagghe, G.; Van Paemel, M.; Tirry, L.; Fujita, T. *Pest Manag. Sci.* **2001**, *57*, 858-865.
27. Smagghe, G.; Nakagawa, Y.; Carton, B.; Mourad, A. K.; Fujita, T.; Tirry, L. *Arch. Insect Biochem. Physiol.* **1999**, *41*, 42-53.

28. Nakagawa, Y.; Smagghe, G.; Tirry, L.; Fujita, T. *Pest Manag. Sci.* **2002**, *58*, 131-138.
29. Smagghe, G.; Degheele, D. *Pestic. Biochem. Physiol.* **1993**, *46*, 149-160.
30. Smagghe, G.; Degheele, D. *Pestic. Biochem. Physiol.* **1994**, *49*, 224-234.
31. Minakuchi, C.; Nakagawa, Y.; Kiuchi, M.; Tomita, S.; Kamimura, M. *Insect Biochem. Mol. Biol.* **2002**, *32*, 999-1008.
32. Minakuchi, C.; Nakagawa, Y.; Kiuchi, M.; Seino, A.; Tomita, S.; Kamimura, M. *Insect Biochem. Mol. Biol.* **2003**, *33*, 41-49.
33. Grebe, M.; Przibilla, S.; Henrich, V. C.; Spindler-Barth, M. *Biol. Chem.* **2003**, *384*, 105-116.
34. Nakagawa, Y.; Soya, Y.; Nakai, K.; Oikawa, N.; Nishimura, K.; Ueno, T.; Fujita, T.; Kurihara, N. *Pestic. Sci.* **1995**, *43*, 339-345.
35. Nakagawa, Y.; Nishimura, K.; Oikawa, N.; Kurihara, N.; Ueno, T. *Steroids* **1995**, *60*, 401-405.
36. Shimizu, B.; Nakagawa, Y.; Hattori, K.; Nishimura, K.; Kurihara, N.; Ueno, T. *Steroids* **1997**, *62*, 638-642.
37. Nakagawa, Y.; Hattori, K.; Minakuchi, C.; Kugimiya, S.; Ueno, T. *Steroids* **2000**, *65*, 117-123.
38. Watanabe, B.; Nakagawa, Y.; Miyagawa, H. *J. Pesticide Sci.* **2003**, *28*, 188-193.
39. Oikawa, N.; Nakagawa, Y.; Nishimura, K.; Ueno, T.; Fujita, T. *Pestic. Sci.* **1994**, *41*, 139-148.
40. Oikawa, N.; Nakagawa, Y.; Nishimura, K.; Ueno, T.; Fujita, T. *Pestic. Biochem. Physiol.* **1994**, *48*, 135-144.
41. Blackford, M.; Dinan, L. *Insect Biochem. Mol. Biol.* **1997**, *27*, 167-177.
42. Takahashi, M.; Kikuchi, K.; Tomita, S.; Imanishi, S.; Nakahara, Y.; Kiuchi, M.; Kamimura, M. *Comp. Biochem. Physiol. B Biochem. Mol. Biol.* **2003**, *135*, 431-437.

Natural Products

Chapter 19

Secondary Metabolites with Diverse Activities toward Phytopathogenic Zoospores of *Aphanomyces cochlioides* in Host and Nonhost Plants

Satoshi Tahara[1] and Md. Tofazzal Islam[1,2]

[1]Laboratory of Ecological Chemistry, Graduate School of Agriculture, Hokkaido University, Kita-Ku, Sapporo 060–8589, Japan
[2]Permanent address: School of Agriculture and Rural Development, Bangladesh Open University, Gazipur-1705, Bangladesh

Our knowledge on molecular basis of life cycle development of oomycete phytopathogens has been compiled in last two decades, which is indicating that the inter-relationships between host plant and phytopathogen are elaborately regulated by chemical signaling substances. In contrast, chemical weapons in nonhost plants which can affect/interfere the life cycle development of oomycete phytopathogens, are supposed to be responsible for their incompatibility to Oomycetes. These facts suggest us some cues for biorational regulation of soilborne oosporogenic and zoosporogenic pathogens which are often difficult to eradicate. The research trend of oomycete pathogens based on ecological chemistry has been briefly reviewed with special reference to *Aphanomyces cochlioides*.

Introduction

Oomycetes (Peronosporomycetes in the new classification) are phylogenetic relatives of brown algae that cause many destructive diseases of plants, as well as several animal and human diseases (*1*). They are mostly water or soil inhabiting organisms (*2*). The motile zoospores which are propelled by flagella are an important means of pathogen distribution and often the key infectious stage of the phytopathogenic Oomycetes. Disease caused by these Oomycetes can be multi-cyclic, resulting in severe epidemics that can destroy whole crops within a single season (*3*). Our knowledge of their biology is limited, since their physiology differs from that of fungi, many fungicides are ineffective against Oomycetes (*4, 5*). New approaches are needed to identify novel targets and to develop biorational control measures to minimize the economic impact of these phytopathogens.

Zoospores have high-affinity receptor based recognition systems for locating hosts by chemotaxis (*6, 7, 8*). They can accumulate at the potential infection sites of host roots by chemotaxis, after which they undergo a series of morphological changes before penetrating the root tissues (*9*). Recent reports suggest that all these key pre-infection events of zoosporic pathogens are triggered by host-specific chemical signals (*10, 11*). A few host-derived zoospore chemical signals have been identified (Table 1). All of these mediate chemotaxis at micromolar to nanomolar levels.

In contrast to susceptible plants, non-susceptible plants may possess some chemical means to ward-off zoosporic phyopathogens (*7*). Surveys of non-susceptible plants using the *Aphanomyces cochlioides* zoospore bioassay revealed that some of the non-susceptible plants possess diverse chemical weapons those can directly affect the viability or regulate some steps of the life cycle development of this Oomycetes. However, very restricted chemical factors disturbing the life cycle of such organisms have been known at present (Table 1), even though understanding well about such chemical factors seems to be very significant to establish new techniques for control of those soilborne oomycete phytopathogens still problematic in agriculture. Here we briefly review our research results concerning the activities of chemical signals in host plants regulating the life cycle development of oomycete phytopathogens. In addition, the diverse activities of some nonhost zoospore regulating metabolites (repellents, cytotoxins and inhibitors of zoospore motility) are discussed in relation to the biorational regulation of soilborne oomycete phytopathogens.

Table I. **Diverse Stimuli Triggering Characteristic Behaviors and/or Morphological Changes of *Aphanomyces cochlioides* or Other Peronosporomycetes (Oomycetes) Zoospores**

(1) Host-specific signal substances triggering life cycle development
  a) *Aphanomyces raphani*   3-indolecarbaldehyde (**1**)*
  b) *A. euteiches*   prunetin (5,4'-dihydroxy-7-methoxyisoflavone, **2**)*
  c) *A. cochlioides*   cochliophilin A (5-hydroxy-6,7-methylenedioxyflavone, **5**)
         N-trans-feruloyl-4-O-methyldopamine (**6**)
  d) *Phytophthora sojae*   daidzein (7,4'-dihydroxyisoflavone, **3**)
         genistein (5,7,4'-trihydroxyisoflavone, **4**)
  Activities: i) Host-specific attraction, ii) Encystment induction
       iii) Germination of cystospores, and iv) Hyphal chemotropism

(2) Non-specific (general) attractants
  a) amino acids
  b) carbohydrates
  c) alcohols
  d) aldehydes (valeraldehyde, isovaleraldehyde)

(3) Inorganic and physical stimuli
  a) low molecular weight cations (*e.g.* H^+, K^+) and electric signals
  b) cyst formation by mechanical stimulation
  c) thigmotropism or contact chemical sense
    (hyphal recognition of surface structure where mycelium is growing)

(4) Repellent against *Aphanomyces cochlioides* zoospores
  a) flavonoid compounds (**8, 13**, 8-prenylnaringenin)
  b) environmental pollutant (**9**)
  c) synthetic and natural estrogens (**10, 11**)
  d) phenylpropanoids (lignans)

(5) Zoosporicidal substances in nonhost plants
  a) antimicrobial substances (synthetic and natural products)
  b) condensed tannins and anacardic acids (cell lytic)
  c) saponins (*e.g.* avenacin, cell lytic)
  d) lectins (*e.g.* concanavalin A)

(6) Encystment triggering substances in nonhost
  a) nicotinamide (**15**)
  b) 1-linoleoyl-2-lysophosphatidic acid monomethyl ester (**14**) + *N-trans*-feruloyl-tyramine (**7**)
  c) *n*-BuOH and *sec*-BuOH
  d) G-proteins activator (*e.g.* mastoparan)

See, Ref. 6 - 8; * not studied on developmental transition of zoospores.

1. Host-specific signal substances

Host-specific attractants of oomycete phytopathogens known so far are shown in Table 1. For example, indole-3-carbaldehyde (**1**) isolated from cabbage seedlings is a chemoattractant down to concentration of 1 nM for *A. raphani* zoospores *(12)*. Prunetin (**2**) isolated from pea seedlings *(13)* is a potent attractant (down to 10 nM) for zoospores of the *A. euteiches*, while the zoospores of *Phytophthora sojae* are attracted to daidzein (3) and genistein (4) from the roots and root exudates of its host, soybean, at a concentration down to 0.1 nM (*10, 14*). Zoospores of *A. cochlioides* are strongly attracted to host metabolites, cochliophilin A (5) and *N-trans*-feruloyl-4-*O*-methyldopamine (6) at concentrations down to 1 nM and 10 nM, respectively (*15, 16*). Structure and activity relationships of these host-specific zoospore attractants have been reported (*17, 18, 19*).

The question of specificity in chemotaxis is an important one as it relates to the contribution of chemotaxis, and subsequent steps of infection by zoospores, to host selection and host specificity (*8*). Several lines of evidence suggest that compounds such as cochliophilin A (5) and the soybean isoflavones (3, 4) function as host-specific signal compounds. (A) These compounds are restricted to a limited number of plant groups which include the host plants. (B) Some amounts are exuded from the host plant roots. (C) They show highly selective activity to the relevant pathogens; and (D) they are not only attractive for zoospores, but also trigger the zoospore encystment followed by cystospore germination, and cause tropism of the germinated hyphae. For example, a spinach root was found to contain 5.3 µg/g of the host-specific attractant, cochliophilin A (5), and root exudation of 5 was estimated to be 34 ng/plant/day. Cochliophilin A was attractive for the spinach pathogen *A. cochlioides* zoospores at 1×10^{-10} M, but inactive to *A. euteiches* and *A. raphani* pathogenic to pea and cabbage, respectively at 1×10^{-5} M.

In *Phytophthora* (*Ph.*) and *Pythium* (*Py.*) species with restricted host ranges, zoospores display a strong attraction to their respective host root exudates (*6,7*). Specific sensitivity of zoospores to the host-derived compounds has also been found in other pathogenic Oomycetes. Kerwin *et al.* (*20*) observed that zoospores of *Py. marinum* encyst on the surfaces of red algae (its hosts) but not on green or brown algae (nonhost). Galactose or anhydrogalactose contents in the surface of red algae were found to be responsible for such a specific response.

2. Diverse activity of secondary metabolites in nonhost plants

In recent years, there has been renewed interest in examining interactions between nonhost plants and Oomycetes (*21*). The molecular basis of nonhost resistance remains one of the major unknowns in the study of plant-microbe interactions. Performed barriers and compounds such as saponins and tannins are ubiquitous in plants and play important roles in nonhost resistance in filamentous fungi (*22, 23*). To assess the potential role of secondary compounds in nonhost resistance we have initiated a screening procedure of root extracts of 200 nonhost plants of *A. cochlioides*. Nearly half of the extracts had direct effects on motility and viability of *A. cochlioides* zoospores. None of the plant extracts showed attractant and subsequent differentiation activities as shown by cochliophilin A (**5**) in earlier experiments (*24*). In addition nonhost extracts exhibited some deleterious activities for example, repellent, stimulant, halting, lysis *etc.* against zoospores (Table 1). These results suggest that many nonhost plants might use secondary metabolites to directly defend themselves from oomycete phytopathogens (*25*). Isolation of different nonhost defense factors (chemical weapons) against Oomycetes may give some new interesting agents for controlling those soilborne phytopathogens.

2-1) Repellent activity of estrogenic compounds toward *Aphanomyces* zoospores

During a survey of physiologically active constituents in 100 species used in Chinese herbal medicines, the constituents of *Dalbergia odorifera* (Leguminosae) repelled *Aphanomyces* zoospores from Chromosorb W AW particles coated with a 500 ppm solution of the acetone extracts. One of the repellent compounds was subsequently identified as medicarpin (**8**). The repellent activity of medicarpin was reinforced by the addition of the isoflavones formononetin and claussequinone which are also components of *D. odorifera* root extracts. Since many isoflavones also elicit estrogenic activity, we postulated that other estrogenic agents might function as zoospore repellents.

Bisphenol A (BPA, **9**), a reputed xenoestrogen (*26*), exhibited potent repellent activity against the zoospores of *A. cochlioides*. Following this finding, we tested a number of androgenic and estrogenic compounds (*e.g.* testosterone and its fungal metabolites, and progesterone, estradiols, diethylstilbestrol, estrone, estriol, pregnenolone, dienestrol *etc.*) on the motility behavior of *A. cochlioides* zoospores (*27*). Interestingly, most of the synthetic and natural estrogenic compounds, *e.g.* diethylstilbestrol (**11**), estriol (**10**), and 17β-estradiol (**11**), exhibited potent repellent activity toward the motile zoospores of *A. cochlioides* at a level of 1 µg/ml or less by the "particle method" [Chromosorb W AW particles (60 - 80 mesh) coated with each test solution were dropped into a zoospore suspension and the responses of zoospores were observed under a microscope]. Furthermore, a phytoestrogen, 8-prenylnaringenin recently identified in drinking beer (*28*) exhibited potent repellent activity toward *A. cochlioides* zoospores. Surprisingly, *O*-methylation of DES (**10**) yielding dimethyl ether (**12**) functioned as a potent attractant. Moreover, the zoospores attracted by **12** encysted and then germinated in the petri dish. This represents the first report of repellent behavior of estrogenic compounds toward oomycete zoospores.

Endocrine disrupters, bisphenol A (**8**), diethylstilbestrol (**10**) *etc.*, pose a potential threat to human and animal health, the detection of these compounds in the environment remains problematic due to the absence of rapid and sensitive bioassays. The particle bioassay method is very simple and convenient to identify compounds affecting zoospore motility. It may also be a useful technique for the identification of repellants from natural sources or as a prescreening technique to identify estrogen-like compounds from environmental samples.

The mechanism of repulsion or negative chemotaxis in zoospores is not yet known, but in bacteria both activities are mediated by the modification of common receptor and signal transduction pathways (29). Notably, while several isoflavones with structural similarity to daidzein functioned as chemoattractants for *Ph. sojae*, many related flavonoids had repellent activities (19). The importance of flavonoids in the case of *Ph. sojae*, and estrogenic substances in the case of *A. cochlioides*, in deterring infection of nonhost plants, certainly warrants additional investigation.

2-2) Motility inhibitory and cell lytic substances against *Aphanomyces* zoospores

The saponins such as avenacin A_1 which is released from oat roots are known to be lytic to a broad range of oomycete zoospores. Our screening assays revealed that crude extracts of stem bark from an Anacardiaceae plant, *Lannea coromandelica* had similar properties. A solution of 200 ppm of the crude extract was found to inhibit the motility of the zoospores and subsequently cause them to lyse (31).

In course of chromatographic studies, five dihydroflavonols were identified and their structures were elucidated by spectroscopic methods. Two of them were new natural products, but not active against zoospores (32). Characterization of zoosporicidal principle in *Lannea* extracts were done by MALDI-TOF-MS as a linear and angular-type condensed tannin which was confirmed by parallel experiments using commercial polyflavonoid tannins in analytical and biological assays (31). Like *Lannea* extracts, commercially available Quebracho and Mimosa tannins showed identical halting and characteristic lytic activities against *A. cochlioides* zoospores. Similar effects were observed with anacardic acid, a product from Ginkgo fruits (33) and the synthetic fungicide fluazinam. Polyflavonoidal tannins are found as a general defense agent in many plants. Their roles against herbivores and microorganisms are well-known, but the zoosporicidal effects of polyflavonoidal tannins have not been previously reported. These observations suggest that such compounds might constitute a new group of natural agents that could be utilized as part of a biorational control of other oomycete phytopathogens.

To gain more insight into the zoosporicidal activity of *Lannea* and commercial tannin extracts, we studied the morphological changes of zoospores by scanning electron microscopy (SEM). The morphological changes (fragmentation of cellular materials and formation of unique structures) of zoospores by polyflavonoid tannins observed in this experiment are similar to some features of apoptosis (*34*). Scanning electron microscopic observation visualized that both *Lannea* and commercial tannins caused lysis of cell membrane followed by fragmentation of cellular materials. The mode of action of these compounds appeared to be different from that of detergent action of saponins on membrane (*30*).

Lannea stem bark contains high proportion of polyflavonoid tannins (*ca.* 13% of dry weight) and, thus raises a possibility of using those naturally occurring compounds as a zoosporicidal agent. To the best of our knowledge, this is the first report of zoosporicidal activity of natural polyflavonoid tannins against an oomycete phytopathogen. Further studies on the zoosporicidal mode-of-action of polyflavonoid tannins and their effects on other phytopathogenic Oomycetes are needed to determine their potential as an organic source of zoosporicidal agents.

2-3) Complex mixtures affecting zoospore motility

Root extracts of fifty herbs were assayed to identify compounds affecting zoospore motility. EtOAc extracts from the roots of the common purslane *Portulaca oleracea* (Portulacaceae) were potent inhibitors of zoospore motility. Further fractionation of these extracts and characterization of the fractions using the particle bioassay identified two active factors, *N-trans*-feruloytyramine (**7**, a stimulant), and 1-linoleoyl-2-lysophosphatidic acid monomethyl ester (**14**, a repellent). Mixtures of the two factors were sufficient to duplicate the inhibition of motility caused by the crude extracts. Moreover, Chromosorb W AW particles coated with an excess of the stimulant (**7**) and either commercially available 1-oleoyl-2-lysophosphatidic acid (100 ppm), or the chemically derivatized monomethyl ester (10 ppm), effectively inhibited zoospore motility. Based on the bioassay results, we conclude that a mixture of the stimulant, *N-trans*-feruloyltyramine (**7**), and the repellent, 1-linoleoyl-2-lysophosphatidic acid monomethyl ester (**14**), in *Portulaca* root are responsible for inhibiting the motility of zoospores of *A. cochlioides* (*35*).

Under the microscope, the treated zoospores were first seen to become stationary, and then settle at the bottom of the petri dish where they encysted to give cystospores. These cystospores germinated within 1-2 h, although germination would not normally be expected in the absence of a host plant. This is the first report on the inhibition of zoospore motility as a result of the interaction of a zoospore stimulant (*N-trans*-feruloyltyramine, **7**) and a repellent (1-linoleoyl-2-lysophosphatidic acid monomethyl ester, **14**).

2-4) Regulation of developmental transition by *Amaranthus gangeticus* metabolites

On our preliminary screening of nonhost plant extracts, *Amaranthus gangeticus*, one of edible Amaranthaceae plants in Bangladesh and some other tropical countries, rich in vitamin A and dietary fibers, showed both potent attractant and motility inhibitory activities towards *A. cochlioides* zoospores. Chromatographic separation of *Am. gangeticus* constituents revealed that the taxis and subsequent motility inhibition of zoospores were regulated by the cumulative effects of two chemically different factors. The attractant was identified as *N-trans*-feruloyl-4-*O*-methyldopamine (**6**), and the motility-inhibiting factor as nicotinamide (**15**) (*36*). In particle assays compound **6** showed attractant activity up to 1×10^{-8} M concentration in a dose dependent manner without halting motility of zoospores even at a very high concentration (1×10^{-5} M) (*37*). The direct application of compound **6** as a homogeneous solution had no effect on the motility of zoospores. On the other hand, nicotinamide (**15**) showed immediate halting activity followed by encystment in both particle (1×10^{-5} M) and homogeneous solution methods (MIC 2×10^{-8} M). Interestingly, cysts formed inducibly by **15** regenerated zoospores (85-90%) instead of germination within 2-3 h in homogeneous solution method or only 20-30 min (*ca* 90% of cysts around the particle) in particle method. However, concomitant application of compounds **6** (1×10^{-6} M) and **15** (1×10^{-5} M) showed encystment of zoospores followed by germination of cystospores nearly 90% within 30-35 min (particle method). As discussed above, nicotinamide (**15**) repeatedly compels zoospores a cycle from one zoospore stage to the next zoospore stage *via* immature cystospores morphologically different from mature cysts which stage is essential for following cyst germination, and finally exhausts the zoospores themselves (Scheme 1).

SEM observation revealed that **15** induced cysts regenerate zoospores leaving their smooth cyst coats, whereas cysts produced by concomitant application of compounds **6** and **15** germinated *via* the mature stage as shown by cochliophilin A (**5**). When an excised root of a 6 days old seedling of *Am. gangeticus* was immersed into a zoospore suspension in a small petri dish, all zoospores around the root tip were immediately halted. The halted spores encysted and then regenerated zoospores after 3 h indicating the possibility of exudation of nicotinamide (**15**) predominantly from the root of *Am. gangeticus*. On the other hand, zoospores were specifically attracted to the root tip of *Celosia cristata* (a host; Amaranthaceae), aggregated and encysted to form a mass of cystospores within 30 min followed by cystospore germination within 40 min instead of zoospore regeneration. However, different levels of motility inhibition followed by regeneration were commonly observed in case of all nonhost species in Amaranthaceae. Semiquantitative TLC examination of root extracts and zoospore bioassay revealed that all incompatible Amaranthaceae roots contain high proportions of nicotinamide (15).

In contrast, compatible *C. cristata* roots contain high proportion of *N*-*trans*-feruloyl-4-*O*-methyldopamine (6) with low amount of nicotinamide (15). It appeared from the results that the ratio of **6** and **15** in the root exudates of Amaranthaceae might determine the compatibility of pathogen to host, whilst further detailed analytical data are awaited.

Zoospores that fail to locate the host or other nutrient sources encyst before the endogenous nutrient reserves are depleted, and the cystospore regenerates a further zoospore. Such kind of zoospore regeneration seems to be an adaptation to increase the possibility of establishing infection. Immature cystospores of *A. cochlioides* induced by mechanical stimulation (vortexing) as well by *n*-BuOH (*38*) or nicotinamide (**15**) regenerated further zoospores in high yields. As shown in Scheme 1, the direction of zoospore development, to zoospore regeneration or hyphal germination, is regulated by host and nonhost plant chemicals.

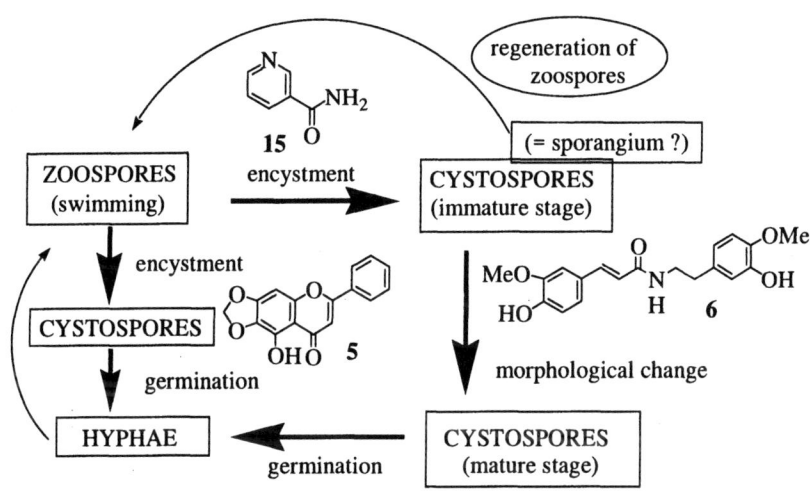

Scheme 1. *Differentiation of* Aphanomyces cochlioides *Zoospores in the Presence of Cochliophilin A (5), Nicotinamide (15) and* N-trans-*Feruloyl-4-O-methyldopamine (6)*

3. Putative signal transduction pathways

Our knowledge about the system of zoospore signal transduction pathway is still not quantitative, but qualitative. However, it is our confidence that the zoospore is an excellent system to study the molecular regulation of development in oomycete pathogens, because they show remarkable behavioral and morphological changes in response to exogenous signals within 10 - 60 min.

G-proteins are believed to be key components of signal transduction pathways in chemotaxis of many other motile cells (*39*). Mastoparan is commonly used as a diagnostics for the participation of G-proteins in both animal and plant signal transduction pathways (*40, 41*). Interestingly, the hetrotrimeric G-protein activator, mastoparan induced zoospore encystment followed by germination of the cystospores at micromolar concentration (*7, 11*). The synthetic peptide analogue Mas 17, predicted not form an amphipathic helix at lipid interface because of the replacement of Leu-6 by Lys, is totally devoid of agonist activity. The concomitant application of mastoparan and the host-specific attractant cochliophilin A (*5*) appeared to enhance further encystment of zoospores and rapid germination of cystospores. In addition, chemicals interfering with phospholipase C activity (neomycin) and Ca^{2+} influx/release (EGTA and loperamide) suppressed cochliophilin A and mastoparan induced encystment and germination. These results suggest that the zoospore differentiation by host-specific cochliophilin A (*5*) may be mediated by G-protein-coupled receptors to activate both phosphoinositide and Ca^{2+} second messenger pathways. Changes of Ca^{2+} fluxes during differentiation of zoospores have been observed by early investigators (*42, 43*). To the best of our knowledge, this is the first indication that G-protein mediated signaling mechanism is involved in oomycete zoospores (*7*). Further research on the quantification of lipid metabolism and PLC activity may open the black box of the signal transduction pathways. Since the components of the pathway represent attractive targets for developing alternative disease control methods, agricultural practice may benefit from such kind of research results in the long term.

Our preliminary experiments to detect the putative receptor responsible for the host-specific attractant cochliophilin A (*5*), by using a probe 5,7-dihydroxyflavone incorporated with 7-*O*-biotinyl and 4'-azido groups show that the receptors are located on the cell membrane (*44*). Detailed characterization of the cochliophilin A receptor would be a starting point to unravel the signal transduction pathway in zoospore chemotaxis and differentiation by host-specific plant signals.

Conclusions

Our screening tests to find plant secondary metabolites interacting with the life cycle development of oomycete zoospores revealed that both host and nonhost metabolites can regulate some key steps in the life cycle of oomycete phytopathogens. The results suggested us to design following projects on phytopathogenic Oomycetes from fundamental and practical viewpoints.

1) Identification and characterization of the receptors corresponding to the host-specific signal substances essential for the pathogenic zoospores to establish infection.
2) Elucidation of intracellular signal transduction processes from the signal reception to metabolic, behavioral, and morphological changes in the zoospores.
3) Establishment of a biorational regulation method for oomycete phytopathogens by using biomimics which can disturb their life cycle development.

Acknowledgements

We are thankful to Professor R. Yokosawa, Health Science University of Hokkaido for his kind gift of microbial strains, and Professor Paul Morris, Bowling Green State University for his scholastic review and linguistic improvement of the manuscript. The financial support from MEXT (to S.T., A(2):No. 14206013) and a Postdoctoral Fellowship from the JSPS (to M.T.I) are also very much appreciated.

References

1. Money, N. P.: *Nature* **2001**, *411*, 644-645.
2. Sparrow, F. K.: *Aquatic Phycomycetes,* 2nd edn., University of MichiganPress, Ann Arbor, Michigan, **1960**.
3. Gow, N. A. R.; Campbell, T. A.; Morris, B. M.; Osborne, M. C.; Reid, B.; Shephered, S. J.; West, P. V. in "Microbial Signaling and Communication", 57th Symp. Soc. Gen. Microbiol. Symp., England, R.; Hobbs, G.; Bainton, N.; Robertson, D. McL., Eds.; Cambridge University Press: U. K., **1999**, pp. 285-305.
4. Tyler, B. M.: *Trends Genet.* **2001**, *17*, 611-614.
5. Govers, F.: *Nature* **2001**, *411*, 633.
6. Deacon, J. W.: *New Phytol.* **1996**, *133*, 135-145.
7. Islam, M. T.; Tahara, S.: *Biosci. Biotechnol. Biochem.* **2001**, *65*, 1933-1948.
8. Tyler, B. M.: *Annu. Rev. Phytopathol.* **2002**, *40*, 137-167.
9. Islam, M. T., Ito, T.; Tahara, S.: *J. Gen. Plant Pathol.* **2002**, *68*, 111-117.
10. Morris, P. F.; Bone, E.; Tyler, B. M.: *Plant Physiol.* **1998**, *117*, 1171-1178.
11. Islam, M. T.; Ito, T.; Tahara, S.: *Plant Soil* **2003**, *255*, 131-142.

12. Yokosawa, R.; Kuninaga, S.: *Ann. Phytopathol. Soc.*, Jpn. **1979**, *45*, 339-343.
13. Yokosawa, R.; Kuninaga, S.; Sekizaki, H.: *Ann. Phytopathol. Soc.*, Jpn. **1986**, *52*, 809-816.
14. Morris, P. F.; Ward, E. W. B.: *Physiol. Mol. Plant Pathol.* **1992**, *40*, 17-22.
15. Horio, T.; Kawabata, Y.; Takayama, T.; Tahara, S.; Kawabata, J.; Fukushi, Y.; Nishimura, H.; Mizutani, J.: *Experientia* **1992**, *48*, 410-414.
16. Horio, T.; Yoshida, K.; Kikuchi, H.; Kawabata, J.; Mizutani, J.: *Phytochemistry* **1993**, *33*, 807-808.
17. Sekizaki, H.; Yokosawa, R.; Chinen, C.; Adachi, H.; Yamane, Y.: 1993. *Biol. Pharm. Bull.* **1993**, *16*, 698-701.
18. Kikuchi, H.; Horio, T.; Kawabata, J.; Koyama, N.; Fukushi, Y.; Mizutani, J.; Tahara, S.: *Biosci. Biotechnol. Biochem.* **1995**, *59*, 2033-2035.
19. Tyler, B. M.; Wu, M. H.; Wang, J. M.; Cheung, W.; Morris, P. F.: *App. Environ. Microbiol.***1996**, *62*, 2811-2817.
20. Kerwin, J. L.; Johnson, L. M.; Whisler, H. C.; Tuiniga, A. R.: *Can. J. Bot.* **1992**, *70*, 1017-1024.
21. Kamoun, S.: 2001. *Curr. Opin. Plant Biol.* **2001**, *4*, 295-300.
22. Osburn, A.: *Trends Plant Sci.* **1996**, *1*, 4-9.
23. Osburn, A.: *Plant Cell* **1996**, *8*, 1821-1831.
24. Islam, M. T.: *Ph D Thesis*, Grad. Sch. Agric., Hokkaido Univ., Japan, **2002**.
25. Islam, M. T.; Tahara, S.: in "Proceedings, International Symposium on Utilization of Natural Products in Developing Countries: Trends and Needs"; Mansingh, A.; Young, R. E.; Yee, T.; Delgoda, R.; Robinson, D. E.; Morrison, E.; Lowe, H. Eds.; Nat. Prod. Inst., Univ, West Ind., Kingston July, **2000**, pp. 210-218.
26. Takai, Y.; Tsutsumi, O.; Ikezuki, Y.; Hiroi, H.; Osuga, Y.; Momoeda, M.; Yano, T.; Taketani, Y.: *Biochem. Biophys. Res. Commun.* **2000**, *270*, 918-921.
27. Islam, M. T.; Tahara, S.: *Z. Naturforsch.* **2001**, *56c*, 253-261.
28. Takamura-Enya, T.; Ishihara, J.; Tahara, S.; Goto, S.; Totsuka, Y.; Sugimura, T.; Wakabayashi, K.: *Food Chem. Toxicol.* **2003**, *41*, 543-550.
29. Mason, M. D.: *Adv. Microb. Physiol.* **1992**, *33*, 277-346.
30. Deacon, J. W.; Mitchell, R. T.: *Trans. Br. Mycol. Soc.* **1985**, *84*, 479-487.
31. Islam, M. T.; Ito, T.; Sakasai, M.; Tahara, S.: *J. Agric. Food Chem.* **2002**, *50*, 6697-6703.
32. Islam, M. T.; Tahara, S.: *Phytochemistry* **2000**, *54*, 901-907.
33. Begum, P.; Hashidoko, Y.; Islam, M. T.; Ogawa, Y.; Tahara, S.: *Z. Naturforsch.* **2002**, *57c*, 874-882.
34. Shiokawa, D.; Murata, H.; Tanuma, S.: *FEBS Lett.* **1997**, *413*, 99-103.
35. Mizutani, M.; Hashidoko, Y.; Tahara, S.: *FEBS Lett.***1998**, *438*, 236-240.
36. Islam, M. T.; Hashidoko, Y.; Ito, T.; Tahara, S.: *J. Pestic. Sci.* **2004**, *29*, 6-14.

37. Shimai, T.; Islam, M. T.; Fukushi, Y.; Hashidoko, Y.; Tahara, S.: *Z. Naturforsch.* **2002**, *57c*, 323-331.
38. Latijnhouwers, M.; Munnik, T.; Govers, F.: *Mol. Plant-Microbe Interact.* **2002**, *15*, 939-946.
39. van Es, S.; Devreotes, P. N.: *Cell. Mol. Life Sci.* **1999**, *55*, 1341-1351.
40. Munnik, T.; van Himbergen, J. A. J.; ter Riet, B.; Braun, F.-J.; Irvine, R. F.; van den Ende, H.; Musgrave, A.: *Planta* **1998**, *207*, 133-145.
41. Pingret, J. L.; Journet, E. P.; Barker, D. G.: *Plant Cell* **1998**, *10*, 659-671.
42. Connolly, M. S.; Williams, N.; Heckman, C. A.; Morris, P. F.: *Fungal Genet. Biol.* **1999**, *28*, 6-11.
43. Xu, C.; Morris, P. F.: *Mycologia* **1998**, *90*, 269-275.
44. Sakihama, Y.; Shimai, T.; Ito, T.; Tahara, S.: 3rd PPCPS, Hawaii, June, **2003**, P-217.

Chapter 20

Nematicidal Compounds from the Fungi

Yasuo Kimura[1], Miyako Kusano[1,2], and Satoshi Nakahara[1]

[1]Department of Agriculture, Tottori University, Tottori, Japan
[2]Umea Plant Science Center, The Swedish University, Umea, Sweden

Fungal metabolites represent a vast repository of materials and compounds with evolved biological activity, including nematicidal effect. Some of these compounds can be used directly or as templates for nematicides. Plants such as a burdok (*Arctium lappa* L), a carrot (*Daucus carota* L), a radish (*Raphanus sativus* L.) and so on are parasitized by the nematode, *Pratylenchus penetrans*, and show necrosis and successively breakdown of root cortex cells.

In this study, we searched the fungal metabolites for new nematicidal compounds to control the root-lesion nematode, *Pratylenchus penetrans*, and succeeded in the discovery of some fungi. In the course of screening for these nematicidal agents for potential development, we found the presence of nematicidal compounds in the culture metabolites of *Penicillium* cf. *simplicissimum* (Oudemans) Thom, the Fungus *Aspergillus* sp.,and *Penicillium bilaiae* Chalabuda.

Materials and Methods

Melting point (mp) data were generated using an YAMAKO MP-S3 instrument. Optical rotation values were determinedwith an HORIBA SEPA-200 instrument. The CD spectra were measured with a JASCO J 720 spectropolarimeter. The IR and UV spectra were recorded with JASCO FT/IR 530 and JASCO FT/IR 700 and SHIMADZU UV 2200 instruments, respectively. The NMR spectra were obtained fromJEOL-JNM-270 (at 270 MHz for ^1H and 67.5 MHz for ^{13}C), JEOL-JNM-ALPHA400 (at 400 MHz for ^1H and 100 MHz for ^{13}C), JEOL-JNM-ECP500 (at

500 MHz for ^{1}H and 125 MHz for ^{13}C), and JEOL-JNM-ALPHA600 (at 600 MHz for ^{1}H and 150 MHz for ^{13}C). The MS spectra were recorded with HITACHI M-80B and JEOL-JMS-SX 102 apparatus. Column chromatography was performed on silica gel of 200 mesh (Wakogel C-200). Gel permeation chromatography was performed on Pharmacia Sephadex LH-20. Analytical TLC and preparative TLC were performed on Merck pre-coated silica gel 60 F_{254} and Merck Kieselgel 60 GF_{254} (10 g silica gel spread on 20 x 20 x 0.05 cm glass plates), respectively.

1. Bioassay for nematicidal activity.

The root lesion nematode, *Pratylenchus penetrans*, was cultured for about two weeks on a slant of alfalfa grown in the Krusberg medium,[1, 2] separated from the callus by the Baermann funnel technique[3] and counted under a microscope (x 40). An aqueous suspension containing a defined number of nematodes (ca. 2000 nematodes /ml) was prepared by appropriate dilution. Test compounds and extracts were dissolved in methanol and added into the nematode suspension (up to 3% volume of the suspension). The nematode suspension (pH 6-7) with defined concentration of test compound was transferred to 12 or 24 well plates. After keeping for 3 days at 24^0C, the nematodes in the well plates were counted under a microscope (x 40). The nematicidal activity was calculated as follows: Mortality (%) = (B-A) B^{-1} x 100[4] ;where A = the number of living nematodes after being treated with a test compound, and B = the number of living nematodes in the control wells (3% methanol in distilled water).

2. Bioassay for nematicidal activity toward free-living nematode

Nematicidal activities were determined in a microwell plate assay[5] with the free-living nematode, *Caenorabditis elegans*. Worms were cultivated on agar plates.[6] A suspension of adults and L_4 larvae (over 90%) from a 4 days old culture was diluted with M9 buffer [KH_2SO_4 3 g/l, Na_2HPO_4 6 g, NaCl 5 g, $MgSO_4$ (1 mol/l) 1 ml] for the preparation containing a defined number nematodes (ca. 2000 nematodes /ml). Tested compounds and extracts were added in methanolic solutions up to 3% of the final volume (0.2 ml). The nematode suspension (0.1 ml) thus obtained was added to 24 well plates, containing a defined amount of the test compound (pH 6-7). After keeping for 1 day at 18^0C, the nematodes in the well plates were counted under a microscope (x 40).

3. Bioassay for the growth of lettuce seedlings.[7]

Lettuce seeds were grown under light (ca. 2000 lux) at 24^0C on a petri dish (150 x 25 mm) laid with a filter paper containing deionized water. After 1 day, twelve seedlings were selected for uniformity (radicles; 2 mm) and transferred to a minipetri dish (35 x 15 mm) laid with a filter paper containing deionized water (1 ml) and a defined amount of the test compound. The petri dishes were kept at 24°C for 4 days under continuous light (ca. 2000 lux). The length of the hypocotyls and roots treated with the alkaloids were measured and the mean value of the length was compared with an untreated control.

4. *Bioassay for the growth of rice seedlings.*[8]

The rice seeds (*Oriza sativa* L.) were sterilized with 75% ethanol for 30 sec, immersed in sodium hypochlorite solution (antiformin) for 2hr, rinsed under running water for 3 hr and transferred to a petri dish (150 x 25 mm) containing deionized water. After 3 days under the light (ca. 2000 lux) at 30°C, seven seedlings were selected for uniformity (radicles; 2-3 mm) and transferred into a test tube laid with filter paper containing deionized water (1 ml) and a defined amount of the test compound. The test tubes (140 x 23 mm) were sealed with a sheet of polyethylene film and incubated at 30°C for 7 days under continuous light (ca. 2000 lux). The length of total, second leaf sheath and primary root after treatment with the compounds were measured, and the mean value of the length was compared with an untreated control.

5. *Fermentation.*

Erlenmeyer flasks (500 ml) containing medium (250 ml) made from malt extract (30 g/l), glucose (20 g/l) and peptone (3 g/l), were inoculated with spores of the fungi previously grown on solid potato dextrose agar. The culture broth was grown without shaking at 24°C for 21 days.

6. *Extraction and isolation .*

The isolation of compounds from the culture broth is shown in Figs. After 21days, the culture broth was filtered to separate the mycelium from the broth. The mycelial mats were dried for 1 week and then extracted three times with acetone.

Results and Discussion

Based on the results mentioned below, the chemistry and biological activities of nematicidal compounds are discussed.

1) Compounds from *Penicillium* cf. *simplicissimum* (Oudemans) Thom

Bioassay-guided fractionation led to the isolation of three active compounds.peniprequinolone (1), penigequinolones A sand B (2a, 2b) [9] and 3-methoxy-4,6-dihydroxy-4-(4,-methoxyphenyl)quinolinone (4)[10] . The rerlated but inactive comound (3)[10]

```
Mycelial mats (malt extract medium)
    │ Extraction with acetone
Acetone extract
    │ silica gel column chromatography (CC)
    │ eluted with benzene-acetone
┌───────────────────┬───────────────────┬───────────────────┐
Fraction (Acetone 10%)  Fraction (Acetone 20%)  Fraction (Acetone 20%)
│ LH-20              │ LH-20              │ silica gel CC
│                    │                    │ PTLC
│ silica gel CC      │ PTLC               │
│ preparative TLC (PTLC) │                │
Compound 1          Compound 4, 2a, 2b   Compound 3
```

Fig. 1 Isolation Procedure of the Compounds from *Penicillium* cf. *simplicissimum*

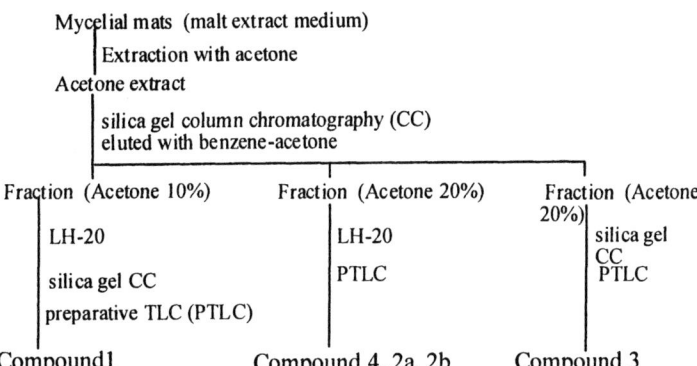

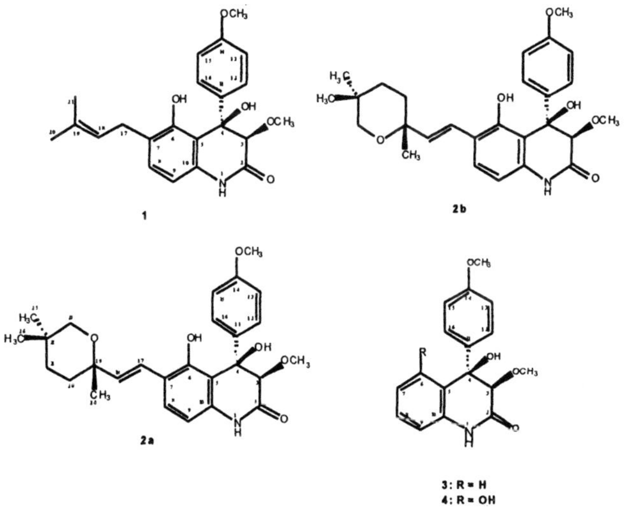

Fig. 2 Structures of Isolaterd Compounds

Table 1. Nematocidal activities of aspyrone, acetylaspyrone and alkaloids 1-5 towards *Pratylenchus penetrans*.

Compound	Nematicidal activities LD_{50}	LD_{90}
1	1000	> 1000
2	100	1000
3	> 100	
4	1000	
Aspyrone	100-300	1000
Acetylaspyrone	100	300

$LD_{50}(LD_{90})$: Concentrations (mg/l) causing over 50% (resp. 90%) immotility of the nematodes after 3 days.

The biological activities of **1**, **2**, and **3** were examined by bioassay methods involving nematoda, together with lettuce, and rice seedlings. Compound **4** was examined using nematoda because of low yield. Compounds **1**, **2**, and **4** showed nematicidal activity toward *P. penetrans* by 82.4%, 69.2%, and 57.7% at the concentration of 1000 mg/l, respectively (Fig. 8). However, **3** hardly showed any effect at the tested concentrations (1-1000 mg/l). The effect could not be changed by removing **2** from the medium after 3 days of incubation with penigequinolones **2** (1000 mg/l) and washing of the nematoda with the water. Eighty percent of the nematoda were dead after 1 day whereas the nematoda of the control were still alive.[11] This fact indicated that alkaloid **2** showed the nematicidal activity within 3 days of incubation. Aspyrone[12] has been reported as a nematicidal compound produced by *Aspergillus melleus*, whose LD_{50} and LD_{90} values are shown in Table 1 as a positive control, together with those of acetylaspyrone.

The selectivity of **2** toward *P. penetrans* was demonstrated by its relative low activity at the tested concentrations (1-1000 mg/l) agai
nst the free-living nematode *C. elegans*. As for rice seedlings, **1** and **2** accelerated only the root growth of the seedlings in proportion to their concentrations from 100 mg/l to 300 mg/l, while **3** showed no activity against the growth. From these results, the presence of a phenolic hydroxyl group at C-5 and/or a tetrahydropyran ring may be necessary to exhibit the nematicidal activity toward the root lesion nematode *P. penetrans*. Compound **2** were non-toxic against the free living nematoda *C. elegans* and the hypocotyl and primary root elongation toward rice and lettuce seedlings, making them potentially useful nematicides.

2) Compounds from Unidentified *Aspergillus* sp.

The EtOAc extract of the culture filtrate of *Aspergillus* sp. was fractionated by silica gel column chromatography and further separated by preparative TLC to afford three active compounds **1, 2** and **3**. The acetone extract of the mycelial mats was dissolved in EtOAc and partitioned with NaHCO$_3$ saturated solution. The remaining EtOAc extract was fractionated by silica gel column chromatography and further fractionated by Sephadex LH-20 column chromatography to afford an active compound **4**.

```
                    Culture broth (malt extract medium)
                                  |
        ┌─────────────────────────┴─────────────────────────┐
     Filtrate                                         Mycelial mats
     Extraction with EtOAc                            Extraction with acetone
     silica gel column chromatography (CC)            Acetone extract
     eluted with hexane-EtOAc-acetone                   silica gel CC
        |                                                eluted with hexane-aceton
                                                       Fraction (Acetone 50%)
  ┌─────┴──────────────┐                                 LH-20
Fraction (EtOAc 40%)  Fraction (Acetone 10%)             PTLC
  silica gel CC                                       Compound 4
  LH-20                 PTLC
  recrystallization     LH-20
  Compound 2            Compound 1, 3
```

Fig. 3 Isolation Procedure of the Compounds from *Aspergillus* sp.

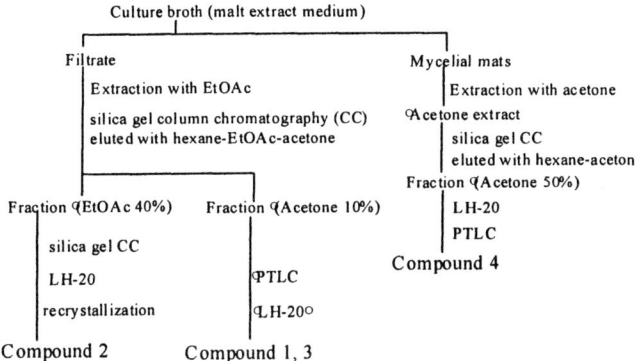

Fig. 4 Structures of Isolaterd Compounds

α,β-Dehydrocurvularin (**2**), 8-β-hydroxy-7-oxocurvularin (**3**) and 7-oxocurvularin (**4**) were identified by comparing the physicochemical properties with those reported.[13-21] The structure of **1** was determined by spectroscopic data.

The biological activities of **1-4** were examined using nematodes, lettuce and rice bioassay methods, but **3** was not examined with rice due to its low yield. Nematicidal activities of **1-4** against *P. penetrans* were shown in Table 2. **1** and **2** promoted nematicidal activity in proportion to its concentration from 1 mg/L

Table 2. Nematicidal activities of compounds 1-4
and aspyrone against *Pratylenchus penetrans*.

Compound	Nematicidal activities (LD_{50}^*)
1	300-1000
2	3-10
3	1000
4	>1000
Aspyrone	100-300

$LD_{50}(LD_{90})$: Concentrations (mg/l) causing over 50%
(resp. 90%) immotility of the nematodes after 3 days.

to 1000 mg/L. 1, 2, 3 and 4 had nematicidal activities against *P. penetrans* by 35%, 80%, 33%, and 23% at a concentration of 300 mg/L, respectively. Furthermore, 1 and 2 had the activities by 87% and 88% at a concentration of 1000 mg/L, respectively. Compound 1 has less nematicidal activity than that of aspyrone as a positive control, but 2 has more effective nematicidal activity than that of aspyrone. In contrast to the growth inhibition of *P. penetrans*, all compounds had hardly any effect on the free-living nematode *C. elegans* at the concentrations tested (1-1000 mg/L). With lettuce seedlings, 2 completely inhibited the root growth at a concentration of 300 mg/L. However, 1 and 3 showed no inhibitory activity, and 4 accelerated the growth at the same concentration. With rice seedlings, 2 completely inhibited the primary root growth at a concentration of 300 mg/L. However, 1 accelerated the primary root growth in proportion to its concentration from 3 mg/L to 300 mg/L, and 4 accelerated the primary root growth to 167 % of control at a concentration of 300 mg/L. From these results, 1 has less nematicidal activity than that of 2, but 1 was nontoxic to plant growth, making it potentially useful nematicides.

3) Compounds from *Penicillium bilaiae* Chalabuda

Bioassay-guided fractionation led to the isolation of active compounds.

Compound 1 was obtained as colorless needles, and its molecular formula was determined to be $C_{12}H_{12}O_4$ by MS and elemental analysis. The IR absorption band at 2120 cm^{-1} and two signals at δ 74.6 and 80.8 ppm in the ^{13}C NMR spectrum indicated the presence of two acetylenic carbons. A band at 1692 cm^{-1} and one ^{13}C NMR signal at δ 166.7 ppm indicated the presence of a carbonyl carbon. A band at 3460 cm^{-1} and one proton signal at δ 4.89 ppm in the ^{1}H NMR spectrum indicated the presence of a D$_2$O exchangeable hydroxyl group. The ^{13}C and ^{1}H NMR spectra of 1 indicated the presence of one methoxy, one *O*-substituted aliphatic methylene, one *O*-substituted aliphatic methine, and one

1,4-di-substituted phenyl group. In the ^{13}C-^{1}H COLOC spectrum of 1, a cross peak between the methoxy proton signal at δ 3.84 ppm and the carbonyl carbon was observed, and this carbonyl carbon had a cross peak with 2-H. The partial structure corresponding to C-1' to C-4' was derived from their coupling constants in the ^{1}H NMR spectrum, and a NOE between 1'-H and 3-H was observed. Thus, the planar structure of 1 was established.

The absolute stereochemistry of C-2' was determined using a modification of Mosher's method.[22] 1 was treated with (R)-(-)- and (S)-(+)-2-methoxy-2-trifluoromethyl-2-phenylacetyl chlorides (MTPA Cls) to afford the C-2'-(S) and (R) MTPA esters of 1. Positive $\Delta\delta$ values (δ_S-δ_R) in the ^{1}H NMR spectra were observed for 2' and 4'-H, while negative $\Delta\delta$ values were located at 1', 2, 3 and 6-H. These results revealed the absolute configuration of C-2' to be R.

```
Culture filtrate (malt extract medium)
        |
        | extracted with EtOAc (PH2 with 2N-HCl)
EtOAc extracts
        |
        | silica gel chromatography
 ┌──────────────────┬──────────────────────┬──────────────────────┐
EtOAc 35% fraction   Acetone 40% fraction    Acetone 90% fraction
silica gel            silica gel              silica gel
 chromatography        chromatography          chromatography
recrystallization     PTLC                    LH-20 chromatography
                      recrystallization
        |                   |                       |
Compound 3           Compounds 1, 2          Compound 4
```

Fig. 5 Isolation Procedure of the Compounds from *Penicillium bilaiae* Chalabuda

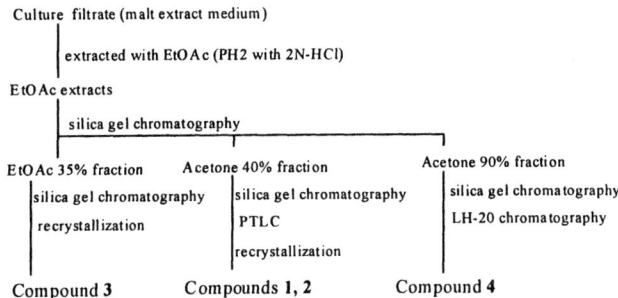

Fig. 6 Structures of Isolaterd Compounds

p-Hydroxyacetophenone (2), 6-methoxycarbonylpicolinic acid (3) and 2,6-pyridinedicarboxylic acid (4), were identified by comparing the physicochemical properties with those reported. [23, 24] Methylation of 3 and 4 with diazomethane gave the same dimethyl-2, 6-pyridinedicarboxylate.

Table 3. Nematicidal activities of compounds 1-4 and aspyrone agaainst *Pratylenchus penetrans*

Compound	Nematicidal activities (LD_{50})
1	10-30
2	10-30
3	100-300
4	<1
Aspyrone	100-300

$LD_{50}(LD_{90})$: Concentrations (mg/l) causing over 50% (resp. 90%) immotility of the nematodes after 3 days.

Nematicidal activities of **1-4** against the root-lesion nematode *Pratylenchus penetrans* were shown Table 3. Compounds **1-4** promoted nematicidal activity proportional to its concentration from 1 mg/L to 1000 mg/L and nematicidal activities by 83%, 100%, 67%, and 99% at concentrations of 1000 mg/L, respectively. In addition, **1, 2** and **4** have more effective nematicidal activities than that of aspyrone.

References

1) Kursberg, L. R., Studies on the culturing and parasitism of plant-parasitic nematodes, in particular *Ditylenchus dipsaci* and *Aphelenchoides ritzemabosi* on alfalfa tissues. *Nematologica*, **6**, 181-200 (1961).
2) In "Experimental method for soil microorganisms (in Japanese)", eds. Dojo Biseibutu kenkyukai Youkendo press, Tokyo, pp.146-157 (1975).
3) Saigusa, T., In "Nematodes (in Japanese)" 2nd ed., Rural Culture Association press, Tokyo, pp.107-113 (1993).
4) Kimura, Y., Mori, M., Hyeon, S., Suzuki, A., and Mitui, Y., A rapid and simple method for assay of nematicidal activity and its measuring the activities. *Agric. Biol. Chem.*, **45**, 249-251 (1981).
5) Stadler, M. and Anke, H., Lachnumon and lachnumol, a new metabolites with nematicidal and antimicrobial activities from the ascomycete *Lachunum papyraceum* (KARST.) KARST.
6) Wood, W. B., In "The nematode *Caenorabditis elegans*", Cold Sprong Harbour press, pp. 587-593 (1988).
7) Kimura, Y., Tani, K., Kojima, A., Sotoma, G., Okada, K., and Shimada, A., Cyclo-(L-tryptophyl-L-phenylalanyl), a plant growth regulator produced by the fungus *Penillium* sp. *Phytochemistry*, **41**, 665-669 (1996).
8) Kimura, Y., Suzuki, A., Tamura, S., Mori, K., Oda, M. and Matsui, M., Biological activity of pestalotins on the elongation growth of rice seedlings. *Plant Cell Physiol.*, **18**, 1177-1179 (1977)Kimura, Y., Nakahara, S., and Fujioka, S., Aspyrone, a nematicidal compound isolated from the fungus, *Aspergillus melleus*. *Biosci. Biotechnol. Biochem.*, **60**, 1375-1376 (1996).

9) Kusano, M., Koshino, H., Uzawa, J., Fujioka, S., Kawano, T., Kimura, Y., Nematicidal alkaloids and related compounds produced by the fungus *Penicillium* cf. *simplicissimum*. *Biosci. Biotechnol. Biochem.*, **64**, 2559-2568 (2000).
10) Mayer, A., Anke, H., and Sterner, O., Omphalotin, a new cyclic peptide with potent nematicidal activity from *Omphalotus olearius* 1. Fermentation and biological activity. *Natural product Lett.*, **10**, 25-32 (1997).
11) Kimura, Y., Nakahara, S., and Fujioka, S., Aspyrone, a nematicidal compound isolated from the fungus, *Aspergillus melleus*. *Biosci. Biotech. Biochem.*, **60**, 1375-1376 (1996).
12) Munro, H. D., Musgrave, O. C., and Templeton, R., Curvularin. Part V. The compound $C_{16}H_{18}O_5$, ___-dehydrocurvularin. *J. Chem. Soc.* (C), 947-948 (1967).
13) Hyeon, S., Ozaki, A., Suzuki, A., Tamura, S., Isolation of ___-dehydrocurvularin and __-hydroxycurvularin from *Alternaria tomato* as sporulation-suppressing factors. *Agr. Biol. Chem.*, **40**, 1663-1664 (1976).
14) Robeson, D. J. and Strobel, G. A., ___-Dehydrocurvularin and curvularin from *Alternaria cinerariae*. *Z. Naturforsch.*, **36c**, 1081-1083 (1981).
15) Robeson, D. J. and Strobel, G. A., The identification of a major phytotoxic component from *Alternaria macrospora* as -dehydrocurvularin. *J. Nat. Prod.*, **48**, 139-141 (1985).
16) Kobayashi, A., Hino, T., Yata, S., Itoh, T. J., Sato, H., and Kawazu, K., Unique spindle poisons. Curvularin and its derivatives, isolated from *Penicillium* species. *Agric. Biol. Chem.*, **52**, 3119-3123 (1988).
17) Arai, K., Rawlings, B. J., Yoshizawa,Y., and Vederas, J. C., Biosynthesis of antibiotics A26771B by *Penicillium turbatum* and dehydrocurvularin by *Alternaria cinerariae*: Comparison of stereochemistry of polyketide and fatty acid enoyl thiol ester reductases. *J. Am. Chem. Soc.*, **111**, 3391-3399 (1989).
18) Lai, S., Shizuri, Y., Yamamura, S., Kawai, K., Terada, Y., and Furukawa, H., Novel curvularin-type metabolites of a hybrid strain ME 0005 derived from *Penicillium citreoviride* B. IFO 6200 and 4692.*Tetrahedron Lett.*, **30**, 2241-2244 (1989).
19) Lai, S., Shizuri, Y., Yamamura, S., Kawai, K., Terada, Y., and Furukawa, H., New metabolites of two hybrid strains ME 0004 and 0005 derived from *Penicillium citreoviride* B. IFO 6200 and 4692. *Chemistry Lett.*, 589-592 (1990).
20) Ghisalberti, E. L. and Rowland, C. Y., 6-Chlorodehydrocurvularin, a new metabolite from *Cochliobolus spicifer*. *J. Nat. Prod.*, **56**, 2175-2177 (1993).
21) Otani, I., Kusumi, T., Kashman, Y., Kakisawa, H., High-field FT NMR application of Mosher's method. The absolute configurations of marine terpenoids. *J. Am. Chem. Soc.*, **113**, 4092-4096 (1991).
22) DE Pascual-T., J., Bellido, I. S., Gonzalez, M. S., Muriel, M. R., Hernandez, J. M., p-Hydroxyacetophenone derivatives from *Artemisia campestris* sp. Glutinosa. *Phytochemistry*, **19**, 2781-2782 (1980).
23) Hirayama, F., Konno, K., Shirahama, H., Matsumoto, T., 4-Aminopyridine-2,3-dicarboxylic acid from *Clitocybe acromelalga*. *Phytochemistry*, **28**, 1133-1135

Chapter 21

Bioorganic Chemistry on Sex Pheromones Secreted by Lepidopteran Insects and Their Application for Plant Protection

Tetsu Ando

Graduate School of Bio-Applications and Systems Engineering (BASE), Tokyo University of Agriculture and Technology (TUAT), Koganei, Tokyo 184–8588, Japan (antetsu@cc.tuat.ac.jp)

Lepidopteran sex pheromones have been identified from more than 500 species. The pheromones in the most predominant group (Type I) are composed of unsaturated C_{10} - C_{18} straight-chain compounds with a terminal functional group, such as bombykol produced by the silkworm moth. In addition to them, females in some evolved families produce C_{17} - C_{23} polyunsaturated hydrocarbons and the epoxy derivatives, constituting a second major group (Type II). While some synthetic pheromones have already been utilized for plant protection on the basis of their strong attractive activities for male moths, many bioorganic chemical studies are currently underway on this exciting topic. This paper addresses recent research conducted mainly in the Chemical Ecology Laboratory in TUAT and explores the future of pheromone studies and potential applications.

The sex pheromones of Lepidoptera are usually produced by female moths and attract males. The pheromone is a main factor for reproductive isolation, so it must be species-specific. Lepidoptera is one of the biggest insect groups which has been established for over 100 million years since the Mesozoic era. To date, lepidopteran sex pheromones have been identified from nearly 540 species. Additionally, sex attractants of another 1,240 species have been found by field tests with synthetic pheromones and their related compounds (*1, 2*).

Although the chemical structures of the pheromones are very simple, the variety of the chemical structures and blending of multiple components cause their diversity. About 75 % of these compounds consist of unsaturated fatty alcohols and their derivatives with a C_{10} to C_{18} straight chain, such as bombykol (E10,Z12-16:OH) (Figure 1). These Type I compounds with a terminal functional group have been identified from many groups in Lepidoptera. Another 15 % of the natural pheromones consist of polyunsaturated hydrocarbons and their epoxy derivatives with a C_{17} to C_{23} straight chain, such as (Z,Z,Z)-3,6,9-nonadecatriene (Z3,Z6,Z9-19:H) and *cis*-(Z,Z)-3,4-epoxy-6,9-nonadecadiene (epo3,Z6,Z9-19:H) secreted by the Japanese giant looper, *Ascotis selenaria cretacea* (Figure 1). These Type II compounds do not have a terminal functional group and have been identified from the species in highly evolved insect groups such as the family of Geometridae (*2*).

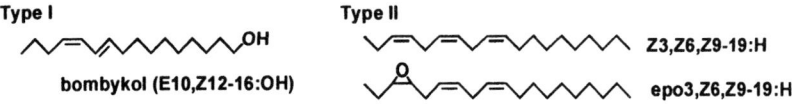

Figure 1. Reprsentative lepidopteran pheromones and their abbreviations.

Studies on lepidopteran pheromones are underway at the Chemical Ecology Laboratory in TUAT. The work is divided into four areas. (A) The systematic synthesis of pheromone compounds and their field evaluation to find new attractants; (B) Development of new analytical techniques and identification of natural pheromones using the synthetic compounds as an authentic sample; (C) Pest control applications for synthetic pheromones; (D) The biosynthesis and olfactory perception system of the sex pheromones. This paper deals with the last three topics (B – D).

Identification of Natural Sex Pheromones

GC-EAD Technique

In addition to the observation of male responses in a flask or a wind tunnel, the activity of pheromone components in the gland extract can be detected using an electrophysiological technique. The electroantennogram

(EAG), a recording of the potential changes measured between the base and tip of an insect antenna as a result of chemical stimulation, has played an important role in the bioassay system (*3*). One advantage of the system is that the EAG recording is easily accomplished with an antenna of an unconditioned male moth in a bright room. Because the activities of several compounds can be successively measured after short intervals, all active components in a pheromone blend are detected by one injection into gas chromatography combined with an EAG detector (GC-EAD) (*4*). Recently, the capability of amplifiers has increased so much that the sensitivity of the EAD against a pheromone component is higher than that of an FID.

Type I Pheromones with a Conjugated Diene System

The persimmon fruit moth, *Stathmopoda masinissa*, is a microlepidopteran species in the family of Oecophoridae and is a well-known harmful pest of persimmon fruits in Japan. The tiny larvae feed on only a small part of the fruits, but the infected fruits drop to the ground before maturity. Since spraying of an insecticide is effective in a very limited term, when the larvae move from the unfolding buds to the fruits for feeding on the core, the sex pheromone would play an important role as a monitoring tool. By GC-EAD analysis of an *n*-hexane extract of the pheromone glands removed from the females in the scotophase, three distinct EAG-active components were detected. GC-MS analysis revealed the EAG-active compounds consisting of an aldehyde, an acetate, and an alcohol with a conjugated diene system in a C_{16} chain (*5*). The mass spectra of the natural components are different from the authentic samples including the diene system at a higher position than the 9-position, which had been synthesized before (*6*); on the other hand, the aldehyde component produced a unique base peak at *m/z* 84 indicating a 4,6-diene structure. However, the contents in the pheromone glands are very low, and the position of the diene system has not been confirmed by a chemical derivatization.

In order to confirm the double-bond positions, four geometrical isomers of 4,6-hexadecadienyl compounds were synthesized in addition to the other positional isomers including the conjugated diene system between the 3- and 10-positions (*7*), and then, their GC-MS data revealed that the natural components are 4,6-dienes with a (*E,Z*)-configuration; *i.e.*, E4,Z6-16:Ald, E4,Z6-16:OAc, and E4,Z6-16:OH. While the base peaks of the alcohol and acetate including the 4,6-dienyl structure have been recorded at *m/z* 79, it has been confirmed that the 4,6-hexadecadienal shows a characteristic base peak at *m/z* 84, which is expected to have a stable 2,3-dihydropyranyl structure ($[C_5H_8O]^+$) probably formed by a cyclization and hydrogen rearrangement after

Table I. Base Peaks in the EI-Mass Spectra of Type I C_{16} Compounds Including a Conjugated Diene System

Compounds	Base peak (m/z)						
	3,5-16	4,6-16	5,7-16	6,8-16 - 10,12-16	11,13-16	12,14-16	13,15-16
Alcohol	67	79	79	67	95 (B)	81 (C)	67 (D)
Acetate	79	79	79	67	95 (B)	81 (C)	67 (D)
Aldehyde	—	84 (A)	80	67	95 (B)	81 (C)	67 (D)

cleavage of the bond between the 5- and 6-positions. The base peaks of all 6,8-, 7,9-, and 8,10-dienes universally appeared at m/z 67, such as 9,11-, 10,12-, and 13,15-dienes, while 5,7-hexadecadienal interestingly showed the base peak at m/z 80 being diagnostic for their double-bond positions (Table I).

This is the second identification from the species in the family of Oecophoridae. Compounds with a conjugated diene system make one of the most important groups among the Type I pheromones. However, their double bonds are usually located in a side of the terminal methyl group such as bombykol, a 10,12-diene with a C_{16} chain, and no 4,6-dienes have been found. Interestingly, 4,6,10-trienes have been identified from the cocoa pod borer moth, *Conopomorpha cramerella*, in the family of Gracillariidae (8), which is taxonomically close to Oecophoridae. Generally, the Type I pheromones are biosynthesized *via* a saturated fatty acyl intermediate (9). The common structure of the 4,6-dienes and 4,6,10-trienes suggests one possibility, namely, that they are produced by a common enzyme(s) catalyzing the dehydrogenation of palmitic acid at the 4- and 6-positions. If this is true, the double bond at the 10-position of the *C. cramerella* pheromone is formed after the desaturation at the 4- and 6-positons. The *S. masinissa* pheromone is valuable not only for its application in IPM programs but also from the biochemical viewpoint.

Type II Pheromones Produce by Highly Evolved Species

The second major group of lepidopteran sex pheromones consists of unbranched C_{17} to C_{23} (Z,Z)-6,9-dienes, (Z,Z,Z)-3,6,9-trienes, and their monoepoxy derivatives, which lack a terminal functional group and are biosynthesized from linoleic acid and linolenic acid. GC-MS is a useful tool to elucidate their chemical structures as well (*10, 11*). On the EI measurement,

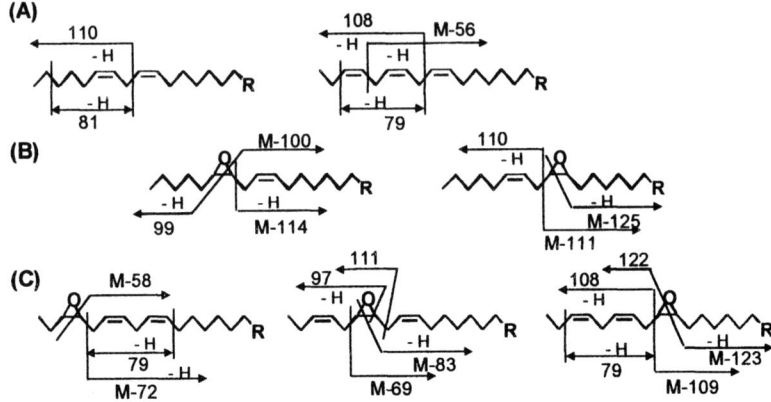

Figure 2. Diagnostic fragment ions in the mass spectra of (A) dienes and trienes, (B) epoxymonoenes, and (C) epoxydienes.

every component produces M^+ and some diagnostic fragment ions listed in Figure 2. Referring to the spectral data of synthetic compounds, we identified epo3,Z6,Z9-19:H and Z3,Z6,Z9-19:H (100:5 - 100:1) from virgin females of *A. s. cretacea* (*12*) and epo6,Z9-19:H and Z3,epo6,Z9-19:H (9:1) from the giant geometrid moth, *Biston robustum* (*13*), as a pheromone component.

Type II pheromones are almost always secreted from the species within the families of Geometridae, Noctuidae, Lymantriidae, and Arctiidae. In spite of the large number of species in these families, their structural diversity is quite limited. Some females in these families may be able to modify these compounds with additional desaturation and/or epoxidation steps, potentially generating a greatly increased number of pheromone components. Recently, a research group of the National Institute of Agrobiological Science in Japan and our research group identified the novel pheromone components from two lymantrid species distributed in Okinawa Islands (Figure 3). Females of the tussock moth, *Orgyia postica*, produce posticlure (Z6,Z9,*t*-epo11-21:H), the first lepidopteran pheromone including a *trans*-epoxy ring (*14*). Females of another tussock moth, *Perina nuda*, produce a diepoxy pheromone (epo3,epo6,Z9-21:H) derived from the corresponding C_{21} triene (Z3,Z6,Z9-

posticlure (Z6,Z9,*t*-epo11-21:H)　　　　　　epo3,epo6,Z9-21:H

Figure 3. Novel sex pheromones produced by tussock female moths.

21:H) (15). The structural determination of posticlure has been accomplished by means of ¹H NMR measurements in addition to GC-MS analysis. The diepoxy pheromone has been identified by referring to the GC-MS data of a series of diepoxy derivatives systematically synthesized (16).

Chiral HPLC Analysis

A chiral GC column has the potential to separate enantiomers of epoxy pheromones in the Type II class, but the applications are very limited because no good column with a universal ability for the resolution has been commercialized. On the other hand, the resolution abilities of chiral HPLC columns have been examined in detail (17). The Chiralpak AD column operated under a normal-phase condition sufficiently separates the two enantiomers of 9,10-epoxydienes, 6,7-epoxymonoenes and 9,10-epoxymonoenes. Another normal-phase column, the Chiralpak AS column, is suitable for the resolution of the 3,4-epoxydienes. The Chiralcel OJ-R column operated under a reversed-phase condition sufficiently accomplishes the enantiomeric separation of 6,7-epoxydienes and 6,7-epoxymonoenes (18). The stereochemistry of each enantiomer separated by chiral HPLC has been studied after methanolysis of the epoxy ring. Examining the ¹H NMR data of esters of the produced methoxyalcohols with (S)- and (R)-α-methoxy-α-(trifluoromethyl)phenylacetic acid by a modified Mosher's method (19), the parent epoxides with shorter Rts have been indicated to be (3S,4R)-, (6S,7R)-, and (9R,10S)-isomers (17).

Utilizing the chiral HPLC columns, we successfully determined the absolute configurations of *cis*-epoxy rings in the natural epoxydienes and epoxymonoenes secreted by three Geometridae species [*A. S. cretacea* (12),

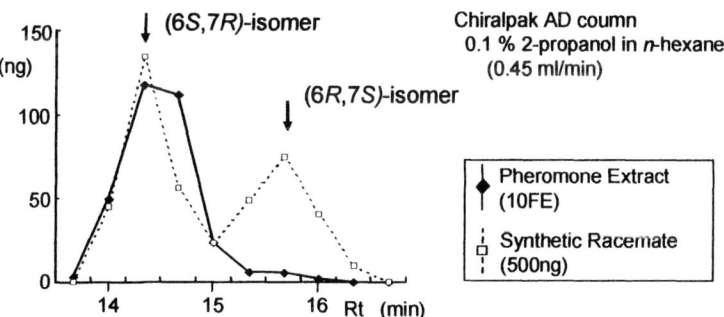

Figure 4. Quantitative GC analyses of chiral HPLC-eluted epo6,Z9-19:H, which were collected every 20 sec.

Menophra atrilineata (*18*), and *B. robustum* (*13*)] and one Noctuidae species [*Oraesia excavata* (*20*)]. Furthermore, the stereochemistry of two *cis*-epoxy rings in epo3,epo6,Z9-21:H of *P. nuda* (*15*) and a *trans*-epoxy ring in postíclure of *O. postica* (*14*) was defined by the chiral HPLC. In addition to the detection with a UV detector or RID, quantitative GC measurements of the chiral HPLC elutants have confirmed the assignments of the stereochemisty as shown for the analysis of a main component of the *B. robustum* pheromone (Figure 4). Recently, the stereochemistry of the epoxydiene produced by *A. S. cretacea* was also recognized by liquid chromatography combined with time-of-flight mass spectrometry (LC-TOF-MS), which was equipped with a chiral HPLC column and conducted by electrospray ionization (ESI) (*21*).

Foresight of Researches for the Identification

The chemical structures of the lepidopteran sex pheromones are not complicated. Furthermore, good analytical systems have been developed, namely GC-EAD for the bioassay and GC-MS for the structure determination. Therefore, pheromone identification seems to be routine work, particularly for the Type I and II components produced by a big insect. In the case of a small insect, such as a leafminer moth, the pheromone content is too low to record a reliable MS spectrum, sometimes lower than a 1-pg/female order. As a result, new challenges are necessary to establish good methods for determining a double-bond position for the Type I pheromones and stereochemistry for the Type II pheromones. Lepidoptera, which is the second largest insect group, includes nearly 150,000 described species. Considering the species diversity, it should be pointed out that the information given here is still rudimentary. This suggests a chance to identify novel compounds, which stimulate the organic chemists investigating natural products.

Application of Synthetic Pheromones

Overview of Mating Disruption

Lepidopteran sex pheromones have been applied in IPM programs as a monitoring tool and a lure for mass trapping. Furthermore, the disruption of the chemical communication between females and males is carried out by permeating a field with synthetic pheromones. Agricultural fields are very large and wind-blown; however, permeation is possible, and several synthetic pheromones are being utilized as a mating disruptant for more than 20 lepidopteran species after the registrations of ' Nomate PBW,' consisting of

Table II. Utilization of Representative Mating Disruptants in the World [a]

Crop	Pest insect (scientific name)	Country	Applied field (ha) 1977	2002
Cotton	pink bollworm moth [b]	USA	30,000	40,000
	(*Pectinophora*	Egypt	328,000	-
	gossypiella)	Israel	8,000	5,000
		Brazil	-	5,000
Apple &	codling moth [c]	USA	13,200	63,000
pear	(*Cydia pomonella*)	Italy	6,800	14,000
Japanese	cherry treeborer [d]	Japan	4,000	3,800
plum	(*Synanthedon hector*)			
Tea	smaller tea tortix [e]	Japan	400	500
	(*Adoxophyes honmai*)			
	Oriental tea tortrix [e]			
	(*Homona magnanima*)			
Forest	gypsy moth [f]	USA	10,000	150,000
	(*Lymantria dispar*)			

a: Information supplied from Shin-Etsu Chemical Co. Ltd. in Japan.
Disruptant; b: Z7,Z11-16:OAc + Z7,E11-16:OAc (1:1, gossyplure),
  c: E8,E10-12:OH (codlelure), d: Z3,Z13-18:OAc + E3,Z13-18:OAc (1:1),
  e: Z11-14:OAc, f: Me2,epo7-18:H (disparlure)

Z7,Z11-16:OAc and Z7,E11-16:OAc for the control of the pink bollworm moth, *Pectinophora gossypiella*, in the USA (1976) and 'Hamaki-con,' containing Z11-14:OAc for the simultaneous control of some leafroller species in Japan (1983). Table II shows the application areas of some representative disruptants. In cotton fields, the synthetic pheromone of *P. gossypiella* is widely used. In Japan, unfortunately, utilization is still limited. On the contrary, in the USA, the pheromone is widely used even in forests for the control of the gypsy moth, *Lymantria dispar*. The pheromone of *L. dispar* is an epoxy compound, but it does not include a double bond. Other chemicals utilized as a disruptant are almost all Type I compounds, and trials of mating disruption by Type II pheromones are very limited.

While details of the mechanism of mating disruption are not clear, it is expected that males are meaninglessly excited by the permeated pheromone and that females are masked by a higher level of the synthetic pheromone or by a modification of the natural mixing ratio. Generally, a synthetic material that mimics a natural pheromone blend is an optimal lure for male attraction. For the disruption of mating, information is inadequate to prove whether or not a copy of the natural pheromone blend is the best. The best disruptants have been selected after many trials and errors.

Disruption by Type II Compounds

A. S. cretacea is a harmful pest insect in tea gardens in Japan. The main pheromone component is epo3,Z6,Z9-19:H and the corresponding hydrocarbon (Z3,Z6,Z9-19:H) is a minor component, which shows a synergistic effect on male attraction (22). This triene is easily synthesized from linolenic acid, and its oxidation with a peracid yields a mixture of three epoxydienes. These positional isomers are separable on medium-pressure liquid chromatography, but a large-scaled separation for the preparation of a disruptant is difficult. If Z3,Z6,Z9-19:H or the epoxydiene mixture (EDM) inhibits the mating communication, the application of Type II can be realized.

The above possibility was examined in a tea garden (23). The orientation of the males to the synthetic pheromone placed in a trap was strongly disrupted by Z3,Z6,Z9-19:H or EDM, which was impregnated in septa and placed around the trap. Based on this result, polyethylene tubes containing Z3,Z6,Z9-19:H or

Table III. Mating Ratio of *A. s. cretacea* Females Tethered in a Tea Garden in Mie Prefecture Which Was Permeated with Z3,Z6,Z9-19:H or EDM (Epoxydiene Mixture) Released from Dispensers

Dispenser		Number of females		Mating ratio
	(No./ha)	Tethered	Mated	(%)
(A) Z3,Z6,Z9-19:H	0 (control)	11	11	100
Tested from Sept.	500	10	6	60
7 to 14, 1999	1000	9	6	67
	3000	10	8	80
	5000	10	4	40
(B) EDM	0 (control)	14	14	100
Tested from Sept.	250	13	3	23
7 to 18, 1999	500	14	4	29
	1000	14	1	7
	3000	12	0	0
	5000	12	0	0

EDM were prepared. Strong inhibition of male orientation to synthetic pheromone traps was achieved in orchards permeated with Z3,Z6,Z9-19:H at dispenser density of 3000 and 5000 tubes/ha and with EDM at every tested dose, 250-5000 tubes/ha. Table III shows the mating ratios of females in a field permeated with the chemicals. In this test of Z3,Z6,Z9-19:H, the mating ratio decreased when the dispenser density increased, but half of the females still mated with males. In the case of EDM, the mating ratio effectively decreased, and mating was perfectly inhibited at 3000 and 5000 tubes/ha. This is the first formulation for the mating disruption of a geometrid pest.

Pheromone Biosynthesis and Perception

Recent studies revealed that the biosynthetic processes for two pheromone types are quite different, as shown in Figure 5. Type I pheromones are biosynthesized, starting from acetyl CoA in a pheromone gland. For example, Z7-12:OAc is synthesized from palmitic acid *via* Δ11-desaturation, chain shortening by β-oxidation, reduction, and acetylation (*24*). The desaturation step is effectively blocked by cyclopropene compounds (*25, 26*). In contrast, the biosynthesis of the Type II pheromone, such as epo3,Z6,Z9-19:H, starts from dietary linolenic acid, and the trienyl intermediate is probably produced in an

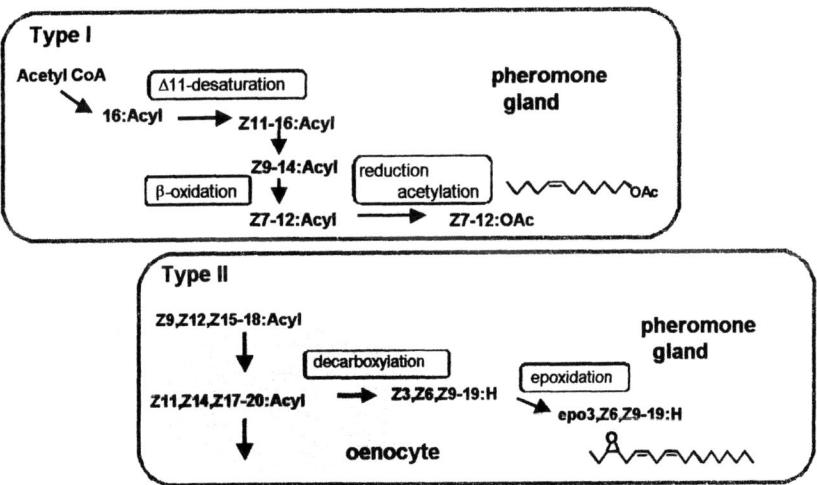

Figure 5. Biosynthetic pathways of sex pheormones (Type I: Z7-12:OAc and Type II: epo3,Z6,Z9-19:H) and propspected sites of their biosynthesis.

oenocyte and moves to a pheromone gland (27). Only the epoxidation proceeds in the pheromone gland (28).

The pheromone titer is synchronized with a light-dark cycle. The photo signal is received at the head and transported to the pheromone gland with the help of the hormone, which activates the pheromone biosynthesis and is called a pheromone biosynthesis-activating neuropeptide (PBAN). PBAN is composed of about 30 amino acids, and the sequence has been determined for those from 10 lepidopteran species (29). On the other hand, pheromone production is terminated after mating. Experiments with *B. mori* showed that the pheromone gland of the mated female maintained its ability to synthesize bombykol but could not produce the pheromone due to suppression of the PBAN secretion from a suboesophageal ganglion *via* a neural signal of the mating (30).

There are many sensilla trichodea on the antenna. A photo-labeling experiment found an interesting protein, which specifically bound the sex pheromone and was called a pheromone-binding protein (PBP). The molecular size of PBP is ca. 15 KDa, and the amino acid sequence has been determined for PBPs from more than 20 species. Their role has been estimated to transport a lipophilic pheromone to a receptor on the dendrite (31). To data, no pheromone receptor of the lepidopteran insect has been identified.

Conclusion

The diversity of the species and their pheromone systems are of interest to many organic chemists, biochemists, and entomologists, who attempt to understand insect evolution chemically and biochemically in detail and to apply their findings for plant protection. Recently, new research has produced results in the fields of the biosynthesis and perception of the pheromones. If they were better understood, these interesting scientific subjects would become potentially important targets for pest control. In addition to the direct application of the sex pheromones, inhibition of these processes seems to be another profitable approach to control pest insects.

Acknowledgements

These studies were supported by the following co-workers, Drs. S. Wakamura of the National Institute of Agrobiological Sciences, N. Arakaki of the Okinawa Prefectural Agricultural Experiment Station, K. Ohtani of the Mie Central Office of Agricultural Extension, M. Kiso and J. Takeuchi of the Agricultural Technology Center in Tokyo, H. Naka and K. Tuchida of Gifu University, Y. Ohmasa of the Ehime Fruit Tree Experiment Station, and G.-Q. Pu of Su Zhou University in China. I am grateful to these co-workers as well as to the students of the Chemical Ecology Laboratory in TUAT who wrote their theses for master's and doctor degrees, and to Drs. K. Ogawa and F. Mochizuki of Shin-Etsu Chemical Co. Ltd. for information on the utilization of mating disruptants.

References

1. Arn, H.; Tóth, M.; Priesner, E. **2000**, http://www-pherolist.slu.se/
2. Ando, T.; Inomata, S.; Yamamoto, M. *Topics Current Chem.* **2003**, in press; http://www.tuat.ac.jp/~antetsu/LepiPheroList.htm
3. Roelofs, W. L.; In *Techniques in Pheromone Research*; Hummel, H. E.; Millar, T. A., Eds.; Springer, **1984**, pp 131-159.
4. Arn, H.; Stadler, E.; Rauscher, S. *Z. Naturforsch.* **1975**, *30c*, 722-725.
5. Naka, H.; Le, V. V.; Inomata, S.; Ando, T.; Kimura, T.; Honda, H.; Tsuchida, K.; Sakurai, H. *J. Chem. Ecol.* **2003**, *29*, 2447-2459.
6. Ando, T.; Ogura, Y.; Uchiyama, M. *Agric. Biol. Chem.* **1988**, *52*, 1415-1423.
7. Nishida, T.; Le, V. V.; Yamazawa, H.; Yoshida, R.; Naka, H.; Tsuchida, K.; Ando, T. *Biosci. Biotechnol. Biochem.* **2003**, *67*, 822-829.
8. Beevor, P. S.; Cork, A.; Hall, D. R.; Nesbitt, B. F.; Day, R. K.; Mumford, J. D. *J. Chem. Ecol.* **1986**, *12*, 1-23.
9. Roelofs, W.; Bjostad, L. *Bioorg. Chem.* **1984**, *12*, 279-298.
10. Ando, T.; Ohsawa, H.; Ueno, T.; Kishi, H.; Okamura, Y.; Hashimoto, S. *J. Chem. Ecol.* **1993**, *19*, 787-798.
11. Ando, T.; Kishi, H.; Akashio, N.; Qin, X.-R.; Saito, N.; Abe, H.; Hashimoto, S. *J. Chem. Ecol.* **1995**, *21*, 299-311.
12. Ando, T.; Ohtani, K.; Yamamoto, M.; Miyamoto, T.; Qin, X.-R.; Witjaksono *J. Chem. Ecol.* **1997**, *23*, 2413-2423.
13. Yamamoto, M.; Kiso, M.; Yamazawa, H.; Takeuchi, J.; Ando, T. *J. Chem. Ecol.* **2000**, *26*, 2579-2590.
14. Wakamura, S.; Arakaki, N.; Yamamoto, M.; Hiradate, S.; Yasui, H.; Yasuda, T.; Ando, T. *Tetrahedron Lett.* **2001**, *42*, 687-689.
15. Wakamura, S.; Arakaki, N.; Yamazawa, H.; Nakajima, N.; Yamamoto, M.; Ando, T.; *J. Chem. Ecol.* **2002**, *28*, 449-467.
16. Yamazawa, H.; Nakajima, N.; Wakamura, S.; Arakaki, N.; Yamamoto, M.; Ando, T. *J. Chem. Ecol.* **2001**, *27*, 2153-2167.
17. Qin, X.-R.; Ando, T.;Yamamoto, M.; Yamashita, M.; Kusano, K.; Abe, H. *J. Chem. Ecol.* **1997**, *23*, 1403-1417.
18. Pu, G.-Q.; Yamamoto, M.; Takeuchi, Y.; Yamazawa, H.; Ando, T. *J. Chem. Ecol.* **1999**, *25*, 1151-1162.
19. Ohtani, I.; Kusumi, T.; Ishitsuka, M. O.; Kakisawa, H. *Tetrahedron Lett.* **1989**, *30*, 3147-3150.

20. Yamamoto, M.; Takeuchi, Y.; Ohmasa, Y.; Yamazawa, H.; Ando, T. *Biomed. Chromatogr.* **1999**, *13*, 410-417.
21. Yamazawa, H.; Yamamoto, M.; Karasawa, K. I.; Pu, G.-Q.; Ando, T. *J. Mass Spectrom.* **2003**, *38*, 328-332.
22. Witjaksono; Ohtani, K.; Yamamoto, M.; Miyamoto, T.; Ando, T. *J. Chem. Ecol.* **1999**, *25*, 1633-1642.
23. Ohtani, K.; Witjaksono; Fukumoto, T.; Mochizuki, F.; Yamamoto, M.; Ando, T. *Entomol. Exp. Appl.* **2001**, *100*, 203-209.
24. Komoda, M.; Inomata, S.; Ono, A.; Watanabe, H.; Ando, T. *Biosci. Biotechnol. Biochem.* **2000**, *64*, 2145-2151.
25. Ando, T.; Ohno, R.; Ikemoto, K.; Yamamoto, M. *J. Agric. Food Chem.* **1996**, *44*, 3350-3354.
26. Ando, T.; Ikemoto, K.; Ohno. R.; Yamamoto, M. *Arch. Insect Biochem. Physiol.* **1998**, *37*, 8-16.
27. Wei, W.; Miyamoto, T.; Endo, M.; Murakawa, T.; Pu, G.-Q.; Ando, T. *Insect Biochem. Mol. Biol.* **2003**, *33*, 397-405.
28. Miyamoto, T.; Yamamoto, M.; Ono, A.; Ohtani, K.; Ando, T. *Insect Biochem. Mol. Biol.* **1999**, *29*, 63-69.
29. Rafaeli, A. *Int. Rev. Cytology* **2002**, *213*, 49-91.
30. Ando, T.; Kasuga, K.; Yajima, Y.; Kataoka, H.; Suzuki, A. *Arch. Insect Biochem. Physiol.* **1996**, *31*, 207-218.
31. Sandler, B. H.; Nikonova, L.; Leal, W. S.; Clardy, J. *Chem. Biol.* **2000**, *7*, 143-151.

Chapter 22

Phytotoxin Produced by *Streptomyces cheloniumii* Causing Potato Russet Scab

Masahiro Natsume[1], Mayumi Komiya[1], Fumie Koyanagi[1], Hiroshi Kawaide[1], Nobuya Tashiro[2], and Hiroshi Abe[1]

[1]Department of Applied Biological Science, Tokyo University of Agriculture and Technology, Fuchu, Tokyo 183–8509, Japan
[2]Saga Fruit Tree Experiment Station, Ogi-cho, Saga 845–0014, Japan

The russet scab phytotoxin was isolated from *Streptomyces cheloniumii* MAFF 304020 and identified as an 18-membered macrolide, FD-891. It induced necrosis in potato slices at 50 μg/disk, which was 1/100th the activity of thaxtomin A. The phytotoxin was produced by other pathogenic strains, *Streptomyces* spp. MAFF 225003, 225005 and MAFF 225006, which indicates that this phytotoxin is a pathotoxin.

Studies of phytotoxins produced by phytopathogenic microorganisms are needed not only to elucidate diseases mechanisms, but also to find new targets for the development of herbicides.

Potato tuber diseases caused by actinomycetes can be classified into two groups: common scab, which is characterized by corky erumpent or pitted symptoms, and russet or netted scab, which shows superficial reticulations (*1*).

The phytotoxin thaxtomin A (**1**) was isolated from *Streptomyces scabies*, the pathogen of common scab, by King *et al.*(*2*) Thaxtomin A production has been shown to be a pathogenicity factor in common scab because there is a positive correlation between the ability to produce thaxtomin A in various *S. scabies* isolates and their pathogenicity (*3, 4*). Other pathogens of common scab, such as *S. acidiscabies* (*3*, *4*, *5*) and *S. turgidiscabies* (*6*), also produce thaxtomin A.

We showed that *S. scabies* produces another phytotoxins, concanamycin A and B (**2**)(*7*), and that their production is specific to *S. scabies* (*5*). We detected concanamycin A and B in common scab lesions. The diversity of symptoms in common scab such as erumpent or pitted type may be attributed to the production of concanamycins, although their contribution to pathogenicity has yet to be clarified.

On the other hand, there have not been any reports with regard to phytotoxins produced by pathogens that induce superficial symptoms, such as russet or netted scab. Oniki *et al.* (*8*) and Suzui *et al.* (9) showed that the pathogen of russet scab is a *Streptomyces* sp. and is distinguished from *Streptomyces* spp. that cause common scab by its negative production of melanin and the spiny structure of its spore surface and they named it *S. cheloniumii*. They also showed that *S. cheloniumii* induced russet scab symptoms and not common scab symptoms (*8*). These results indicate that the pathogen of russet scab produces a phytotoxin other than thaxtomin A and that the phytotoxin is a causal agent for russet scab. We therefore searched for the phytotoxin produced by *S. cheloniumii*.

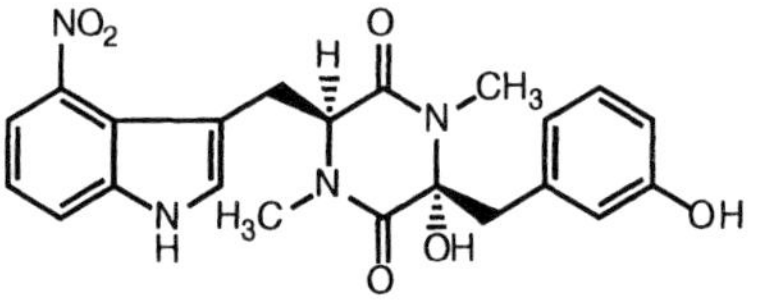

Thaxtomin A (**1**)

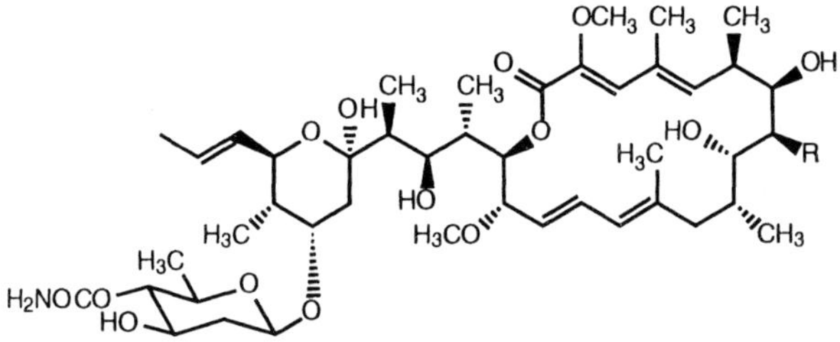

Concanamycins (**2**)
(A: R=C_2H_5, B: R=CH_3)

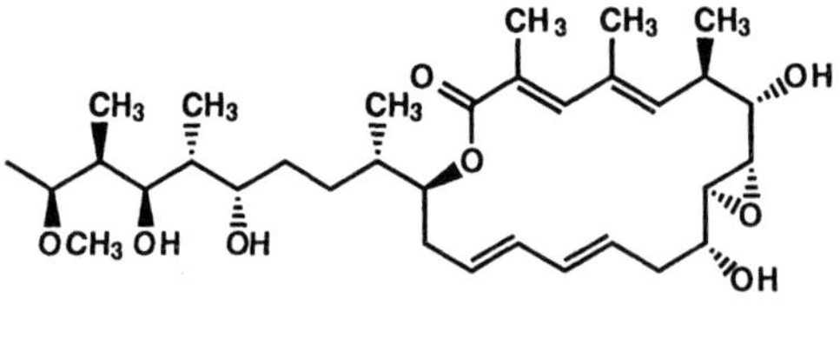

Russet scab toxin (**3**)
(= FD-891)

Figure 1. Structures of phytotoxins produced by Streptomyces species causing potato scab.

We describe here the isolation and identification of a phytotoxin produced by *S. cheloniumii* and by other *Streptomyces* strains causing russet scab.

Materials and Methods

Bacterial strains
The bacterial strains used in this study are listed in Table 1. *S. cheloniumii* MAFF 304020 and *Streptomyces* spp. MAFF 225003 and 225004 were collected in Chiba prefecture, Japan, *Streptomyces* sp. MAFF 225005 in Hokkaido, and *Streptomyces* sp. MAFF 225006 in Ibaraki prefecture. For isolation of the russet scab toxin, *S. cheloniumii* MAFF 304020 was used.

Pathogenicity assay in greenhouse
Each strain was inoculated into a yeast extract-malt extract medium from the agar slant culture and cultivated at 28°C for 5 days on a reciprocal shaker. The cultured material (5 ml) was inoculated in 500 g of soil-bran medium in a 1 liter Erlenmeyer flask and mixed well. The flask was incubated at 23°C for 5 months. The cultured material was mixed with sterilized soil to adjust the population density of pathogens to $5.2 - 8.3 \times 10^4$ cfu/g soil. The soil used for dilution of the infected soil and for the control experiment was adjusted to pH 6.6 with lime. Seed potato (cv. Dansyaku) was planted in the infected soil and cultivated for 3 months under short-day conditions in a greenhouse. Three replicate pots of seed pieces were used for each strain.

Analysis for thaxtomin A, concanamycins and russet scab toxin
Each strain was cultured on an oatmeal agar medium (*10*) at 28°C for 14 days. Thaxtomin A and concanamycins were determined with an HPLC equipped with a photodiode array detector. The analytical procedure was described in detail previously (*5*). Conditions for analysis of the russet scab toxin were the same as those for the concanamycins except for the detection wavelength (266 nm).

Isolation of russet scab toxin produced by S. cheloniumii
S. cheloniumii MAFF 304020 was cultured on an oatmeal agar medium at 28°C for 14 days. The cultured material was macerated with acetone, and the acetone extract was purified as outlined in Figure 2. Tuber slice assays (*4*) were used for fractionation guides.

Results

Symptoms induced by pathogenicity assay
First, it was determined whether or not the symptoms caused by the russet scab pathogens differed from those by common scab pathogens. All of the *S. cheloniumii* MAFF 304020 and *Streptomyces* spp. MAFF 225003, 225004, 225005 and 225006 strains induced superficial symptoms, which were rough in

texture and had shallow cracks such as tortoise shell (Table 1). Lesions caused by *S. scabies*, *S. acidiscabies*, or *S. turgidiscabies* are necrotic and sunken and are entirely different from those caused by russet scab pathogens. There was no clear difference among symptoms caused by *S. scabies*, *S. acidiscabies* and *S. turgidiscabies*.

Table 1. Results of Pathogenicity Assay and Production of Phytotoxins by *Streptomyces* spp.

species	strain	symptoms induced[1]	production[2] of Txt (1)[3]	Con (2)[3]
S. cheloniumii	MAFF 304020	superficial	n.d.[4]	n.d.
Streptomyces sp.	MAFF 225003	superficial	n.d.	n.d.
Streptomyces sp.	MAFF 225004	superficial	n.d.	n.d.
Streptomyces sp.	MAFF 225005	superficial	n.d.	n.d.
Streptomyces sp.	MAFF 225006	superficial	n.d.	n.d.
S. scabies	JCM 7914	necrotic, sunken	2.50	0.086
S. acidiscabies	JCM 7913	necrotic, sunken	5.31	n.d.
S. turgidiscabies	IFO 16080	necrotic, sunken	2.37	n.d.

[1] Results of pathogenicity assay.
[2] unit: μg/ ml cultured material
[3] Txt (**1**): thaxtomin A; Con (**2**): cncanamycins A plus B
[4] n.d. : not detected

Production of thaxtomin A and concanamycins

Production of the known phytotoxins was examined in an agar culture. None of the pathogens that caused russet scab produced thaxtomin A or concanamycins (Table 1). *S. scabies* produced both thaxtomin A and concanamycins, as described previously (*5*). *S. acidiscabies* and *S. turgidiscabies* produced only thaxtomin A. These results indicate that the pathogens of russet scab produce a new phytotoxin.

Isolation and identification of the russet scab phytotoxin

The procedure for purifying the russet scab toxin is outlined in Figure 2. Chloroform extract was fractionated by silica gel column chromatography. Only the 10% MeOH - $CHCl_3$ fraction induced necrosis by the tuber slice assay. This fraction was rechromatographed using a different solvent system. The active fraction was then purified by preparative TLC and separated into five fractions with the guidance of UV absorption. The active fraction with R_f 0.38-0.47 was finally purified by ODS-HPLC and afforded a single peak.

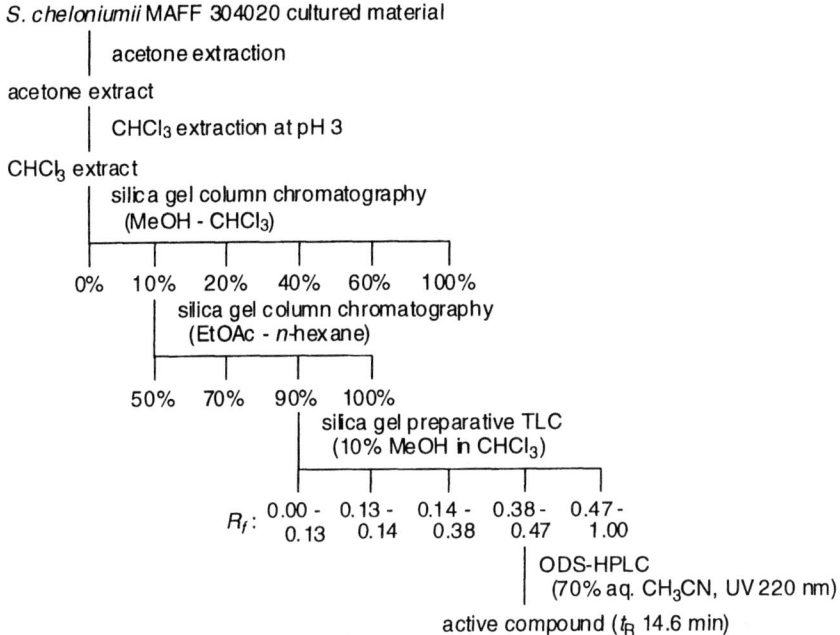

Figure 2. Purification procedure for russet scab toxin.

The phytotoxin showed a quasi molecular ion at m/z 579 $(M + H)^+$ with FAB-MS (matrix: glycerol), the molecular formula of which was determined to be $C_{33}H_{55}O_8$ based on the results of high resolution FAB-MS (m/z 579.3867, calcd. 579.3897). The UV spectrum of the toxin showed maxima at λ_{max} 205 and 267 nm. In the ^{1}H-NMR spectrum (500 MHz, CDCl$_3$), 5 doublet methyl, 2 olefinic or acetyl methyl, 1 methoxy methyl and 6 olefinic proton signals were observed as characteristic peaks.

Searching a database based on the molecular formula and narrowing down candidates from the features of the ^{1}H-NMR spectrum, the russet scab toxin was presumed to be the 18-membered macrolide, FD-891 (3) (11, 12, 13). Comparison of the retention time and UV spectrum in ODS-HPLC and of the ^{1}H-NMR spectrum of the isolated toxin with those of an authentic sample confirmed that the phyototxin was FD-891.

Necrosis-inducing activity of the russet scab toxin
The russet scab toxin induced weak necrosis at 50 μg/disk and clear activity at 100 μg/disk (Figure 3). Thaxtomin A showed similar activity at 1 μg/disk.

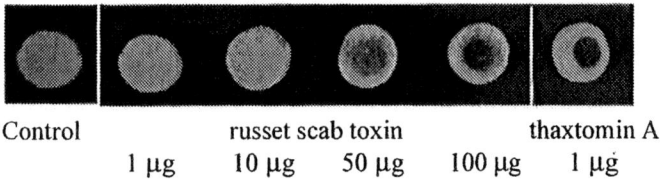

Control	russet scab toxin			thaxtomin A
1 μg	10 μg	50 μg	100 μg	1 μg

Figure 3. Necrosis inducing-activity of the russet scab toxin and thaxtomin A.

Production of the russet scab toxin by other strains
Whether production of the russet scab toxin would be common to other russet scab pathogens was examined in two strains isolated in different regions. The results showed that *Streptomyces* spp. MAFF 225003, 225005 and 225006 produced a phytotoxin with the same retention time and UV spectrum in HPLC analysis.

Discussion

We showed that the russet scab pathogens produced a phytotoxin other than thaxtomin A or concanamycins and isolated it. The phytotoxin was identified as the 18-membered macrolide, FD-891, and its effect in inducing necrosis on potato tuber disks was about 1/100th that of thaxtomin A. The russet scab toxin was produced by the other three pathogenic strains that were isolated in different regions. These results strongly suggest that the phytotoxin is a pathotoxin. To confirm this possibility, we are now trying to detect the phytotoxin in lesions.

FD-891 has been reported as a cytocidal compound for mammal cells ([11]), and this is the first report on its phytotoxicity. FD-891 has the same ring size as concanamycins and both have a quite similar structure. Phytotoxic activity of concanamycins is based on the inhibition of V-ATPase ([14]). FD-891 does not show inhibitory activity in mammal cells ([15]), though the first report described its inhibitory activity ([11]). The action mechanism of the toxin remains an interesting problem.

Acknowledgment

We are grateful to Professors Katsumi Kakinuma and Tadashi Eguchi, Tokyo Institute of Technology for providing authentic FD-891 and its NMR spectra.

References

(1) Loria, R.; Bukhalid, R. A.; Fry, B. A.; King, R. R. *Plant Dis.* **1997**, *81*, 836-846.
(2) King, R. R.; Lawrence, C. H.; Clark, M. C.; Calhoun, L. A. *J. Chem. Soc., Chem. Commun.* **1989**, 849-850.
(3) King, R. R.; Lawrence, C. H.; Clark, M. C. *Am. Potato J.* **1991**, *68*, 675-680.
(4) Loria, R.; Bukhalid, R. A.; Creath, R. A.; Leiner, R. H.; Oliver, M.; Steffens, J. C. *Phytopathology* **1995**, *85*, 537-541.
(5) Natsume, M.; Yamada, A.; Tashiro, N.; Abe, H. *Ann. Phytopathol. Soc. Jpn.* **1998**, *64*, 202-204.
(6) Toth, L.; Akino, S.; Kobayashi, K.; Doi, A.; Tanaka, F.; Ogoshi, A. *Soil Microorganisms* **1998**, *51*, 29-34.
(7) Natsume, M.; Ryu, R.; Abe, H. *Ann. Phytopathol. Soc. Jpn.* **1996**, *62*, 411-413.
(8) Oniki, M.; Suzui, T.; Araki, T.; Sonoda, R.; Chiba, T.; Takeda, T. *Bull. Natl. Inst. Agro-Environ. Sci.* **1986**, *2*, 45-59 (in Japanese with English summary).
(9) Suzui, T.; Miyashita, K.; Tashiro, N. In *Abstracts of Papers, 5th Int. Cong. Plant Pathology*; Kyoto, Japan, 1988; p 177.
(10) Babcock, M. J.; Eckwall, E. C.; Schottel, J. L. *J. Gen. Microbiol.* **1993**, *139*, 1579-1586.
(11) Seki-Asano, M.; Okazaki, T.; Yamagishi, M.; Sakai, N.; Hanada, K.; Mizoue, K. *J. Antibiotics* **1994**, *47*, 1226-1233.
(12) Seki-Asano, M.; Tsuchida, Y.; Hanada, K.; Mizoue, K. *J. Antibiotics* **1994**, *47*, 1234-1241.
(13) Eguchi, T.; Kobayashi, K.; Uekusa, H.; Ohashi, Y.; Mizoue, K.; Matsushima, Y.; Kakinuma, K. *Org. Lett.* **2002**, *4*, 3383-3386.
(14) Dröse, S.; Bindseil, K. U.; Bowman, E. J.; Siebers, A.; Zeeck, A.; Altendorf, K. *Biochemistry* **1993**, *32*, 3902-3906.
(15) Kataoka, T.; Yamada, A.; Bando, M.; Honma, T.; Mizoue, K.; Nagai, K. *Immunology* **2000**, *100*, 170-177.

Chapter 23

Synthesis and Biological Evaluation of Abscisic Acid, Jasmonic Acid, and Its Analogs

Hiromasa Kiyota, T. Oritani, and S. Kuwahara

Graduate School of Agricultural Science, Tohoku University, Sendai 981–8555, Japan

Plant hormones (+)-abscisic acid (ABA) and (–)-jasmonic acid (JA) play important roles in plant growth regulation and environmental stress response. Several analogs of ABA and JA were prepared and their plant growth regulatory activities were investigated. ABA analogs modified at the 2-position decreased the activities. Some JA analogs with 3,7-double bond or 12-fluorine substituent showed the activities comparable to JA. In addition, practical syntheses of ABA, JA and its derivatives were developed.

1. Introduction

Plant hormones play important roles in controlling many aspects of plant growth and development. Among several compounds recognized as plant hormones, only abscisic acid (ABA, **1**) and jasmonic acid (JA, **2**) are not used practically because of their instability and cost. To overcome these points, we have made efforts to develop new synthetic routes and new analogs of these hormones. Here we describe our recent advances of these studies.

2. Synthesis and plant growth regulatory activities of ABA and JA analogs

2-1. ABA analogs[1a]

(+)-abscisic acid
(ABA, **1**)

(+)-β-ionylideneacetic acid
(β-IAA, **2**)

Abscisic acid (ABA, **1**) is a plant hormone with particular activities such as growth promotion or inhibition, hypertrophy, leaf senescence, stomatal control of plant water balance, resistance to environmental stress etc.[1] In spite of these important activities, ABA is not used practically as a plant growth regulator or as a pesticide. One of the two main reasons is ABA's instability in nature. ABA is rapidly hydroxylated to give 8'-hydroxy-ABA, which is cyclized by Michael addition to give biologically inactive phaseic acid (metabolic pathway). On the other hand, 2-position of ABA is easily isomerized under sunlight to give *trans*-ABA, which is also inactive. Another reason is cost. Both microbial production and chemical synthesis are costly.

We developed several ABA analogs with substituents at 2-position to prevent the *Z-E* isomerization.[2] Epoxy-β-ionylideneacetic acid (β-IAA, **2**) was used as the parent compound, because **2** is converted to ABA in plants thus shows comparable activity as ABA does. As shown in Scheme 1, β-ionone was converted to (2*E*)- and (2*Z*)-2-fluoro-epoxy-β-IAA (**3**). Wittig-Horner reaction of β-ionone with triethyl phosphonofluoroacetate gave 2-fluoro-β-IAA. The electron rich double bond was selectively epoxidized, *E*- and *Z*-isomers were separated, and each ester was hydrolyzed to give the fluorine analogs **3***E* and **3***Z*. The stereochemistry was determined by ^{19}F-NMR, compared with the literature value. Interestingly *E*-isomer completely isomerized to *Z*-isomer under mercury lamp (using the same lamp for non-F compound, the ratio was about 1 to 1). This was probably due to the strong intramolecular H-bonding effect. Our aim was to control the stability of this double bond by introducing F atom, however, undesired compound was more stable. Using a similar sequence cyano analog **4** was prepared using ethyl cyanoacetate. However, direct Knoevenagel condensation of β-ionone with malonate failed. Thus, diacid analog **5** was synthesized using selenium chemistry.

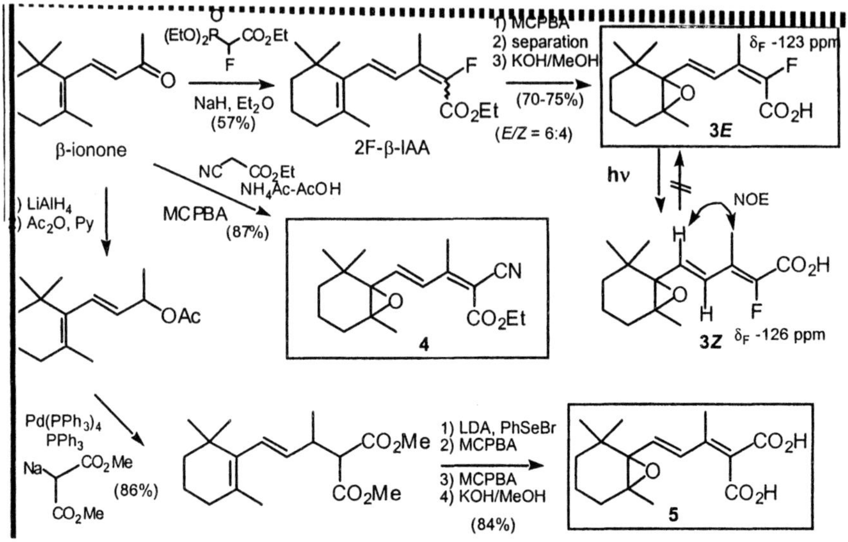

Scheme 1. *Synthetic schemes ABA analogs*

The results of bioassay are shown in Table 1. The synthetic compounds were tested for growth inhibition assay using rice and lettuce. The cyano and diacid analogs were inactive. On the other hand, fluorine analog **3E** showed a comparable activity to the parent compound. We interpreted these results as the steric requirement of 2E-position was severe for the activity and electronic change was not so important. The results that **3Z** showed some activity was curious, because **4Z** would not isomerize to its E-isomer *in vivo*.

Table 1. Plant growth inhibitory activity of ABA analogs

	2	3E	3E	4	5
rice seedlings elongation*	+++	++	++	−	−
lettuce seeds germination†	+++	+++	++	−	+

* The length of the second leaf sheath were measured after 4 d (*Oryza sativa* cv. Satohonami).

† Germination rates were measured after 6 d (*Lactuca sativa* cv. Great Lake).

2-2. JA analogs -1, block of epimerization at the 7-position[3]

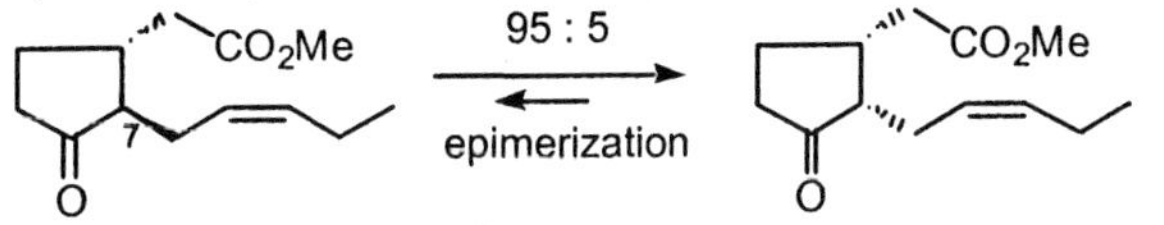

(+)-methyl jasmonate (MJA, **6**) ⇌ (95:5, epimerization) ⇌ (+)-methyl epijasmonate (MepiJA, **7**)

Jasmonic acid (JA) is recently recognized as one of the plant hormones. JA and its methyl ester (MJA, **6**) sustain many plant actions such as plant growth inhibition, senescence promotion, abscission, tuber formation, stress response etc.[1a] Recent studies suggested that really active compound is its 7-epimer, epijasmonic acid (epiJA). Actually, methyl epijasmonate (MepiJA, **7**) is epimerized to MJA as 20:1 mixture at equilibrium. It means, on the other hand, even purified MJA contains a small amount of MepiJA and it might be responsible for the activity. We began the studies to know these points and to prepare more effective analogs. As shown in Scheme 2, analogs with 7-substituents were synthesized to block epimerization. 3,7-Didehydro compound **9**, the root of two side chains was in a same plane as MepiJA, was prepared through 4,5-didehydro-MJA (**8**). Fluoro compounds **10** were prepared according to Taapken's report.[4] The stereochemistry of C-F bonds are supposed from the results of odour evaluation.[5]

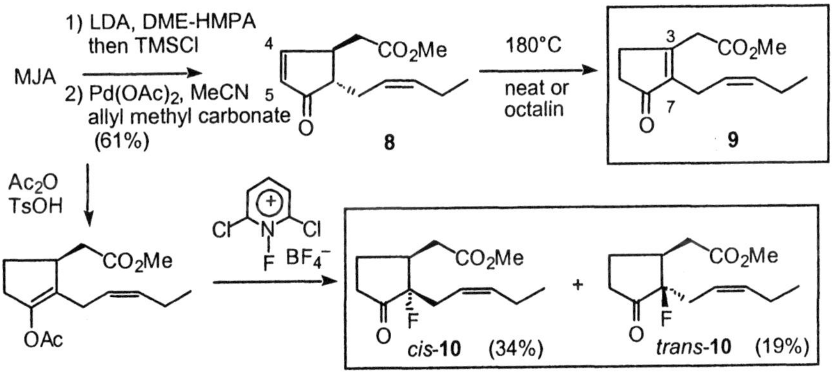

Scheme 2. *Synthesis of MepiJA analogs*

Plant growth inhibitory activity was evaluated (Table 2).[6] The result that 7-F analogs showed little activity coincided with the Taapken's report.[4] This was due to the strong electronegative charge around 7-position. Didehydro analogs showed stronger activity for lettuce seeds. We thought that for lettuce, the coplanarity of two side chains as MepiJA was important for activity.

Table 2. Plant growth regulatory activity of JA analogs-1

	MJA	8	9	cis-10	trans-10
rice seedlings elongation*	++	++	+	+	+
lettuce seeds germination†	++	+++	++++	+	+
radish seeds germination‡	++	+	+	+	+

* The length of the second leaf sheath were measured after 5 d (*Oryza sativa* cv. Sasaminori).

† Germination rates were measured after 5 d (*Lactuca sativa* cv. Great Lake).

‡ Germination rates were measured after 4.5 d (*Raphanus sativus* cv. Sakuranbo).

2-3. JA analogs -2, antimetabolites

Hydroxylation of the 11- or 12-position is known as one of the metabolic pathways of JA.[7] In fact, 11- or 12-hydroxy-JA did not show usual plant growth inhibitory activities as JA.[8] We thought 11- or 12-fluoro derivative could be an antimetabolite with stronger activities. Preparations of the fluoro compounds are summarized in Scheme 3 and the results of bioassays are listed in Table 3. 12-Trifluoro analog **12** showed comparable activity to MJA and 11-fluoro analog **11** had less activity.[9,10] These indicated that ω-trifluoromethyl group in **12** acted as methyl group equivalent, and 11-C–F bond in **11** worked as not C–H bond mimic but C–O(H) group mimic.

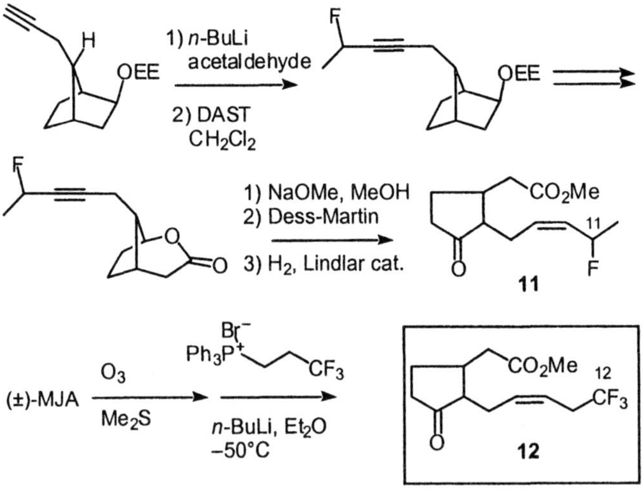

Scheme 3. *Synthesis of 11- or 12-fluoro MJA analogs*

Table 3. Plant growth regulatory activity of JA analogs-2

	MJA	11	12
rice seedlings elongation*	++	+	+++
lettuce seeds germination[†]	++	¶	++
radish seeds germination[‡]	++	+	¶
potato tuber formation[§]	++	¶	++

* The length of the second leaf sheath were measured after 5 d (*Oryza sativa* cv. Sasaminori).
† Germination rates were measured after 5 d (*Lactuca sativa* cv. Great Lake).
‡ Germination rates were measured after 4.5 d (*Raphanus sativus* cv. Sakuranbo).
§ Formation of sessile microtuber was observed (*Solanum tuberosum* L. cv Irish Cobbler).
¶ Not tested.

12-Hydroxyepijasmonic acid (tuberonic acid, TA) was isolated as potato tuber-forming substance.[8] On the other hand, its biosynthetic precursor epiJA, and MJA also showed this activity. We also tested **12** for potato tuberization assay to know whether the 12-hydroxy group of TA was essential or not. The result that **12** showed strong potato tuber-forming activity indicated that the 12-OH group of TA is not essential for the activity, and this hydroxylation is for the transportation from leaves to stocks of potato (transportation form of TA is 12-*O*-glucoside).

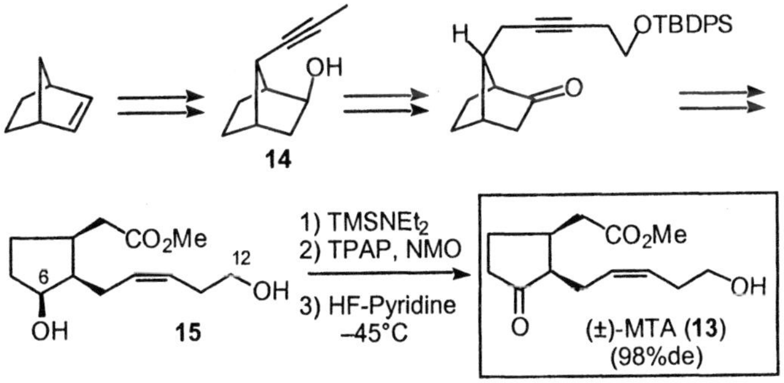

Scheme 4. *Synthesis of (±)-methyl tuberonate.*

We also achieved the synthesis of methyl tuberonate (MTA, **13**) (Scheme 4).[11] Bicyclic alcohol **14** was converted to diol **15**. The key of the synthesis was oxidation of 6-hydroxy group and removal of 12-*O*-protecting group without epimerization at 7-position. 12-OH group of **15** was protected as trimethylsilyl (TMS) ether, and 6-OH group was oxidized with tetrapropylammonium perruthenate (TPAP). Finally, TMS group was removed under mild conditions to give MTA with 98%de.

3. Practical synthesis

3-1. ABA and β-IAA

We have developed practical method to supply ABA and β-IAA by optical resolution of synthetic intermediate. As shown in Scheme 5, optically pure (*S*)-phorenol (**16**), the synthetic intermediate for ABA, was obtained using esterase SNSM-87 (Nagase).[12] As for β-IAA, the intermediate epoxy-β-cyclogeraniol (**17**) was resolved in high selectivity (E = 1600) using lipase P (Amano).[13] Each intermediate could be converted to (+)-ABA[14] or (+)-β-IAA[15] in good yields.

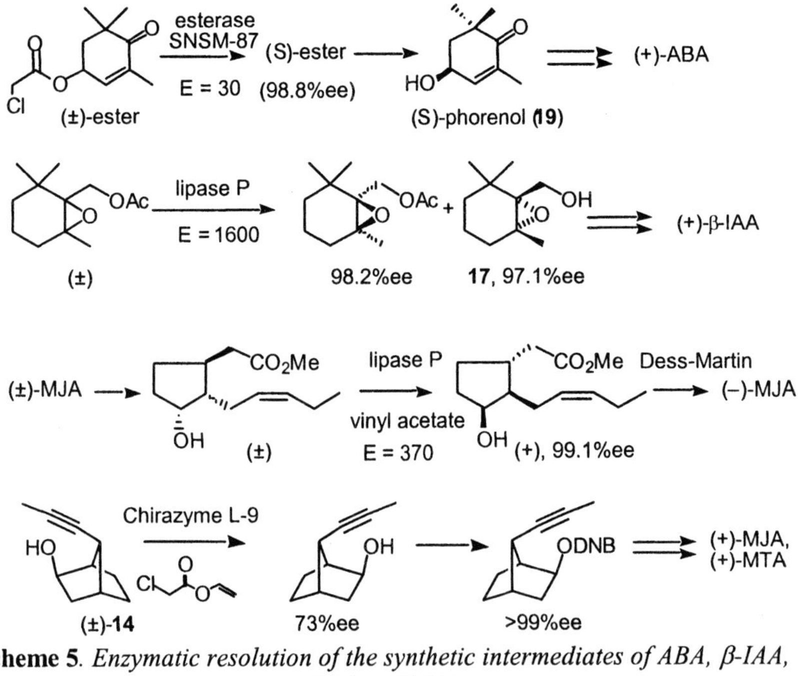

Scheme 5. *Enzymatic resolution of the synthetic intermediates of ABA, β-IAA, MJA and MTA*

3-2. MJA and MTA.

(–)-Methyl jasmonate was prepared from its commercial racemate in overall 40% yield.[16] We also succeeded to resolve a sterically hindered hydroxy compound 14 using Chirazyme L-9 (Roche), which was the key intermediate of our synthesis of (±)-methyl tuberonate.[11]

4. Conclusion

The unique plant hormonal activities of ABA and JA are very important for control of both production of crops and prevention of phytopathogens. The results of our studies in making novel and effective analogs and developing practical synthesis of these hormones will be of great service to these purposes.

---

[1] a) *Plant hormones*; ed. by Davies, P. J., 2nd ed. Kluwer, Netherlands, 1995. b) *Abscisic acid*; ed. by Addicott, F. T., Praeger, New York, 1983. c) Oritani, T.; Kiyota, H. *Nat. Prod. Lett.*, **2003**, *20*, 414-415.

[2] Kiyota, H., Masuda, T., Chiba, J., Oritani, T. *Biosci. Biotechnol. Biochem.*, **1996**, *60*, 1076-1080.

[3] Beale, M. H., Ward, J. L. *Nat. Prod. Lett.*, **1998**, *15*, 533-548.

[4] Taapken, T., Blechert, S., Weiler, E. W., Zenk, M. H. *J. Chem. Soc., Perkin Trans. 1*, **1994**, 1439-1442.

[5] Kiyota, H., Takikawa, S., Kuwahara, S., under submission.

[6] Kiyota, H. Yoneta, Y., Oritani, T. *Phytochemistry*, **1997**, *46*, 983-986.

[7] Sembdner, G., Meyer, A., Miersch, O, Bruckner, C. in *Plant growth substances 1988*; eds. by Pharis, R. P., Wood, S. B., Springer Verlag, Berlin/Heidelberg, pp 374-379, 1990.

[8] a) Koda, Y. *Int. Rev. Cytology*, **1992**, *135*, 155-199. b) Miersch, O., Kramell, R., Parthier, B., Wasternack, C. *Phytochemistry*, **1999**, *50*, 353-361.

[9] Kiyota, H., Saitoh, M., Oritani, T., Yoshihara, T., *Phytochemistry*, **1996**, *42*, 1259-1262.

[10] Kiyota, H., Koike, T., Higashi, E., Satoh, Y., Oritani, T. *J. Pesticide Sci.*, **2001**, *25*, 96-99.

[11] Kiyota, H., Nakashima, D., Oritani, T. *Biosci. Biotechnol. Biochem.*, **1999**, *63*, 2110-2117.

[12] Kiyota, H., Nakabayashi, M., Oritani, T. *Tetrahedron: Asymmetry*, **1999**, *10*, 3811-3817.

[13] Okazaki, R., Kiyota, H., Oritani, T. *Biosci. Biotechnol, Biochem.*, **2000**, *64*, 1444-1447.

[14] Kinzle, F., Mayer, H., Minder, R. E., Thommen, H. *Helv. Chim. Acta*, **1978**, *61*, 2616-2627.
[15] Oritani, T., Yamashita, K. **1983**, *22*, 1909-1912.
[16] Kiyota, H., Higashi, E., Koike, T., Oritani, T. *Tetrahedron: Asymmetry*, **2001**, *12*, 1035-1038.

New Chemistry–Green Chemistry

Chapter 24

Discovery of Pyridalyl: A Novel Compound for Lepidopterous Pest Control

Noriyasu Sakamoto, Shigeru Saito, Taro Hirose, Masaya Suzuki, Sanshiro Matsuo, Keiichi Izumi, Toshio Nagatomi, Hiroshi Ikegami, Kimitoshi Umeda, Kazunori Tsushima, and Noritada Matsuo

Sumitomo Chemical Company Ltd., Agricultural Chemicals Research Laboratory, 2-1, Takatsukasa, 4-Chome, Takarazuka, Hyogo, Japan

In our research to find new insecticides, we found out that 2-(trifluoromethy)-4-phenoxyphenyl 3,3-dichloro-2-propenyl ether, showed weak insecticidal activity against lepidopterous larvae. Optimizations of this lead compound led to the discovery of pyridalyl (PLEO®, S-1812) belonging to a new class of insecticides. Pyridalyl is being developed worldwide and the first market introduction is expected in Japan and some Asian countries in the years between 2004 and 2005. This compound gives very good control of various lepidopterous and thysanopterous pests on cotton and vegetables. Pyridalyl also controls insecticide-resistant strains of lepidopterous pests as well as susceptible strains. It produces unique insecticidal symptoms so that it may have a different mode of action from any other existing insecticides . Moreover, pyridalyl is safer to mammals and various beneficial arthropods and it will provide an important tool in IPM and insecticide resistant management programs.

Introduction

The discovery of insecticides with novel modes of action is vital for managing pest strains that are resistant to existing products. The insecticide market has generally been dominated since the 1970s by the three chemical classes, organophosphates, carbamates and synthetic pyrethroids. As a consequence, insect resistance to them has become a significant problem for farmers. In addition, toxicity issues have put restrictions on their use. Regulatory agencies worldwide have placed a premium on crop protection by chemical pesticides having improved toxicological profiles for non-target organisms. It is very desirable to develop new insecticidal agents, as IPM-compatible products, that are both active against resistance strains and safe to humans and to the environment.

Pyridalyl (PLEO®, S-1812) is an insecticide of a novel chemical class. This compound controls insecticide-resistant strains of lepidopterous pests as well as susceptible strains. It is also safer to mammals and various beneficial arthropods so that it is expected to be a useful material for controlling lepidopterous and thysanopterous pests in IPM and insecticide resistant management programs.

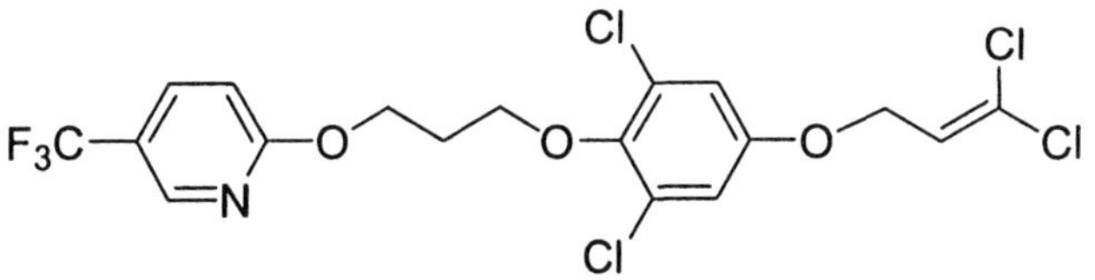

Chemical name	: 2,6-dichloro-4-(3,3-dichloroallyloxy)phenyl 3-[5-(trifluoromethyl)-2-pyridyloxy]propyl ether
Trade name	: PLEO (Japan)
ISO name	: Pyridalyl
Code No.	: S-1812

Figure 1. Pyridalyl: Structure and Nomenclature

This paper reviews the discovery, the structure-activity relationships, insecticidal activity and biological properties of pyridalyl.

Discovery of a lead compound

In our research to find a new insecticide, we focused on two known insecticidal compounds **1** and **2** (Fig. 2), which have a 3,3-dichloro-2-propenyloxy group in compound (*1, 2*). This group attracted much of our attention as an interesting group to generate a new lead compound.

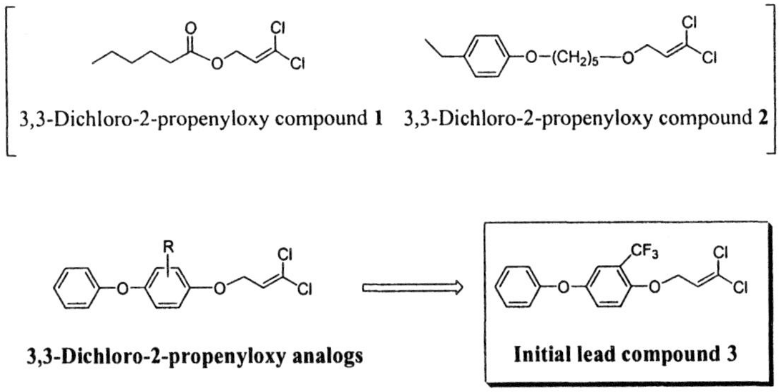

Figure 2. Discovery of the Initial Lead Compound 3

We synthesized several compounds containing a 3,3-dichloro-2-propenyloxy group as initial analogues. These compounds were evaluated in our primary screening, and no compound cleared our standard for secondary screening. However, we noticed that 2-(trifluoromethy)-4-phenoxyphenyl 3,3-dichloro-2-propenyl ether (**3**) was slightly active at 500 ppm against the larvae of *Spodoptera litura* in the artificial diet assay (Fig. 2). Evaluated in a topical assay, **3** was found to be active at 25μg (insect)$^{-1}$ against *S. litura*, while **1** and **2** were inactive in both evaluations (Table 1)(*3,4*). These results prompted us to perform further structural modifications of the compound **3**.

Evolution to find the second lead compound

Our optimization program on the lead structure **3** involved replacement of chlorine atoms on the unsubstituted moiety (**A**) by other halogen atoms and introduction of substituents on the left-side phenyl ring (**B**). Many analogs of **3** were prepared (*5, 7*) and tested against *S. litura* in the artificial diet assay. The structure-activity relationships for these compounds are summerized in Figure 3.

Table 1. Insecticidal Activity of Lead Compound 3 against *S. litura*

Compound[a]	Mortality (%)[b]	
	Artificial diet assay[c]	Topical assay[d]
1	0	0
2	0	0
3	60	90
Control	0	0

[a] See Fig. 2
[b] Mortality was determined six days after treatment.
[c] Artificial diet assay: at 500 ppm
[d] Topical assay: at 25 μg insect^{-1}

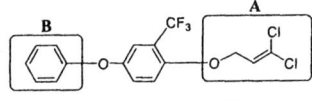

Initial lead compound 3

A: 2-Propenyloxy moiety

B: Unsubstituted phenyl moiety

Compound 5   Compound 4

Figure 3. Structure-activity Relationships for Compound 3 against the Forth-instar Larvae of S. litura in the Artificial Diet Assay

The activity of the 3,3-dibromo-2-propenyloxy derivative was equally, or only slightly less active than that of the initial lead **3**. On the other hand, derivatives containing the 3,3-chlorofluoro-, 3,3-bromofluoro- and 3,3-difluoro-2-propenyloxy moieties should reduced activity. Introduction of various substituents on the unsubstituted phenyl ring led to a marked increase of insecticidal activity. In particular, compound **4** containing a 2,6-dichloro-4-(trifluoromethyl)phenyl group showed the highest activity against *S. litura*.

Then, compound **5**, containing a 3-chloro-5-(trifluoromethyl)pyridyl group as an alternative to the 2,6-dichloro-4-(trifluoromethyl)phenyl group, was prepared, evaluated and found to retain the insecticidal activity. We, therefore, restarted our optimization program using the compound **5** as a second lead.

Optimization from the second lead structure

The second lead **5**, the most favorable proto-type, can be divided into three moieties, two aromatic parts and a linkage as shown in Figure 4. The modifications of these parts were summarized in Table 2-4. Each activity order was determined by evaluations against the fourth-instar larvae of *S. litura* in the artificial diet assay.

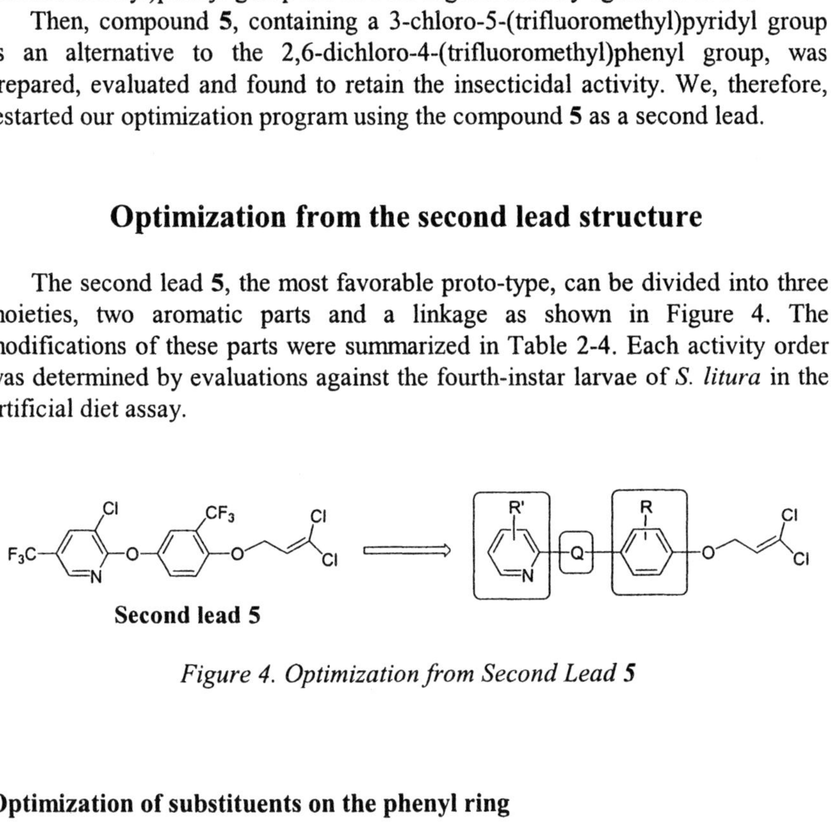

Second lead 5

Figure 4. Optimization from Second Lead 5

Optimization of substituents on the phenyl ring

We first prepared the derivatives **6-17** with the 4-position on the phenyl ring fixed as 3-chloro-5-(trifluoromethyl)-2-pyridyloxy group, to examine the effect of substituents on the phenyl ring (Fig. 4)(*4, 5*). Introduction of a chlorine (**7**) or bromine atom (**8**) at the 2-position on the phenyl ring reduced the insecticidal activity against *S. litura*, while a chlorine (**9**) or bromine atom (**10**) at the 3-position increased the activity. Introduction of a chlorine atom at both the 2- and 5-positions (**11**), or both the 3- and 5-positions (**13**) resulted in a strong increase of the activity. One of these compounds, the compound **13** gave the highest activity (Table 2). Other substituents in the 3- and 5-positions (**14-17**) also increased the activity.

From the results shown in Table 2, it was concluded that the incorporation of the substituent, especially, a chlorine atom at both the 3- and 5-positions on the phenyl ring was important for high activity.

Table 2. Effect of Substituents on the Phenyl Ring on *S. litura* Activity

Compound	Substituents R_1	Activity rating [a,b]
6	H	0
7	2-Cl	0
8	2-Br	0
9	3-Cl	2
10	3-Br	2
11	2,5-Cl$_2$	3
12	2,6-Cl$_2$	1
13	3,5-Cl$_2$	4
14	3-Cl, 5-Br	3
15	3,5-(CH$_3$)$_2$	2
16	3-CH$_3$, 5-CH$_2$CH$_3$	3
17	3,5-(CH$_2$CH$_3$)$_2$	2
5	2-CF$_3$	1

a LC$_{50}$ (ppm); >200: 0, 200-100:1, 100-20: 2, 20-5: 3, 5-1.25: 4, 1.25-0.31: 5
b Artificial diet assay: Mortality was assessed six days after treatment.

Optimization of substituents on the pyridyl ring

Next, we synthesized the derivatives **18-24** with substituents at the 3- and 5-positions on the phenyl ring fixed as chlorine atoms, to examine the effects of substituents on the pyridyl ring (Table 3)(*4, 5*). Introduction of a cyano (**19**) or a nitro group (**20**), being hydrophilic and electron-withdrawing, at the 5-position on the pyridyl ring, drastically reduced the insecticidal activity, whereas induction of a trifluoromethyl group (**21**), which is hydrophobic and electron-withdrawing, at the 5-position on the pyridyl ring resulted in a strong increase of the activity. The further substitution of a chlorine (**13**), bromine (**22**) or trifluoromethyl group (**23**) at the 3-position on the pyridyl ring of the compound **21**, retained the same activity as that of **21**. The introduction of a chlorine atom (**24**), at the 6-position on the pyridyl ring of the compound **23**, reduced the activity.

The results obtained here, showed that a trifluoromethyl group at the 5-position on the pyridyl ring was essential for the development of optimal insecticidal activity.

Table 3. Effect of Substituents on the Pyridyl Ring on *S. litura* Activity

Compound	Substituents R_2	Activity rating [a]
18	H	0
19	5-CN	0
20	5-NO$_2$	0
21	5-CF$_3$	4
13	3-Cl, 5-CF$_3$	4
22	3-Br, 5-CF$_3$	4
23	3, 5-(CF$_3$)$_2$	4
24	3, 5-(CF$_3$)$_2$, 6-Cl	2

[a] See Table 2

Optimization of the linkage moiety between the phenyl and pyridyl rings

We set out to optimize the elongation of the linkage moiety between the phenyl and pyridyl rings, some alkylenedioxy group were introduced (**S-1812, 25-28**)(*4, 7*). The results are shown in Table 4. Introduction of an ethylenedioxy group (**25**) between the two rings, reduced the insecticidal activity compared to the compound **21**, whereas introduction of a 1,3-propylenedioxy group (**S-1812**) and a 1,4-butylenedioxy group (**26**) between the two rings resulted in a marked increase of the activity. Introduction of a 1,5-pentylenedioxy (**27**) and 1,6-hexylenedioxy groups (**28**) between the two rings produced relatively high insecticidal activity.

Table 4. Effect of the Linkage Moiety between Two Rings on *S. litura* Activity

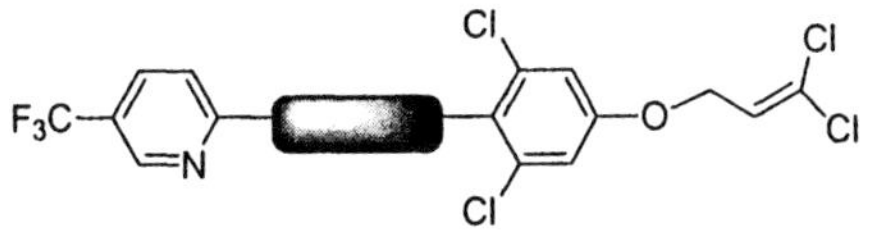

Compound		Activity rating [a]
25	$O(CH_2)_2O$	3
S-1812	$O(CH_2)_3O$	5
26	$O(CH_2)_4O$	5
27	$O(CH_2)_5O$	4
28	$O(CH_2)_6O$	4
21	O	4

[a] See Table 2

Selection of the candidate compound S-1812

Based on overall considerations of insecticidal efficacy and safety to non-target organisms (e.g. predatory insects, mammals, fish), S-1812 (pyridalyl) was finally selected as the most promising compound for advanced pre-commercialization studies.

Biological Properties of S-1812, pyridalyl

Laboratory studies

LC_{50} values of pyridalyl against various lepidopterous larvae were between 0.77 and 4.48 ppm as shown in Table 5. Pyridalyl also showed good activity against thysanopterous insects at 100 ppm (data are not shown). The compound was also found to be highly active against a strain of *Plutella xylostella* that is known to be highly resistant to synthetic pyrethroids, organophosphates and benzoylureas (Table 6). It should be noted that pyridalyl showed little toxicity toward various beneficial arthropods at 100 ppm (Table 7).

Table 5. Insecticidal Activity of Pyridalyl against Lepidopterous Pests

Scientific name	Stage [a]	Test method	DAT [b]	LC_{50} (ppm)
Cnaphalocrosis medinalis	L3	Foliar spray	5	1.55
Helicoverpa armigera	L3	Leaf dip	5	1.36
Helicoverpa zea	L2	Leaf dip	5	3.23
Heliothis virescense	L2	Leaf dip	5	4.29
Mamestra brassicae	L3	Foliar spray	5	1.98
Spodoptera exigua	L3	Leaf dip	5	0.93
Spodoptera litura	L3	Foliar spray	5	0.77
Pieris rapae	L2	Foliar spray	5	3.02
Plutella xylostella	L3	Leaf dip	3	4.48

[a] L2 and L3 means 2nd and 3rd instar larva, respectively.
[b] Days after treatment.
Ref. 8.

Table 6. Insecticidal Activity of Pyridalyl against Insecticide Resistant Strain of *P. xylostella*

Insecticide	Class	LC_{50} (ppm) [a]	
		resistant strain	susceptible strain
Pyridalyl (S-1812)		2.6	4.5
cyfluthrin	synthetic pyrethroid	> 500	3.7
pyrimifos methyl	organophosphate	> 450	12.0
chlorfluazuron	benzoylurea	> 25	3.4

[a] Cabbage leaf dip assay: Mortality was assessed six days after treatment.
Ref. 8.

Table 7. Beneficial Arthropods not Affected by Pyridalyl at 100 ppm

Scientific name	beneficials	Stage	Test method
Trichogramma japonicum	Egg parasitic wasp of lepidoptera	Adult	Foliar spray
Chrisoperla carnea	Predatory Chrysopidae	2-3rd L	Insect dip
Halmonia axyridis	Predatory Coleoptera	2-3rd L	Foliar spray
Orius sauteri	Predatory Hymenoptera	Adult/Nymph	Foliar spray
Phytoseiulus persimilis	Predatory Acarina	Adult	Foliar spray
Apis mellifera	Pollinator	Worker	Direct spray
Bombus terrestris	Pollinator	Worker	Direct spray

Ref. 8.

Field studies

Extensive field trials have been carried out over several years on key target pests such as *P. xylostella* and *H. virescens*, and pyridalyl has proven to be an outstanding insecticide. Pyridalyl shows very good efficacy for control of various lepidopterous and thysanopterous pests on cotton and vegetables without any phytotoxicity in practical treatments which range from 83 to 300g AI ha^{-1}(*8*).

Mode of action

The symptoms in larvae of lepidopterous insects treated with pyridalyl are unique and differ from those obtained with existing insecticides. Treated insects at lethal dose rates lost their vigor gradually and were killed in 2-3 hours. While moribund symptoms such as vomiting or convulsion were not observed in the treated larvae. The biochemical mechanism of insecticidal action is still under investigation but appears to differ from that of any other existing insecticides because the compound shows good control of populations of *H.virescense* or *P. xylostella* which are resistant to various insecticides and the insecticidal symptoms are unique.

Conclusion

Pyridalyl resulted from a major synthetic program based on a series of lead compounds and has proven to be very active against the larvae of some important pests of cotton and vegetable crops. Pyridalyl also controls insecticide-resistant strains of lepidopterous pests as well as susceptible strains. It produces unique insecticidal symptoms so that it may have a different mode of action from any other existing insecticides. Pyridalyl is safer to mammals and various beneficial arthropods so that it will provide an important tool in IPM and insecticide resistant management programs. Pyridalyl will become a safer chemical, and certainly contribute to establishment of sustainable chemistry.

Acknowledgements

We would like to thank the many colleagues within Sumitomo Chemical Co., Ltd. and Valent USA Corporation who have contributed to the discovery and the development of pyridalyl.

References

1. Quistad, B. G.; Cerf, C. D.; Kramer, J. S.; Bergot, B. J.; Schooley, A. D.: *J. Agric. Food Chem.* **1985**, 33:47-50.
2. Piccardi, P.; Massardo, P.; Bettarini, F.; Longoni, A.: *Pestic. Sci.* **1980**, 11:423-431.
3. Sakamoto, N.; Saito, S.; Hirose, T.; Suzuki, M.; Umeda, K.; Tsushima, K.; Matsuo, N. *Abstracts of Papers*, 10th IUPAC International Congress on the Chemistry of Crop Protection, Basel 2002; 1. 254.
4. Sakamoto, N; Saito, S.; Hirose, T.; Suzuki, M.; Matsuo, S.; Izumi, K.; Nagatomi, T.; Ikegami, H.; Umeda, K.; Tsushima, K.; Matsuo, N.: *Pest Manag. Sci.*, **2004**, *60*, 25-34.
5. Sakamoto, N.; Suzuki, M.; Nagatomi, T.; Tsushima, K.; Umeda, K.: (Sumitomo Chemical Co. Ltd.): E.P. Patent 648729, 1995.
6. Sakamoto, N.; Matsuo, S.; Suzuki, M.; Hirose, T.; Tsushima, K.; Umeda, K.: (Sumitomo Chemical Co. Ltd.): E.P. Patent 785923, 1997.
7. Izumi, K.; Ikegami, H.; Suzuki, M.; Sakamoto, N.; Takano, H. (Sumitomo Chemical Co. Ltd.): E.P. Patent 787710, 1997.
8. Saito, S.; Isayama, S.; Sakamoto, N.; Umeda, K.; Kasamatsu, K. *Abstracts of Papers*, Proc Brighton Crop Prot Conf-Pests and Diseases, BCPC, Farnham, Surrey, UK, 2002; 33-38.

Chapter 25

Synthetic Study on Macrocyclic Musks, Mints, and Jasmine Perfumes Utilizing Ti-Claisen and Aldol Reactions

Yoo Tanabe

Department of Chemistry, School of Science and Technology, Kwansei Gakuin University, 2–1 Gakuen, Sanda, Hyogo 669–1337, Japan

We developed powerful Ti- or Zr-Claisen condensation between esters and direct crossed Ti-aldol addition between ketones and ketones or aldehydes. These reactions were successfully applied to the short and practical method for synthesizing some natural macrocyclic musks (civetone and muscone), mints (mintlactone and menthofuran), and a jasmine perfume.

We have been engaged in the exploitation of new original and practical organic reactions directed toward the process chemistry and its application to the syntheses of useful fine chemicals and structurally interesting natural products.

Especially, we focused our attention on the Ti (*or* Zr)-Claisen condensations[1] and related direct aldol additions[2] from a recent standpoint of the environmentally benign (green) organic synthesis.

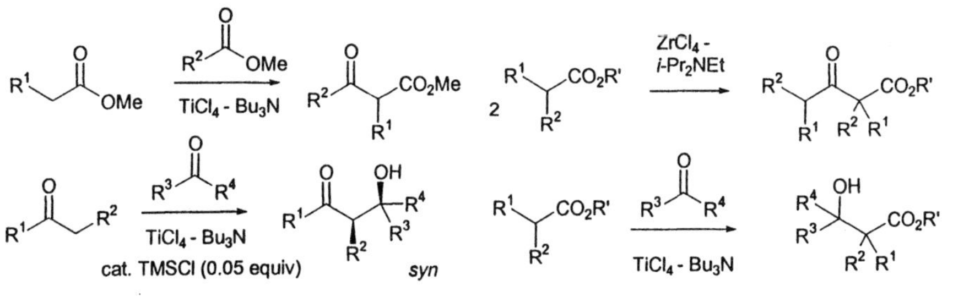

The salient features of these reactions are as follows.
(1) High reaction velocities and yields.
(2) Higher atom-economy and lower cost than the indirect methods using enol silyl ethers and ketene silyl acetals.
(3) Use of readily available and very low toxic metal reagents (e.g., $TiCl_4$, $ZrCl_4$), and use of practical amines (Et_3N, Bu_3N) and solvents (toluene or CH_2Cl_2).
(4) Toleration against basic labile functionalities.
(5) Enhanced reactivity using catalytic TMSCl.

1. Practical synthesis of Z-Civetone (1)[3]

The Ti- and Zr- Claisen and Dieckmann condensations exhibit powerful reactivity to realize C-C bond formations between various carboxylic esters under mild reaction conditions. This characteristic merit would promise a large ring closing with higher concentrations and speed.

Practical synthesis of natural macrocyclic musks, especially Civetone and Muscone, is one of the most important topics in perfume chemistry. Z-Civetone (1) is an attractive ingredient of civet cat and has been one of the most challenging synthetic targets for producing macrocyclic musks due to its unique symmetric 17-membered structure. Several hitherto reported syntheses of 1, however, are limited by the laboratory methods because of their low yields and/or multistep procedures. The Washington treaty claims that harvesting musk constitutes ill treatment of wild animals. An efficient, practical, and stereocontrolled synthesis of natural Z-civetone was performed utilizing a Ti-Dieckmann (intramolecular Ti-Claisen) condensation of dimethyl Z-9-octadecanedioate as the key step. This cyclization reaction to afford the 17-membered β-keto ester has some advantages compared with the traditional basic Dieckmann condensation with respect to higher concentration (100~300 mM), lower reaction temperature (0-5 °C), shorter reaction time (1~3 h), use of environmentally benign (low toxicity and safe) reagents ($TiCl_4$ and Et_3N *or* Bu_3N), and economical reagents and solvents.

2. Efficient short step synthesis of E/Z-Civetone (2)[4]

A laboratory scale synthesis was also performed utilizing intermolecular the Ti-Claisen condensation of methyl 10-decenoate followed by an intramolecular olefin metathesis using the Grubbs' reagent afforded the 17-membered β-keto ester. The overall isolated yield is 74%, which is highest compared with hitherto reported syntheses.

As a further notable extension, a one-pot reaction sequence was performed. Eventually, methyl 9-decenoate was straightforwardly transformed into E/Z-civetone (2) *in a one-pot manner* in 48% yield. It is worth noting that (i) the present ring closing metathesis proceeded using even the titanium-intermediate of β-keto ester, and that (ii) interestingly, the elimination of methoxycarbonyl

group spontaneously took place without any special procedure and/or any addition of reagents.

3. Short step formal synthesis of *R*-Muscone (3)[2b]

R-muscone (3) is an attractive ingredient of musk deer and has been one of the most challenging synthetic targets of perfume industry due to its unique 15-membered structure with β-chiral methyl center. Ti-aldol addition of readily available 2,15-hexadecanedione successfully afforded the 15-membered aldol adduct, which is a key precursor of 3, and was obtained for the first time due to high C-C bond forming ability of the present reaction. The reaction proceeded with a higher concentration (10-50 mM) compared with ring closing metathesis. Stereoselective dehydration of the aldol adduct using $Ti(OR)_4$ afforded *E*-3-methylcyclopentadecenone ($E : Z = 91 : 9$), a key precursor of the Noyori asymmetric hydrogenation using Ru-BINAP, which was developed by the Takasago group (ca. 99% high enantioselectivity).[5] Thus, a formal synthesis of *R*-muscone (3; estimated, 80% ee) was achieved utilizing the intramolecular powerful the aldol addition.

4. Design and facile synthesis of lactone analogs of dihydrojasmone (4) and *cis*-jasmone (5) utilizing Ti-aldol type addition of esters[6]

We developed the present protocol for the aldol-type addition using simple phenyl esters and its application to a design and short step synthesis of the lactone isoster analogs of dihydrojasmone (4) and of *cis*-jasmone (5), which is a representative perfume with jasmine odor. Givordan group reported the synthesis of lactone analog (**4**) using the Reformatsky reaction, which cannot be, however, applied to the synthesis of *cis*-jasmone analog (**5**) bearing Z-double bond.[7] A facile short-step synthesis of **5** was performed utilizing the present Ti-aldol type addition of esters. The analogs were found to possess unique odor for fragrance.

cis-Jasmone   Jasmone lactone analog (**5**)   Dihydrojasmone lactone analog (**4**)

2.5. Efficient one step, general synthesis of 2(5*H*)-furanone (6) and its application to the synthesis of *R*-Mintlactone (7) and *R*-Menthofuran (8)[8]

2(5*H*)-Furanones (α,β-butenolides) comprise an important heterocycle incorporated in natural products and serve as useful synthetic building blocks for lactones and furans. There are many methods available for their synthesis, however, the majority of these are limited to the production of mono or dialkylsubstituted 2(5*H*)-furanones. Previously, we reported the synthesis of trialkylsubstituted 2(5*H*)-furanones using enol silyl ethers and α,α-dimethoxyketones, however, three tedious steps were required.[9]

TiCl$_4$–Bu$_3$N-mediated condensation of ketones with α,α-dimethoxyketones afforded trialkylsubstituted 2(5*H*)-furanones in a one-pot manner, wherein aldol addition and furanone formation occurred sequentially. Its application to straightforward synthesis of (*R*)-mintlactone (**7**) and (*R*)-menthofuran (**8**), two representative natural mint perfumes, was demonstrated. Due to their interesting structure and usefulness, these compounds have been challenging synthetic targets. Our present method of (*R*)-mintlactone seems to be the simplest of the reported methods, because of the one-step synthesis (Reported syntheses required 8-10 steps).

In conclusion, we developed the first Lewis acid-promoted Ti- (or Zr-) Claisen condensation and a related direct aldol additions from a recent standpoint of the green chemistry. These methods had a variety of applications to the practical syntheses of useful fine chemicals including perfumes and 1β-methylcarbapenem.[10]

References
1. a) Tanabe, Y. *Bull. Chem. Soc. Jpn.* **1989**, *62*, 1917. b) Yoshida, Y.; Hayashi, R.; Sumihara, H.; Tanabe, Y. *Tetrahedron Lett.* **1997**, *38*, 8727. c) Yoshida, Y.; Matsumoto, N.; Hamasaki, R.; Tanabe, Y. *Tetrahedron Lett.* **1999**, *40*, 4227. d) Tanabe, Y.; Hamasaki, R.; Funakoshi, S. *Chem. Commun.* **2001**, 1674.
2. a) Evans, D. A.; Urpi, F.; Somers, T. C.; Clark, J. S.; Bilodeau, M. T. *J. Am. Chem. Soc.* **1990**, *113*, 8215. b) Tanabe, Y.; Matsumoto, N.; Higashi, T.; Misaki, T.; Itoh, T.; Yamamoto, M.; Mitarai, K.; Nishii, Y. *Tetrahedron (Symposium)* **2002**, *58*, 8269.
3. Tanabe, Y.; Makita, A.; Funakoshi, S.; Hamasaki, R.; Kawakusu, T. *Adv. Synth. Catal.* **2002**, *344*, 507.
4. Hamasaki, R.; Funakoshi, S.; Misaki, T.; Tanabe, Y. *Tetrahedron* **2000**, *56*, 7423.
5. Ogura, S.; Yamamoto, T. JP 94192161; *Chem. Abstr.* **1995**, *122*, 132836.
6. a) Tanabe, Y.; Matsumoto, N.; Funakoshi, S.; Manta, N. *Synlett*, **2001**, 1959. b) Tanabe, Y. *Japan Pat. Appl.* **2000**, 384994.
7. Mueller, P. M.; Wild, H. J. *Eur. Pat. Appl.* **1992**, 479222.
8. Tanabe, Y.; Mitarai, K.; Higashi, T.; Misaki, T.; Nishii, Y. *Chem. Commun.* **2002**, 2542.
9. Tanabe Y.; Ohno, N. *J. Org. Chem.* **1988**, *53*, 1560.
10. Tanabe, Y.; Manta, N.; Nagase, R.; Misaki, T.; Nishii, Y.; Sunagawa, M.; Sasaki, A. *Adv. Synth. Catal.* **2003**, *345*, 967.

Chapter 26

Oxadiazole Derivatives as Novel Insect-Growth Regulators: Synthesis and Structure–Bioactivity Relationship

Xuhong Qian[1], Song Cao[2], Zhong Li[2], Gonghua Song[2], and Qingchun Huang[2]

[1]State Key Laboratory of Fine Chemistry, Dalian University of Technology, Dalian 116012, China
[2]East China University of Science and Technology, Shanghai 200237, China

Eco-friendly insect-growth regulators, 2,4-dichloro-5-fluorophenyl-oxadiazoles, phenoxymethyl oxadiazoles and oxadiazolyl pyridazinones were designed and synthesized. Some efficient and novel synthetic routes were provided for their intermediates. The target compounds showed effective chronic growth inhibition activities against *Pseudaletia separata* Walker and the antifeedent activities against Asiatic corn borer. QSAR study suggested that the lowest unoccupied orbital energies of these compounds were very important for bioactivities.

Recently, heterocyclic compounds have taken more important role in pesticide chemistry. The number of pesticides derived from five- or six-membered aromatic heterocycles has increased dramatically. Therefore, the use of aromatic heterocycles to design insecticides and insect-growth regulators is the focus of our research.

Eco-friendly insect-growth regulators are a great breakthrough in the research and development of low use rate, highly effective, generally selective insecticides. The typical heterocycles used for insect-growth regulators include pyridine, oxazole, pyridazinone, furan, diazole, thiazole, thiadiazole, pyrimidinone, trazine and tetraazines, etc. However, few are reported using oxadiazoles as insect-growth regulators or insecticide, except closely related thiadiazoles LY-13125.

Benzoylphenylureas as the first generation IGR are inhibitor of insect's chitin synthesis. Dibenzoylhydrazines, as the second generation IGR, are mimics of ecdysone to control insect's molting. The reason for which we concentrate on the oxadiazoles is that there seems to be some correlations in chemical structure among oxadiazoles, benzoylphenylureas, and dibenzoylhydrazines.

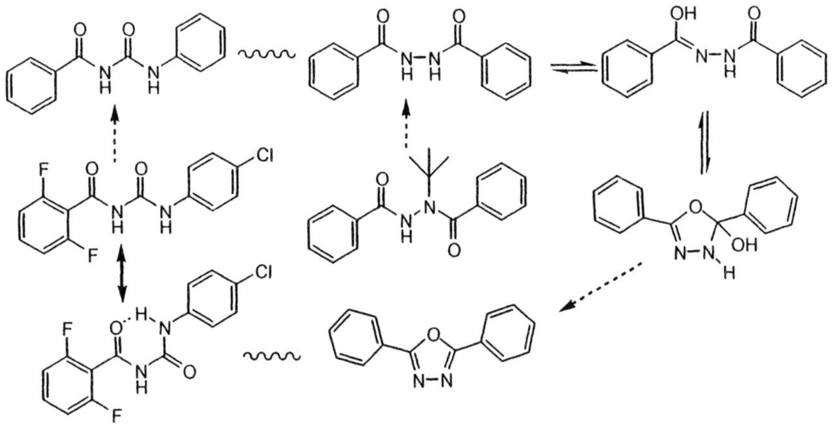

Scheme 1. The evolution of the core structures for benzoylphenylurea, dibenzoylhydrazine and oxadiazole.

2, 5-Bis(2, 4-dichlorophenyl)- 1, 3, 4-oxadiazole (**DCPO**) and analogs are broad-spectrum and effective insecticides or acaricides with potential agricultural uses toward houseflies, faceflies and hornflies*(1, 2)*. In contrast to traditional pesticides, **DCPO** and analogs mainly control the growth and development process of insects by interfering with chitin biosynthesis. Such a process is very similar to the characters of benzoylphenylureas as the inhibitors of chitin synthesis. Their mode of action can be explained as interference with incorporation of [^{14}C] labeled N-acetyl glucosamine active in chitin synthesis*(1, 2)*. In addition, **DCPO** and its analogs also inhibit the synthesis of DNA and protein of insects. However, their limited solubilities in polar solvents make them commercially unattractive, and they tend to have a non-optimal use rate and formulation difficulties. With this in mind, our attention is drawn to changing the physical properties associated with oxadiazoles, while attempting to retain or increase their biological efficacies and promote their solubilities through the modification of substituted oxadiazoles.

1. Novel Oxadiazoles:

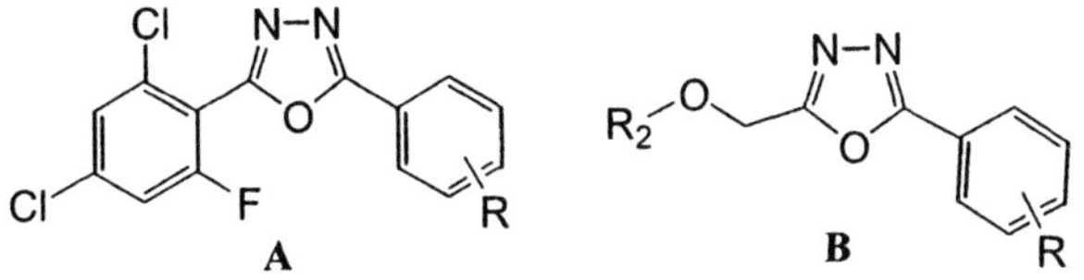

Scheme 2. The target structures A and B for oxadiazoles

It seems that 2-(2, 4-dichlorophenyl)-1, 3, 4-oxadiazole moiety is an essential part for biological activity. However, the planarities and rigidities of 2-(2, 4-dichlorophenyl)- 5-aryl-1, 3, 4-oxadiazoles, might be a main cause for their low solubilities in polar solvents. Therefore, we adopt a strategy by increasing polarity and potential ability for non-bonded interactions in the molecular design. Because electron-negative fluorine atoms have strong hydrogen-bonding capability which has potential to promote bioactivity, the modification of 2,4-dichlorophenyl-1, 3, 4-oxadiazole by introducing fluorine atoms has become our choice to give structure **A**. Previously, we had found that 2-(2,4-dichlorophenyl)-5-*butyl*-1,3,4-oxadiazole also has promising bioactivity*(1)*, implying that one of 2- or 5-position on this oxadiazole ring can be substituted by a flexible alkyl group. Therefore, the introduction of aroxylmethyl group to 2- or 5-position to give structure **B** is an alternative way to improve their bioactivities and polar solubilities. In fact, the ether linkage is frequently used in the structures of some juvenile mimics as insect-growth regulators and pyrethroids as insecticides. Based on the above findings, we design and synthesize two novel types of oxadiazoles **A** and **B***(1-6)*.

The symmetrical **2F-DCPO**, synthesized from the condensation between 4-fluoro-phenoxyacetic acid hydrazide and 2, 4-dichloro-5-fluoro-benzonic acid, is of particular interest, although the expected final product is asymmetrical 2-(2, 4-dichloro-5-fluoro-phenyl)-5-(4-fluoro-phenoxymethyl)-1, 3, 4-oxadiazole. It is believed that an exchange reaction occurs between the acid and the diacylhydrazine, in this case it might be more difficult to transform the asymmetrical hydrazine into the corresponding oxadiazole compared to transform symmetrical hydrazine into **2F-DCPO** (Scheme 3)*(4)*.

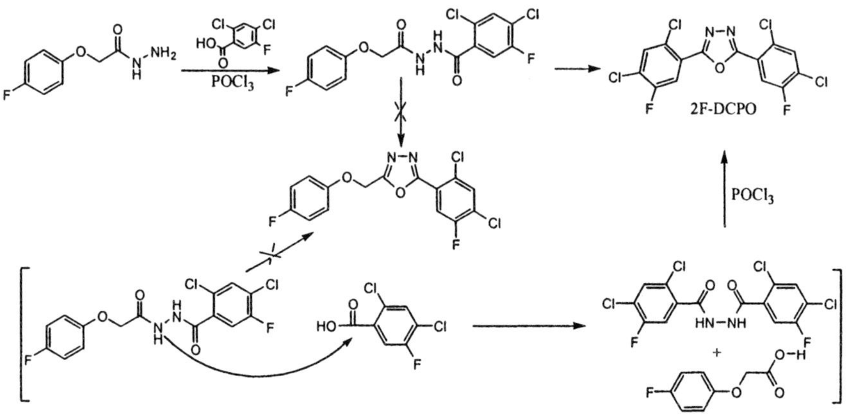

*Scheme 3. Unusual synthesis of symmetrical **2F-DCPO** from an asymmetrical dibenzoylhydrazine*

2-(2, 4-dichloro-5-fluorophenyl)- 5-disubstituted-1, 3, 4-oxadiazoles (**A**), show an enhanced biological activities against *Pseudaletia separata* Walker and favorable polar solubilities. The presence of the 5-fluoro group is important to increase bioactivity. It is observed that the introduction of one or two fluoro groups to **DCPO** would enhance the bioactivity while more than two fluoro groups would decrease the bioactivity.

The data also show that when logP increases bioactivity increases: **2F-DCPO** (LC_{50}, 1.77 ppm; logP: 4.76); **F-DCPO** (LC_{50}, 5.22 ppm; logP: 4.62); **DCPO** (LC_{50}, 10.13 ppm; logP: 4.58). When log P is less than 3.5, only lower or no bioactivity is observed.

The presence of *o*-chloro and -bromo groups contributes to higher activity, while that of *o*-fluoro group decreases the activity.

The activities of **2F-DCPO** against other insects are also determined: *Nilaparvata lugens* Stal in 3rd instar (LC_{50}=25.16ppm), *Chilo suppressalis* Walker (250ppm, 72h, lethal rate 60%), *Lipaphis erysimi* Kaltenbach (125ppm, 8d, lethal rate 60%), *Plutella xylostella* Linnaeus and *Prodenia litura* Fabricius 2nd instar (125ppm, lethal rate 46%).

2-(Substituted phenyl)-5-(substituted phenoxymethyl)-1, 3, 4-oxadiazole (**B**) also shows insect-growth regulator's activity against 2nd instar larvae of *Pseudaletia separata* Walker.

Based on Cerius2 program, the QSAR study for 9 compounds with LC_{50} values suggests that there is a good correlation among their bioactivities against *Pseudaletia separata* Walker, hydrophilicities and the lowest unoccupied orbital enrergies.

$$LC_{50} = 1799.44 - 309.865 \log P - 344.139 E_{LUMO} \quad n=9 \quad r^2 = 0.9749$$

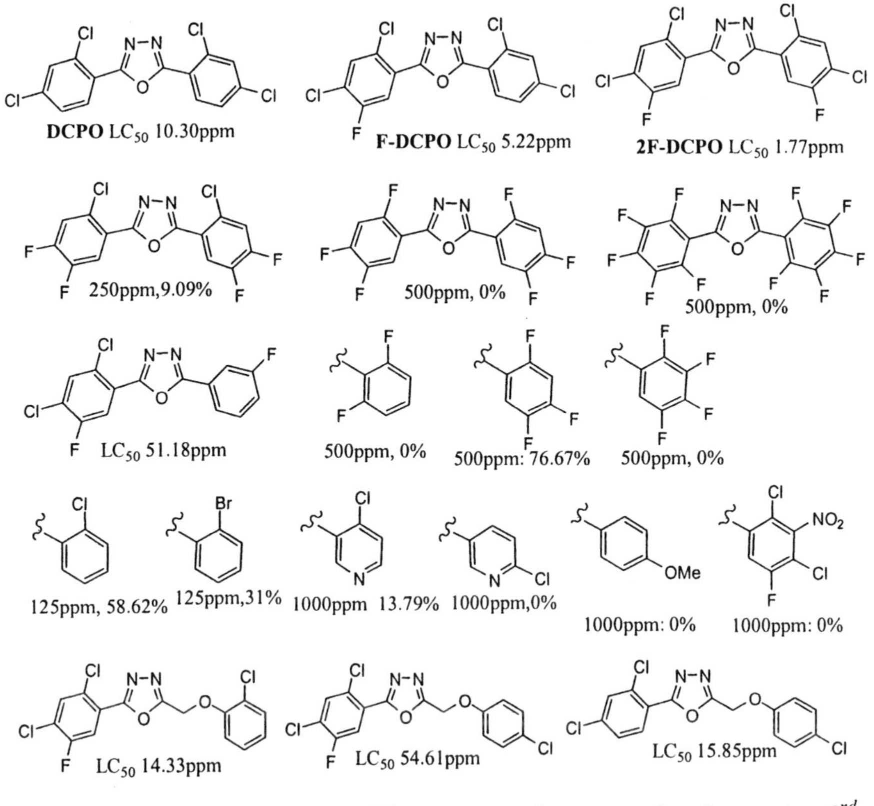

DCPO LC_{50} 10.30ppm

F-DCPO LC_{50} 5.22ppm

2F-DCPO LC_{50} 1.77ppm

250ppm, 9.09%

500ppm, 0%

500ppm, 0%

LC_{50} 51.18ppm

500ppm, 0%

500ppm: 76.67%

500ppm, 0%

125ppm, 58.62%

125ppm, 31%

1000ppm 13.79%

1000ppm, 0%

1000ppm: 0%

1000ppm: 0%

LC_{50} 14.33ppm

LC_{50} 54.61ppm

LC_{50} 15.85ppm

Scheme 4. The structures and bioactivities of some oxadiazoles against 2nd instar larvae of P. separata

2. Oxadiazolyl Pyridazinones and Analogs:

The above results also suggest that 2-aryl-5-aroxymethyl-1, 3, 4-oxadiazole has very good bioactivity*(4-6)* worth further investigating. We think that the introduction of a heterocycle containing two heteroatoms, such as pyridazinone moiety, might increase their bioactivities and change their insecticidal spectra. It has been previously well-known that 2-*tert*-butyl-4-chloro-5-(arylmethoxyl)-3(*2H*)-pyridazinone derivatives, the juvenoids such as dihydropyridazinones, norflurazon, pyridaben, NC-170, NC-184 and NC-196, possess considerable bioactivities to nematodes, acarid, fungi and insect*(7)*. Therefore, asymmetrical oxadiazoles containing pyridazinone group are synthesized and evaluated for their toxic and antifeedant activities against insects of different orders, especially diptera, lepidoptera and homoptera*(8, 9)*.

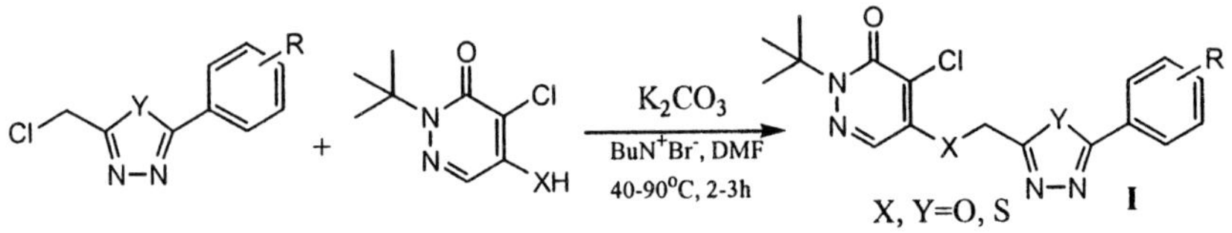

Scheme 5. The synthetic route for oxadiazolyl pyridazinones

Besides compounds **I**, some other similar oxadiazoles are designed and prepared for comparison, namely (**II**) containing pyridazinone moiety with "-SCH$_2$-", (**III**) containing pyridyl with "-OCH$_2$-" and (**IV**) containing guanidinium with "-CH$_2$-" as molecular linkage.

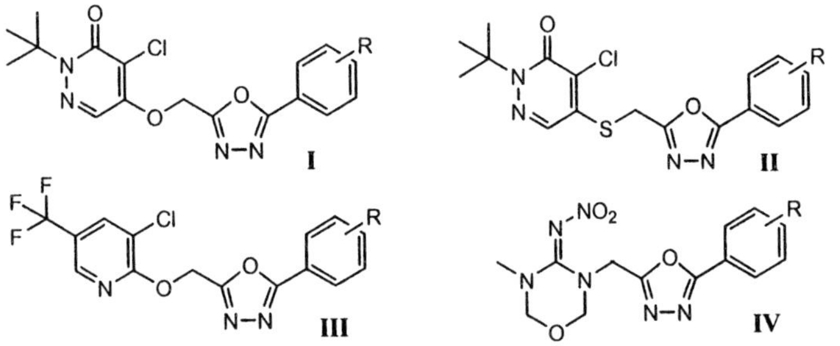

Scheme 6. The target structures of oxadiazolyl pyridazinones

During the synthesis of oxadiazolyl pyridazinones, by treating 2-*tert*-butyl-4-chloro-5-(ethoxycarbonylmethoxy)-3(2H)-pyridazinone (**C$_1$**) with hydrazine hydrate in ethanol at reflux for 6-8 hours (Scheme 7), we obtain an unexpected product **E$_1$** in good yield, instead of the anticipated product of the corresponding acetylhydrazine. Alazawe had previously reported that methoxyl group at 6-position of the pyridazinone could be replaced by the hydrazino group. For this reason, we assume that the key step of this reaction is the formation of intermediate **D**. In fact, we also found that 2-(un)substituted-4, 5-

dichloropyridazinones (**C₂**) can also give 2-(un)substituted-4-amino-3-pyridazinones (**E₂**) under the similar conditions*(10)*.

The known methods lead to amino pyridazinones, which are useful intermediates for making agricultural and pharmaceutical chemicals, involving Raney-Ni cleavage of the hydrazino pyridazinone*(11)*, direct amination of the pyridazinones*(12-14)*, substitution of chloropyridazinone with ammonia at enhanced pressure*(15)*, and the dechlorination of chloropyridazinone performed in the presence of palladium on charcoal*(14)*. Therefore, we believe that a novel and convenient synthetic route is provided for the synthesis of aminopyridazinones, which are not easily accessible by traditional methods.

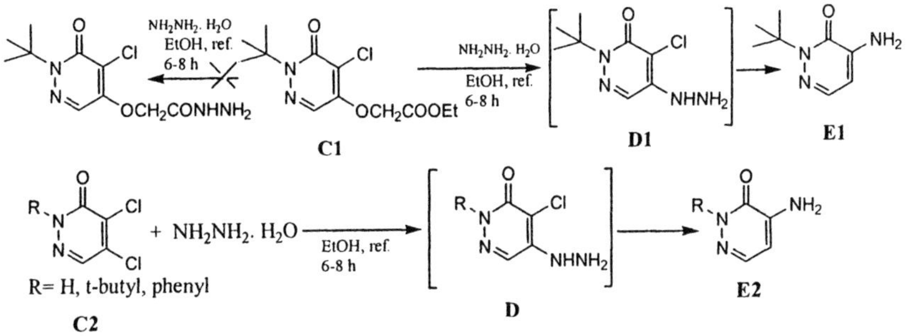

Scheme 7. A novel synthetic route for aminopyridazinones

All final compounds are evaluated for their chronic growth inhibition activities against second-instar armyworm larvae of *Pseudaletia separata* (Lepidoptera: Noctuidae). **Type I** compounds possess considerable inhibition activity to its weight gain, especially **I-2** and **I-3** with the mean effective concentration (EC_{50}) of 14 μM and 22 μM, respectively. The electron-donating groups, e.g. $-CH_3$, $-C_2H_5$, and $-OCH_3$, decrease the bioactivities of **I-7**, **I-8~I-11**.

For type **II**, their data of EC_{50} (>1mM) and EC_{90} (>10mM) show that they are not satisfactory agents for inhibiting the weight gain of the armyworm larvae. Type **III** also show some inhibitory activity, although they are inferior to that of type **I**. However, **IV**, of which the guanidinium moiety is connected with oxadiazole by a "-CH_2-" linkage, is almost ineffective agent against the armyworm larvae even used with high concentration up to 1000 μM.

The QSAR analysis shows that their energies E_{LUMO} of the lowest molecular orbit are important to improve their bioactivities. The multiple regression analysis clearly reveals that the inhibitory activity is positively correlated with E_{LUMO}, but negatively with E_{HOMO}.

$$pEC_{50} = 7.7104\ (\pm 0.0605) + 0.5370\ (\pm 0.0700)\ E_{LUMO}$$

$n=11$, $R^2 = 0.8672$, $SD = 0.1029$, $F = 58.7918$

$pEC_{50} = -11.5831(\pm 12.4628) +0.3922(\pm 0.1140)E_{LUMO} -2.0733(\pm 1.3392) E_{HOMO}$

$n=11$, $R^2 = 0.8978$, $SD = 0.0958$, $F = 35.1558$

Moreover, their mode of action is confirmed as novel insect-growth regulators with ecdysonergic activity according to the symptoms of larval poisoning. The experiments also indicate that **I** does not result in the larvicidal activity, but mainly the inhibition of larval weight gain

Table 1 Chronic growth inhibitory activity of oxadiazolyl pyridazinones and analogs on the weight gain of 2nd instars larvae of *P. separata*.

Compds	R	EC_{50}(mM)	Compds	R	EC_{50}(mM)
I-1	H	0.065	II-3	p-F	2.057
I-2	p-Cl	0.014	II-4	m-F, o,p-2Cl	2.837
I-3	p-F	0.022	II-5	p-C_2H_5	4.313
I-4	m-F	0.029	III-1	H	2.236
I-5	o-F	0.061	III-2	p-Cl	0.478
I-6	m-F, o,p-2Cl	0.039	III-3	p-F	0.581
I-7	p-C_2H_5	0.059	III-4	m-F, o,p-2Cl	0.729
I-8	m-CH_3	0.574	III-5	p-C_2H_5	1.396
I-9	p-CH_3	0.076	IV-1	H	>10
I-10	m,m-$2CH_3$	0.785	IV-2	p-Cl	>10
I-11	p-CH_3O	0.085	IV-3	p-F	>10
I-12	p-NO_2	1.053	IV-4	m-F, o,p-2Cl	>10
II-1	H	2.238	IV-5	p-C_2H_5	>10
II-2	p-Cl	1.066	DCPO	o,p-2Cl	0.029

I, **II** and **III** are consist of planar oxadiazole moiety and planar pyridazinone or pyridyl moiety through a molecular linkage "-OCH_2-" or "-SCH_2-", except unplanar guanidinium moiety (**IV**, no bioactivity). The positive torsion angle between oxadiazole and pyridazinone moieties of type **II** might be the main reason for causing its bioactivity much lower than that of type **I** which has negative torsion angle because of its contrary configuration.

I also shows antifeedent activities against Asiatic corn borer. For the rest of compounds, the deterrency indexes are not determined as their bioactivities are weak at concentrations as high as 500 mg/kg. **I-1** and **I-9** are the most and least active compounds, respectively, and **I-1** is almost as active as **Azadirachtin** at 500 mg/kg. **I-1**, **I-4** and **Azadirachtin** (> 10 mg/kg) significantly reduce the weight gain in choice-diet bioassays. The substituent effects are not generalized except for the following aspects: halogens are more favorable for bioactivity than electron-donating alkyl and alkoxyl groups. For example, a *m*-methyl (**I-8**) or 3,5-dimethyl (**I-10**) group shows reduced bioactivities at 500 mg/kg, whereas **I-4** (m-F) is more active; Bulky electron-withdrawing groups, such as NO_2 displays unfavorable bioactivity. The replacement of the oxo by mercapto bridge (**II**), also leads to a loss of bioactivity, which suggests that the oxo bridge is critical.

Conclusion

2-(2,4-Dichloro-5-fluorophenyl)-oxadiazoles and phenoxymethyl derivatives show enhanced biological activity against *Pseudaletia separata* Walker and favorable polar solubility. Oxadiazolyl pyridazinones show effective chronic growth inhibition activities against the second-instars armyworm larvae of *Pseudaletia separata* (Lepidoptera: Noctuidae) and the antifeedent activities against Asiatic corn borer.

Acknowledgements

National Natural Science Foundation of China, the National Key Project for Basic Research (2003CB114400) and National 863 High-tech Project partially support this research.

References

1. Idoux, J. P.; Gibbs-Rein; Gupton, J. T. *J. Chem. Eng. Data*, **1988**, 33, 385.
2. Qian, X.; Zhang, R. *J. Chem. Tech. And Biotech.* **1996**, 67, 124.

3. Zhang, R.; Qian, X. *J. Fluorine Chem.*, **1998**, 93, 39.
4. Shi, W.; Qian, X.; Song, G.; Zhang, R.; Li, R. *J. Fluorine Chem.* **2000**, 106, 173.
5. Shi, W.; Qian, X.; Zhang, R.; Song, G. *J. Agri. Food Chem.* **2001**, 49, 124,.
6. Cao, S. ; Qian, X.; Song, G.; Huang, Q. *J. Fluorine Chem.* **2002**, 117, 63.
7. Miyake, T.; Haruyama, H.; Ogura, T.; Mitsui, T.; Sakurai, A. *J. Pestic. Sci. (Int. Ed.)*, **1992**, 17, 75.
8. Cao, S.; Qian, X.; Song, G.; Cai, B.; Jiang, Z. *J. Agri. Food Chem.*, **2003**, 51, 152.
9. Huang, Q.; Qian, X.; Song, G.; Cao, S. *Pest. Management Sci.* **2003**, 59, 933-939,.
10. Cao, S.; Qian, X.; Song, G. etc., *Chemitry Letters*, **2001**, 54.
11. Osner, W. M. ; Castle, R. N. *J. Pharm. Sci.*, **1963**, 52, 539.
12. Singh, B. *Heterocycles*, **1984**, 22, 1801.
13. Coates, W. J.; McKillop, A. *Heterocycles*, **1989**, 29, 1077.
14. Coates, W. J.; McKillop, A. *Heterocycles*, **1993**, 35, 1313.
15. Konecny, V.; Kovac, S.; Varkonda, S. *Collect. Czech. Chem. Commun.*, **1985**, 50, 493.

Chapter 27

Natural Products as Green Pesticides

Denise C. Manker

AgraQuest, Inc., 1530 Drew Avenue, Davis, CA 95616

Our ability to provide adequate pest control to increase crop yields and reduce land requirements cannot keep pace with the growing demand of the world's population using conventional agriculture. The costs associated with the rapid development of modern pesticides include environmental contamination, unpredicted human health consequences and deleterious effects on wildlife and other non-targeted organisms in the food chain. This has driven the creation of the new field of green chemistry. This field includes the design of new processes and methods that reduce the use and generation of hazardous substances. Natural products are an excellent alternative to synthetic pesticides as a means to reduce negative impacts to human health and the environment. The move toward green chemistry processes and the continuing need for developing new crop protection tools with novel modes of action makes discovery and commercialization of natural products as green pesticides an attractive and profitable pursuit that is commanding attention.

With increasing world population comes the need to exact the highest yields possible in agriculture. This has led to higher chemical inputs to control insect pests, plant diseases and to enrich overworked soil. Consequently, this has placed an undue burden on the planet's ecosystems. The balance between the need to provide sustenance and the need to leave adequate resources for future generations has led to the sustainable agriculture movement. The realization that sustainable industrial processes must be compatible with environmental concerns has developed into the philosophy of Green Chemistry. These two movements share many goals and ideals. Natural product biopesticides represent a natural overlap between sustainable agriculture and green chemistry. The importance of the contribution of natural products to sustainable agriculture has been recognized several times by the US EPA's annual Green Chemistry Challenge Award (*1*). Over the years, a number of reviews have discussed the status of biopesticides in the pesticide market (*2,3,4,5*). An evaluation of how biopesticides can be applied to green chemistry goals is discussed in this overview.

There are a number of areas in which natural product biopesticides can meet the goals of green chemistry. An increased measure of safety can be addressed when lowering the potential of fire or explosion during the manufacture of crop protection products. In addition, addressing human health concerns by lessening dependence on carcinogens or endocrine disrupters is an important goal of green chemistry. Environmental impact is a concern that is prevalent at various levels of industry and government. Achieving sustainable processes that will allow production of pesticides without compromising the ability of future generations to provide for themselves runs parallel to reducing the waste created in the process. Lastly, reducing the use of hazardous substances benefits the established balances between management, labor and regulatory authorities.

Safety

There have been numerous unfortunate events that have lead to an increased awareness of the importance of safety when producing agricultural crop protection agents. The worst industrial accident in history occurred in Bhopal, India in 1984 at a plant producing the carbamate pesticide, carbaryl. Volatile methyl isocyanate gas that escaped from the plant in one night resulted in the immediate death of over 2,500 people (*6*). Current estimates of mortality resulting from that accident are between 15,000 and 20,000 (*7*). Many synthetic processes in chemical production can result in fire, explosions or other unforeseen events such as those recorded in Bhopal.

Natural products offer many advantages over synthetic processes in the area of safety. Fermentation derived biopesticides largely involve water-based methods which do not require high pressures or temperatures with attendant reduction in risk and to human health from fire. Manufacturing using fermentation is relatively low risk.

Human Health

There is an immediate need in agriculture for safer alternatives to synthetic pesticides. Many existing pesticides face tolerance reassessment under EPA's Food Quality Protection Act (FQPA). Some pesticides cause immediate reactions such as the contact dermatitis reported in banana workers in Panama (8). The most frequent reactions were to chlorothalonil, thiabendazole, imazalil and aluminum hydroxide. However, there are also concerns about the long-term effects of pesticides in the more susceptible segments of the population, such as children. In a study evaluating the presence of the acutely toxic organophosphate pesticides in the urine of children ages 2-5, it was determined that those fed a largely organic diet of juices, fresh fruits and vegetables had significantly lower levels of OP's in their urine compared to children fed conventional produce (9). The median total OP concentration was six times higher for children fed conventional diets than for children with organic diets.

The recognition that past generations of agricultural chemicals have had serious drawbacks in the arena of human health has resulted in an opportunity for natural products as green pesticides. While not all natural products are non-toxic, there is a greater selectivity that may apply to certain target organisms such as insects or plant pathogens. Target specificity and high specific activity are two advantages that natural product possess over the more general biocides that can be produced synthetically.

Environmental impacts

With the publication of Rachel Carson's Silent Spring in 1962 (10), the general public became very aware of the unintended consequences of pesticides distributed by the chemical industry. DDT was used widely as an insecticide to control disease vectors such as mosquitoes. The devastation to food chain caused by the egg thinning properties of chlorinated hydrocarbons raised the public awareness of the hazards of chemical pesticides. The discovery of pesticides in run off water and in drinking water has lead to a demand for safer crop protection agents. The long-lived nature of many of the highly halogenated synthetic compounds offers longer crop protection but results in persistent environmental impact. Several fungicides that are frequently used in current agricultural applications also have deleterious effects on non-target organisms. Chlorothalonil is very toxic to fish, shrimp, frogs, beneficial microbes and earthworms. In addition, the major break down product of chlorothalonil is thirty times more acutely toxic that the fungicide itself. The fumigant methyl bromide has additional deleterious effects on the environment as one of the most effective ozone-depleting compounds known. Synthetic pesticides can also enter the atmosphere and be transported to areas away from agricultural land.

Hamers *et al.* (*11*) report that rainwater samples collected in the Netherlands contained 2000 times more than the maximum permissible concentration of dichlorvos set for surface water.

While not all natural products are non-toxic, all are highly biodegradable. This offers great advantages in lessening the environmental impacts in crop protection. In sustainable agriculture, it is very advantageous to maintain healthy populations of predatory insects such as ladybirds, laceswings and predatory wasps. Many synthetic agrochemicals have unintended consequences for this population leading to a greater need to use crop protection agents. Biopesticides that are very target-specific maintain natural predator populations.

Reduction of waste

Production of waste in industrial chemical processes has been a growing problem since the beginning of the revolution of synthetic agrochemicals. A number of EPA superfund sites that are mandated to remediate heavily contaminated soils have been created as the result of agricultural activities. The clean up of these sites has required large amounts of money and time. Fermentation derived products have an advantage in that the by-products are largely biodegradable. In some cases, such as in the production of Bt, the by-products can actually be used as fertilizers. Since biopesticide production does not generally require the use of organic solvents or other substances hazardous to the environment, material by-products can be released with reduced concern for impact at the downstream sites. This saves cost to production and ameliorates certain hazardous waste problems.

Hazardous substances

Reducing the use of hazardous substances has positive impacts on the quality of human life and the environment. Toxic intermediates required for some synthetic processes require special containment facilities and result in toxic by-products. Some processes require expensive facilities able to handle highly dangerous reagents. These problems are absent in the fermentation processes used in production of biopesticides.

Driving Forces Behind the Development of Biopesticides

There has been a world wide, downward trend in the sales of traditional chemical pesticides. In 1995, the global pesticide market was estimated to be $30 billion while the most recent estimate is closer to $28 billion. In the face of this shrinking global market, the biopesticide market has enjoyed growth of greater than 20% compounded annual growth over the last decade. This figure may not be so surprising when one looks at the rate of growth in the organic market. The consumer demand for safer foods has resulted in a US market of over $12 billion with more than 20% compounded annual growth. Positive public perception of organic foods has provided an excellent opportunity for safer crop protection products such as natural product biopesticides. While biopesticides still only comprise 1% of the global pesticide market, the trends suggest that their market share will continue to grow.

Another factor favoring the development of biopesticides is the cost and regulatory hurdles encountered for new products. It is estimated that a new synthetic chemical requires over $185 million and 10 years development time before it reaches the market (*12*). The cost for developing a biopesticide is dramatically lower than that of a synthetic chemical and is estimated at $6 million. The creation of the Biopesticide and Pollution Prevention Department within the EPA has streamlined the process for registration of biopesticides that have low toxicological effects on non-target organisms. This has resulted in a much shorter time for development, estimated at 3 years. In the current economic climate, companies have a great need to get products to market quickly and efficiently. Natural product biopesticides are one way to achieve this goal.

The rapid development of resistance to traditional pesticides is another excellent reason to provide biopesticides as one of the tools available in an integrated pest management system. Many of the new lower risk synthetic pesticides have a single site mode of action. In some cases, resistance has been observed within a few seasons. In 1988, Georghiou and Lagunes reported that over 500 species of insects were known to be resistant to one or more insecticides (*13*). Rotation with products such as biopesticides that have unique or multiple modes of action will extend the longevity of currently available pesticides. From these examples it is clear that there are distinct advantages to pursuing the development of natural products as green pesticides. Several successful examples of natural product pesticides currently on the market that meet the criteria of green chemistries are given below.

Examples of Natural Products as Green Pesticides

Bacillus thuringiensis

Biopesticides are largely considered to be newcomers to the agricultural chemical industry. However, *Bacillus thuringiensis* (Bt) has a much longer history than many synthetic products (*14*). In 1901, the cause of a disease responsible for the collapse of the silk worm industry was identified by Japanese biologist Shigetane Ishiwatari. The bacterium he isolated was as the causal agent was Bt. Ten years later, German scientist Ernst Berliner found the same bacterium in a dead Mediterranean flour moth and he named it after the town where the moth originated, Thuringia. He also observed a crystal present within spores of Bt but did not link it to the activity. The first commercial preparation of Bt for use in agriculture was in France in 1938 where a spore based product was distributed as Sporine for use against flour moths. The identification of the crystal's role in the insecticidal activity was not determined until 1956 and this led to a renewed interest in research on Bt. Bt was registered by the US EPA in 1961 and it has been used for more than four decades. The largest segment of the biopesticide market is made up by Bt based products. Its use has increased in the face of deregulation of other existing products with unfavorable toxicological profiles and increasing development of insect resistance. In 1997, Bt was ranked fourth by acreage of fruit crops treated by all insecticides (2). While many newly developed synthetic products show resistance in less than five years, the complexity of the action of Bt has lead to the longevity of its use.

Bt is an excellent example of a natural product that can serve as a green pesticide. This biopesticide is manufactured through liquid fermentation, thus obviating the need for processes that are prone to conflagration or explosion. The specific activity of the various subspecies of Bt results in overall higher safety to non-target organisms, including humans. The protoxin of Bt becomes activated only in the highly alkaline gut of the insects and is unaffected by the acidic mammalian gut. There is further specificity within the subspecies of Bt. The bipyramidal crystal endotoxin present in Bt *kurstaki* is active specifically on lepidopteran species while Bt *israelensis* produces an endotoxin active on mosquitoes. Bt *tenebrionis* produces a flat rhomboid crystal with activity specific to Coleopteran beetles. This results in a suite of insecticides that are environmentally safe and do not harm beneficial predatory insects present in agricultural fields. Further, the fermentation process does not require the use of organic solvents or other hazardous materials, thus the waste stream is non-toxic and can be easily discarded and even used as fertilizer in some instances. The application of this technology in developing countries can help address the mounting worldwide problem of accumulated or abandoned, obsolete pesticides in developing countries. The FAO estimates that at least 100,000 tons of obsolete pesticides are stockpiled including large stocks of organochlorine compounds such as DDT, dieldrin and organophosphates including parathion and dichlorvos (15). The relatively simple Bt fermentation process with the ability to substitute simple biomass for chemical feedstock has been carried out with success in a number of countries including Mexico, South Korea, Nigeria, Brazil and India (16). The use of Bt to reduce reliance on more toxic alternatives will save lives and help protect the environment in the developing world where much of our remaining natural resources exist.

Spinosad

A mixture of two naturally occurring compounds produced from the fermentation of *Saccharopolyspora spinosa* is the basis of the insecticide spinosad. The strain was discovered during a screening program carried out by Eli Lilly assessing fermentations in a miniaturized mosquito larvicide bioassay (*17*). During the research and development phase of the project, Dow Chemical and Eli Lilly merged to form the plant sciences division Dow Elanco. When Dow Chemical assumed full ownership of the joint venture, DowElanco emerged as Dow AgroSciences and maintained the work on development of Spinosad. The activity of this new species of actinomycete was traced to a family of novel macrocyclic lactones called spinosyns (Figure 1). These unique structures also exhibited high potency against lepidopteran pests such as tobacco budworm, *Heliothis virescens*. Spinosyn A has been shown to be more active than pyrethroids, one of the most active natural insecticides, and exceeds the activity of some of the organophosphate compounds such as carbamate and cyclodiene (*18*). A major research effort was undertaken by Lilly Research Laboratories and then by Dow AgroSciences to determine if more potent analogs could be produced. Although hundreds of semi-synthetic derivatives of the spinosyns were made during this effort, the original natural products found from the isolated strain, spinosyn A and D, remain some of the most active compounds known in this family.

The activity of spinosyns A and D appears to be a novel neurotoxic effect causing hyperexcitation followed by disruption of the insect central nervous system. The disruption of the insect nicotinic and gamma-aminobutyric acid receptor mechanisms has been shown to effect new binding sites resulting in an additive resistance management benefit for this natural product (*19*). Extensive analysis of the environmental fate of the spinosyns has shown that they rapidly degrade by photolysis in water (half life less than 1 day) and soil (half life of 9-10 days). The leaching potential of Spinosad is very low and it is not considered to pose a threat to groundwater (*17*).

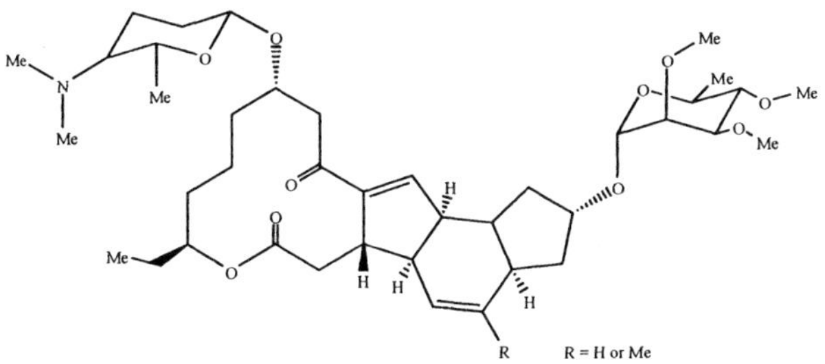

Figure 1. Structure of Spinosyn A (R=H) and Spinosyn D (R=Me).

Spinosad has relatively low toxicity to mammals and birds and has slight to moderate toxicity towards aquatic species. Chronic mammalian toxicology tests have demonstrated that spinosad is not carcinogenic, teratogenic, mutagenic or neurotoxic. The low mammalian toxicology of spinosad means that minimal safety precautions are necessary for agricultural workers with regard to pre-harvest and re-entry intervals. Spinosad's low activity on beneficial insects allows it to be incorporated into Integrated Pest Management programs that rely on natural predators and parasites. This fermentation-derived process was awarded the US EPA Presidential Green Challenge Award in 1999 for achieving many of the goals of safer chemical products and processes (*1*).

Messenger®

Utilizing plants natural defensive response to pathogen attack is the basis for Eden Biosciences unique crop protection product. Messenger® is based on a new class of nontoxic, naturally occurring proteins known as harpins. Harpins were first described in 1992 as elicitors of a hypersensitive response produced by *Erwinia amylovora*, the causal agent of fire blight of pears and apples (*20*). The elicitor was described as an acidic heat-stable cell-envelope associated protein with a molecular weight of 44 kDa. Harpin induces resistance in a variety of plants against a broad array of pathogens. In addition, the activation of the systemic acquired resistance (SAR) pathways has also been linked to increased plant vigor and yield. When applied to crops, harpin increases plant biomass, photosynthesis, nutrient uptake and root development resulting in the observed increase in crop yield and quality.

Investigation of the mechanism of action of the harpin protein has demonstrated that other known elicitor molecules, such as jasmonates and ethylene, operate via alternative resistance signaling pathways (*21*). As is true of most proteins, harpin is rapidly degraded by UV and natural microorganisms and does not bioaccumulate or pose a risk of groundwater contamination. Extensive safety evaluation has shown that Messenger® has virtually no adverse effects on mammals, birds, honeybees, plants, fish, aquatic invertebrates and algae.

Messenger® is produced in a water-based fermentation process that does not use solvents or toxic reagents. The process requires low energy inputs and does not result in hazardous chemical wastes. The by-products of the fermentation are biodegradable and present a minimal waste problem. Carriers used in the formulation of Messenger® are food-grade substances providing an end-use product that is environmentally safe with no non-target toxicity. This unique technology was recognized through receipt of the US EPA Presidential Green Chemistry Award in 2001 (*1*). Messenger® represents one of the first examples of a natural product based on elicitation of plant defenses to help fight off pests and increase yield. It is anticipated that other products based on this concept will emerge in future.

Serenade®

Serenade® Biofungicide is based on a naturally occurring strain of *Bacillus subtilis* QST-713, discovered in a California orchard by AgraQuest scientists. Serenade® is not toxic to beneficial and non-target organisms such as trout, quail, lady beetles, lacewings, parasitic wasps, earthworms and honey bees. Serenade® is exempt from tolerance because there are no synthetic chemical residues, and it is safe to workers and ground water. Serenade's wettable granule formulation is listed with the Organic Materials Review Institute (OMRI) for use in organic agriculture and will continue to be listed under the National Organic Standards which were enacted in the United States in October 2002. The safe toxicology profile of Serenade® also allows for a 4 hour re-entry period and a zero day pre-harvest interval. This provides economic advantages for agriculture in situations where there is heavy disease pressure close to harvest time.

Because of Serenade®'s novel, complex mode of action, environmental friendliness and broad spectrum control, it is well suited for use in integrated pest management (IPM) programs that utilize many tools such as cultural practices, classical biological control and other fungicides. Serenade® can be applied right up until harvest, which provides needed pre and post harvest protection when there is weather conducive to disease development around harvest time.

Serenade® works through a complex mode of action that is manifested both by physiological action of the bacteria and through action of secondary metabolites produced by the bacteria. Serenade® prevents plant diseases by first covering the leaf surface and physically preventing attachment and penetration of the pathogens. In addition, Serenade® produces three groups of lipopeptides (iturins, agrastatins/plipastatins, and surfactins) that act in concert to destroy germ tubes and mycelium. The iturins and plipastatins have been reported to have antifungal properties. Strain QST-713 is the first strain reported to produce iturins, plipastatins and surfactins and two new compounds, the agrastatins, with a novel cyclic peptide moiety. The surfactins have no activity on their own but low levels (25 ppm or less) cause significant inhibition of spores and germ tubes in combination with the iturins or the agrastatin/plipastatin group. In addition, the agrastatins and iturins exhibit synergism towards inhibition of plant pathogen spores.

The molecular structure of each of the three groups of lipopeptides are composed of a cyclic peptide core and a fatty acid side chain. The core consists of a cyclic octapeptide with seven α-amino acids and one β-amino acid with an aliphatic side chain. The molecular weights of the iturin A series range from 1043 to 1085 Da and are dependent upon the number of methylene groups in the aliphatic side chain. The plipastatins are a family of acylated decapeptides that differ in amino acid composition and the nature of the fatty acid side chain. The molecular weight range of these compounds is from 1467 to 1509 Da. In the course of our analysis of secondary metabolites produced by QST713, we found an addition to the plipastatin family where the isoleucine moiety was replaced by a valine (22). Two of these compounds named agrastatins 1 and 2 have been described (Figure 2). We also identified three compounds with the plipastatin amino acid configuration with shorter acyl side chains. The plipastatins have been reported to inhibit phospholipase A2 (23). Plipastatin A1 and B1 have been reported to inhibit the growth of *Alternaria mali*, *Botrytis cinerea* and *Pyricularia oryzae in vitro* and to protect rice seedlings from *Helminthosporium* infections (24).

The third group of lipopeptides identified from QST713 is the surfactins. The structures of the surfactins consist of seven α-amino acids and one β-hydroxy fatty acid ranging in molecular weight from 1008 to 1036 Da.

A detailed study at AgraQuest of the effects of the individual groups of lipopeptides on pathogen spore germination compared to mixtures of the groups provided a better understanding of the effectiveness of this fermentation product. Morphological differences were observed on treatment of spores with different lipopeptide groups. The iturin group resulted in inhibition of spore

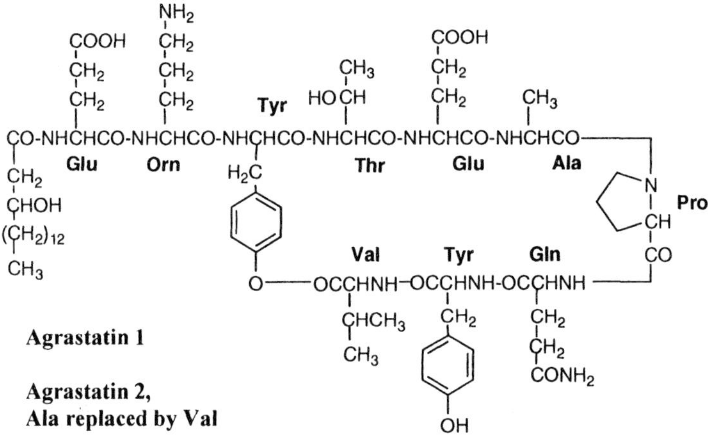

Figure 2. Structures of Agrastatin 1 and 2.

germination where the spores appeared to be normal but had a reduced rate of germination dependent upon the concentration of iturins present. The iturins were most effective on *Botrytis cinerea* spores giving an EC_{50} as low as 15ppm (50% spores were not germinated). The EC_{50} of *Monilinia fructicola* spores occurred at 30 ppm and for *Alternaria brassicicola* the level required was 25ppm. Exposure of the spores to the iturin/plipastatin group resulted in abnormal spore appearance where the spore had a large bubble-like growth replacing the normal appresorium. This effect was most notable with *A. brassicicola* spores where the EC_{50} was 5 ppm.

Investigation of the effects of combining the groups of lipopeptides gave further explanation for the excellent efficacy observed with Serenade®. Addition of as little as 1ppm agrastatin/plipastatin to 10ppm iturin gave a large reduction in spore germination down to approximately 5% for *M. fructicola*. This also resulted in the abnormal spore appearance. The surfactin group was found to have no effect on spores at the highest rate tested (250ppm). Addition of 25ppm surfactin to 20ppm iturin reduced spore germination from 85% to less than 5%. Surfactin at 25ppm added to agrastatin/plipastatin reduced germination from 100% to 10%.

Utilizing a microbial fermentation process allows for production of a fungicide without the use of solvents or non-renewable resources. The feedstock includes agricultural materials such as protein from soybeans, and starch and sugars from plant based sources. This allows for reduction of petroleum-based feedstocks. The products of the fermentation process including the biomass, feedstock which is incorporated into the spray dried product, carbon dioxide and water are non-hazardous. Synthetic production processes of several alternative fungicides are under review from the Clean Air Act by the EPA. Production of an agricultural fungicide without creating solvent-containing waste materials offers a clear environmental advantage over synthetically derived fungicides. Part of the mission of the Office of Industrial Technologies in the U.S. Department of Energy is to deploy energy-efficient, renewable and pollution prevention technologies. Their Renewable Bioproducts Vision includes a goal of using 10% of basic chemical building blocks from plant-derived sources (*25*). Production of Serenade® as an agricultural fungicide is fulfilling this mandate by utilizing plant-derived feedstocks, producing non-hazardous by-products and a highly effective natural product pesticide. Development of Serenade® was recognized through receipt of the US EPA Presidential Green Chemistry Award in 2003.

References

1. Presidential Green Chemistry Challenge Award Recipients. EPA744-K-02-002. US Government Printing Office: Washington D.C., 2002.
2. Marrone, P. *Outlook on Agriculture*. **1999**, *28*, 149.
3. Copping, L. G.; Menn, J. J. *Pest Manag. Sci.* **2000**, *56*, 651.

4. Warrior, P. *Pest Manag. Sci.* **2000**, *56*, 681.
5. *Biologically Active Natural Products*: Agrochemicals; Cutler, H.G.; Cutler, S.J., Eds.; Boca Raton, FL, 1999.
6. Dhara, V.R.; Dhara, R. *Arch. Environ. Health.* **2002**, *57*, 391.
7. Lappierre, D.; Moro, J. *It was Five Past Midnight in Bhopal.* Full Circle Publishing, New Delhi, India, 2001.
8. Penagos, H.G. *Int. J. Occup. Environ. Health.* **2002**, *8*, 14.
9. Curl, C.L.; Fenske, R.A.; Elgethun, K. *Environ. Health Perspect.* **2003**, *111*, 377.
10. Carson, R. *Silent Spring*, Houghton Mifflin, Boston, MA, 1962.
11. Hamers, R.; Smit, M.G.D.; Murk, A.J.; Koeman, J.H. *Chemosphere.* **2002**, *45*, 609.
12. *The Cost of New Agrochemical Product Discovery, Development and Registration in 1995 and 2000*, URL http://www.ecpa.be /library/reports/ PhillipsMcDoug-4-03.pdf
13. Georghiou, G. P.; Lagunes, A., *The Occurrence of Resistance to Pesticides: Cases of Resistance Reported Worldwide through 1988*. FAO. Rome, 1988.
14. Frankenhuyzen, K.V. In *Bacillus thuringiensis*, An Environmental Biopesticide: Theory and Practice; Entwhistle, P.F.; Cory, J.S.; Bailey, M.J.; Higgs, S., Eds.; Wiley, West Sussex, England, 1993, pp 1-23.
15. *Prevention and Disposal of Unwanted Pesticide Stocks in Africa and the Near East*; Report W8419; Plant Production and Protection Division, Food and Agriculture Organization of the United Nations: Rome, 1994.
16. Salama, H.S.; Morris, O.N. In *Bacillus thuringiensis*, An Environmental Biopesticide: Theory and Practice; Entwhistle, P.F.; Cory, J.S.; Bailey, M.J.; Higgs, S., Eds.; Wiley, West Sussex, England, 1993, pp 237-249.
17. Thompson, G.D.; Dutton, R.; Sparks, T.C. *Pest Manag. Sci.* **2000**, *56*, 696.
18. Sparks, T.C.; Thompson, G.D.; Kirst, H.A.; Hertlein, M.B.; Larson, L.L.; Worden, T.V.; Thibault, S.T. *J. Econ. Enotomol.* **1998**, *91*, 1277.
19. Salgado, V.L. *Pestic. Biochem. Physiol.* **1998**, *60*, 91..
20. Wei, Z.M.; Laby, R.J.; Zumoff, C.H.; Bauer, D.W.; He, S.Y.; Collmer, A.; Beer, S.V. *Science.* **1992**, *257*, 85.
21. Dong, G.; Delaney, T.P.; Bauer, D.; Beer, S.V. *Plant J.* **1999**, *20*, 207.
22. Heins, S.D.; Manker, D.C.; Jimenez, D.R.; McCoy, R.J.; Marrone, P.G.; Orjala, J.E. US Patent 6,103,228, 2000.
23. Umezawa, H.; Aoyagi, T.; Nishikiori, T.; Okuyama, A.; Yamagishi, Y.; Hamada, M.; Takeuchi, T. *J. Antibiotics.* **1986**, *39*, 737.
24. Yamada, S.; Takayama, Y.; Yamanaka, M.; Ko, K.; Yamaguchi, I. *J. Pest. Sci.*, **1990**, *15*, 95.
25. www.oit.doe.gov

Chapter 28

ELISA and Liquid Chromatography/Mass Spectrometry/Mass Spectrometry Methods for Sulfentrazone and Its Acid Metabolite in Groundwater Samples

Audrey W. Chen

Agricultural Products Group, FMC Corporation, P.O. Box 8, Princeton, NJ 08543

Sulfentrazone represents a new class of herbicides (aryl triazolinones). It inhibits the protoporphyrinogen oxidase (PPO) in the plant chlorophyll biosynthetic pathway. The major metabolite identified in the water is sulfentrazone-3-carboxylic acid (SCA). An ELISA (enzyme-linked immunosorbent assay) test kit has been used as an analytical screening tool for the water samples (including well, lysimeter, and other source water) from three groundwater studies. Compared to the conventional chemical assay, ELISA uses minimum chemicals (solvents and reagents) and is simpler and time- and cost-effective. LC/MS/MS (liquid chromatograph equipped with triple quadruple mass spectrometry) was used for the confirmation of all positive and partial negative residues. Method development and comparisons of precision, accuracy, cross-reactivity, recovery, and false negatives and positives for ELISA and LC/MS/MS are discussed herein.

Introduction

Sulfentrazone is a broad-spectrum, pre-emergent herbicide that provides good control over broadleaf weeds, grasses and sedges in crop fields and turf. The residue of interest in water includes the parent sulfentrazone and SCA (Figures 1 and 2).

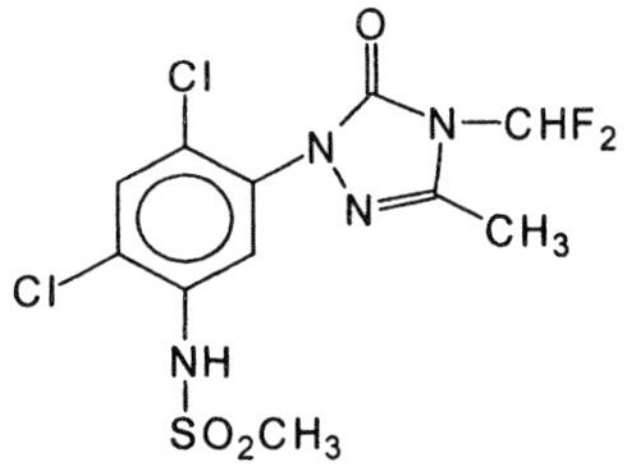

Chemical name: N-[2,4-dichloro-5-[4-(difluoromethyl)-4,5-dihydro-3-methyl-5-oxo-1H-1,2,4-triazol-1-yl]phenyl] methanesulfonamide
Trade name: Authority™ and Spartan®
Common name: Sulfentrazone
CAS No.: 122836-35-5

Figure 1. Sulfentrazone Structure and Nomenclature

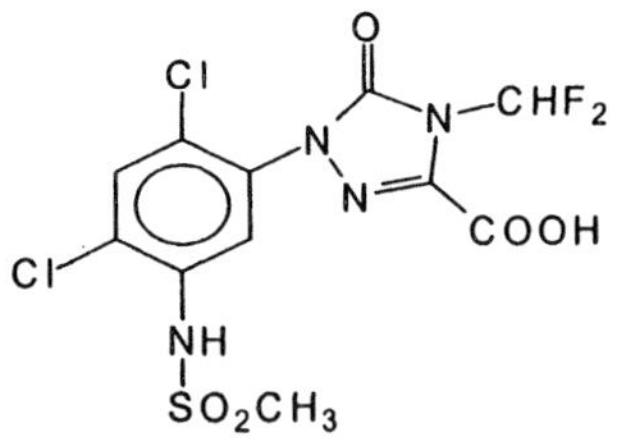

Chemical name: 1-[2,4-dichloro-5-(N-(methyl-sulfonyl)aminuteo)phenyl]-4-difluoromethyl-4,5-dihydro-5-oxo-1H-1,2,4-triazole-3-carboxylic acid
CAS No.: 134391-01-8

Figure 2. Sulfentrazone-3-carboxylic acid (SCA) Structure and Nomenclature

The ELISA (enzyme-linked immunosorbent assay) test kit for sulfentrazone and SCA residues in groundwater is developed with limit of quantitation (LOQ) and limit of detection (LOD) at 0.1 and 0.05 ppb, respectively. Polyclonal antibodies were generated by immunizing with a sulfentrazone crop metabolite (3-hydroxymethyl sulfentrazone, HMS). In order to provides better sensitivity and reproducible results, SCA is decarboxylated and converted to 3-desmethyl sulfentrazone (DMS) in acidic conditions with heat (i.e., reflux) prior to LC/MS/MS measurement. The LOQ and LOD for the analytes of interest by LC/MS/MS are at 0.1 and 0.02 ppb, respectively. The structures and chemical names for HMS and DMS are presented in Figures 3 and 4.

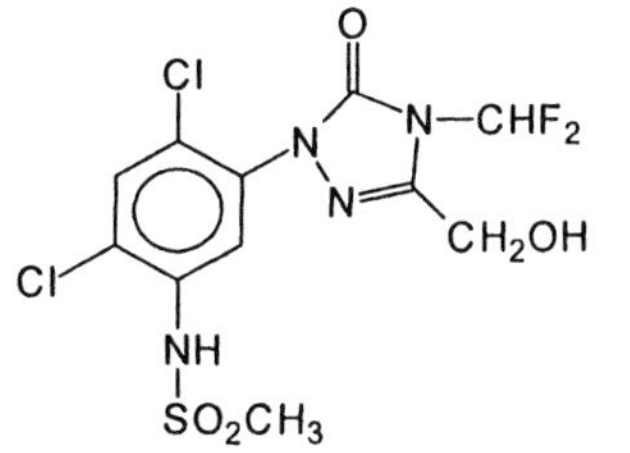

Chemical name: N-[2,4-dichloro-5-[4-(difluoromethyl)-4,5-dihydro-3-hydroxymethyl-5-oxo-1H-1,2,4-triazol-1-yl]phenyl]-methanesulfonamide
CAS No.: 134390-99-1

Figure 3. HMS Structure and Nomenclature

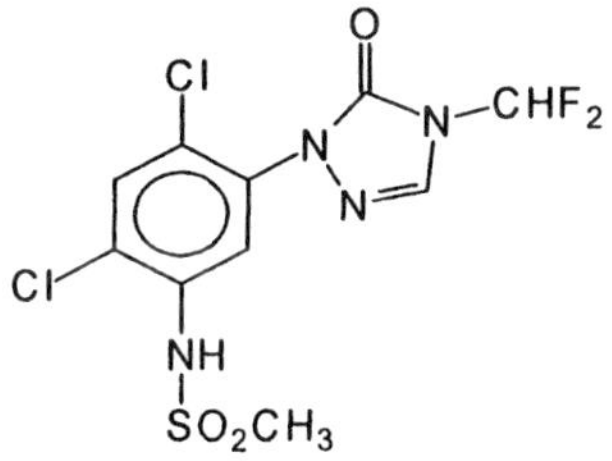

Chemical name: N-[2,4-dichloro-5-[4-(difluoromethyl)-4,5-dihydro-5-oxo-1H-1,2,4-triazol-1-yl]phenyl]-methanesulfonamide
CAS No.: 134391-02-9

Figure 4. DMS Structure and Nomenclature

ELISA Method

Three immunogens, sulfentrazone conjugated with keyhole limpet hemocyanin (KLH), HMS with KLH, and SCA with ovalbumin (OVA) have been tested to detect sulfentrazone and SCA in water. Only the HMS conjugated with KLH would react with both analytes of interest with 100% inhibition (recognition and reactivity) for sulfentrazone and 50% inhibition for SCA (Table 1). Therefore, LOD and LOQ are 0.05 and 0.1 ppb and 0.1 and 0.2 ppb for sulfentrazone and SCA, respectively.

Table I. Comparison of Immunogen and Reactivity

Immunogen	Recognition and Reactivity (%)	
	Sulfentrazone	SCA
Sulfentrazone-KLH	100	0
HMS-KLH[a]	100	50
SCA-OVA	0	100

a. The only immunogen that reacts with both analytes.

HMS

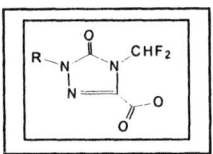

Charge of O (-H): -0.70
Rotational Barrier (N-C-C-O): 2.6 kcal/mol

SCA

Charges of O (=C) and O (-H): -0.55 and -0.68
Rotational Barrier (N-C-C-O): 7.3 kcal/mol

Figure 5. N-C-C-O bonds in HMS and SCA

The charge densities of "O" in HMS (-0.70) and SCA (-0.68) are similar based on the high-level quantum mechanical calculations. However, the C-C bond of N-C-C-O in HMS can rotate freely at room temperature (< 3 kcal/mol), but such rotation is hindered in SCA. The degree of freedom of N-C-C-O in HMS might be the reason why only HMS-KLH could react with both analytes of interest. The Calculated charge density and rotational energy were provided by FMC Computational Sciences Team. The N-C-C-O bond angles in HMS and SCA are demonstrated below.

The competitive inhibition ELISA 96-well format used in this program was developed by Strategic Diagnostics Inc. (SDI). The antibody will bind sulfentrazone either immobilized to the wells of the plate or the free analytes in sample solution. When sulfentrazone is present in the water sample and added to the plate with antibody, it competes with the immobilized sulfentrazone for a limited number of antibody binding sites. Since the same number of sulfentrazone molecules are immobolized on every test well and each well receives the same concentration of antibody (same number of antibody binding sites), a sample containing a low concentration of sulfentrazone will allow more antibody to bind to the immobilized sulfentrazone. The antibody bound to the plate can then be marked by a labeled secondary antibody. This marker, in the presence of substrate, results in a yellow solution and the concentration of analyte can then be measured by a spectrophotometric plate reader. If a sample contains a high concentration of sulfentrazone, less antibody will be available to bind to the immobilized sulfentrazone and ultimately result in a lighter or even colorless solution. This ELISA kit is sensitive and selective and allows reliable and rapid screening for a total concentration of sulfentrazone and SCA. The kit has been tested on several agrochemicals with similar chemical structures (e.g., carbofuran, carfentrazone-ethyl, alachlor, chlorosulfron, atrazine, trifluralin, metribuzin, imazethapyr, terbufos and chlorimuron ethyl) and none of them reacted with the kit [IC_{50} (concentration produces 50% inhibition of maximum signal) > 1000 ppb].

LC/MS/MS Method

The water sample is acidified (pH=1) and boiled one hour under reflux to convert SCA to DMS. The analytes of interest are then concentrated and separated using C_8 and silica gel solid phase extraction (SPE) cartridges prior to the instrument analysis. Micromass Quattro LC triple quadrupole mass spectrometer is used for analyte concentration determination. The molecular ions monitored are 385 (MS) and 199 (MS/MS) for sulfentrazone and 371 (MS) and 201 (MS/MS) for DMS. The LOD and LOQ are 0.02 and 0.1 ppb for both sulfentrazone and SCA (analyzed as DMS) in water. Comparing to ELISA, which determines a total concentration of the two analytes, LC/MS/MS measures individual analyte and provides better sensitivity than ELISA. Therefore, LC/MS/MS is commonly used for the confirmation of analyte residues. The example of chromatograms for sulfentrazone and DMS by LC/MS/MS are presented in Figure 6.

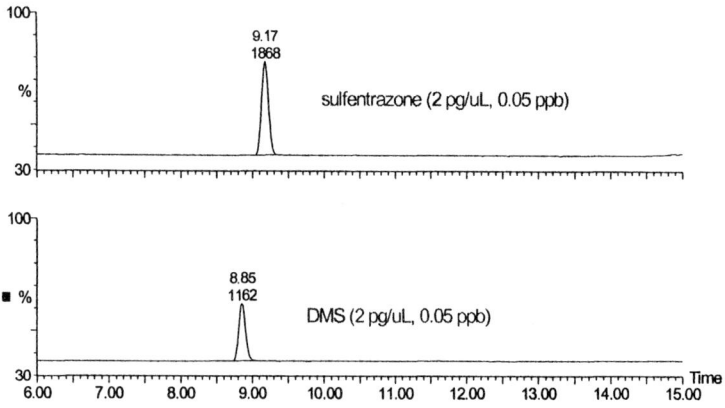

Figure 6. sulfentrazone and DMS at 0.05 ppb by LC/MS/MS

Method Recoveries of Sulfentrazone and SCA in Water

Water samples from the three groundwater studies (A, B and C) were first screened for residues using ELISA method. The samples were then analyzed using LC/MS/MS when the residues were found at detectable levels or higher ($\geq$ 0.05 ppb). Also, a representative number of water samples with negative responses from ELISA were analyzed by LC/MS/MS. In general, a set of 30 (triplicate), 40 (duplicate) or 80 (singular) water samples can be analyzed on each ELISA 96-well plate and a set of 10-12 samples can be analyzed for conventional LC/MS/MS method. Both methods require about a day to perform a set of samples. When residues were expected to be detected in most of the water samples collected at certain times, only LC/MS/MS method was used for residue determination. Method recoveries of sulfentrazone and SCA in water from Study A by both methods are provided in Table 2. Both methods provide reliable and reproducible results based on the respective accuracy (average recoveries) and precision (standard deviations).

Table 2. Method Recovery (%) of Sulfentrazone and SCA in Water from Study A

Analyte	No. of Analysis	Fortification Level (ppb)	Recovery Range	Average ± Std. Dev.
ELISA (Contract Lab)[a]				
Sulfentrazone	18	0.06-2.0	79-159	120±24
ELISA (FMC)				
Sulfentrazone	25	0.1	65-115	85±13
SCA	25	0.2	66-124	96±13
LC/MSD[b] (FMC)				
Sulfentrazone	154	0.1-25.0	62-138	91±12
SCA	154	0.1-25.0	61-140	87±14

a. From the ELISA kit validation report by SDI.
b. Including LC/MS and LC/MS/MS

Result Comparison of ELISA and LC/MS/MS for Sulfentrazone and SCA in Water

For Study A there were 521 water samples analyzed by both ELISA and LC/MS/MS methods. These samples were initially screened using ELISA and resulted in 19 positive samples (residues ≥ 0.05 ppb). These 19 samples along with 128 negative samples (no detectable resiudes) were further analyzed by LC/MS/MS for confirmation. Of the 19 positive samples by ELISA, 10 samples contained positive residues (≥ 0.02 ppb) by LC/MS/MS. The remaining 9 positive samples by ELISA were considered false-positives. Of the 128 negative samples by ELISA, 7 samples contained positive residues (≥ 0.02 ppb) by LC/MS/MS. These samples represented false-negatives by ELISA method. The negative and positive results using both methods are presented in Table 3.

Table 3. Comparison of Positive and Negative Residues by ELISA and LC/MS/MS in Study A

Event	DAT[a]	ELISA Negative	ELISA Positive	LC/MSD Negative	LC/MSD Positive
1	-1	54	3	11	0
5	7	17	1	10	0
6	14	45	0	16	0
7	30	52	2	10	0
8	60	53	2	18	1
9	90	52	3	12	4
10	120	52	3	15	2
11	150	52	2	18	8
12	180	33	2	9	1
13	210	23	1	11	1
15	270	17	0	1	0
Total		502	19	131	17

a. Days After Treatment

The overall residue results by ELISA and LC/MS/MS for Study A correlated rather well, except for the 7 false-negative samples by ELISA. The residue levels for those 7 samples were between the LC/MS/MS LOD (0.02 ppb) and LOQ (0.1 ppb) and therefore, could not be detected by ELISA. The overall negative and positive results by both methods for three groundwater studies are also provided in Table 4.

Table 4. Comparison of Positive and Negative Residues by ELISA and LC/MS/MS in Three Groundwater Studies

Study No.	No. of Samples Analyzed	False Positive	False Negative
A (95th)[a]	502	9	7
B (85th)	283	10	11
C (75th)	330	21	1
Total	1115	40	19

a. Percentile of soil vulnerability

A total of 1115 water samples were analyzed by ELISA for three groundwater studies and only 40 of them were false-positives (3.6%) and 19 samples (1.7%) were false-negatives after confirmation by LC/MS/MS. Since all the positive residues from ELISA would require confirmation by LC/MS/MS, the false-positives from ELISA are not of concern. In addition, the residues of the false-negatives from ELISA are insignificant because these rsidues are less than 0.05 ppb.

Conclusion

A total of ~15 work days would be needed to analyze the 1115 water samples by ELISA based on the 80 singular sample on each 96-well plate. Whereas for LC/MS/MS, a total of ~115 work days would be required to analyze the same amount of samples based on the 10-12 samples per assay set. In addition, ELISA uses and generates minimum chemicals and waste(solvents and reagents) and is simpler and time-and cost-effective compared to the conventional chemical method. The ELISA kit for sulfentrazone and SCA is, therefore, a reliable and effective analytical tool for screening water samples. It is particularly useful in the beginning and at the end of each groundwater study, when detectable residues are not expected.

Acknowledgements

The author gratefully thanks D. Baffuto, J. Carroll, J.F. Culligan, R. Jones, D.J. Letinski, E.M. McCoy, R.T. Morris and M. Xiong for their help with sample preparation, analysis and report and L.R. Young for calculation of charge density and rotational energy.

Chapter 29

Synthesis and Acaricidal Activity of Novel 2-Substituted-3-trifluoromethylquinoxalines

Yoshitaka Fukushima[1], Naoki Ishii[1], Tetsuya Imai[1], Makio Usui[1], and Noriharu Ken Umetsu[2]

[1]Naruto Research Center, Otsuka Chemical Company, Ltd., Naruto, Tokushima 772–8601, Japan
[2]Otsuka Chemical Holdings Company, Ltd, Osaka 554–0021, Japan

Recently fluorine-containing compounds have attained importance as a target for useful agrochemicals. The 2-substituted-3-trifluoromethylquinoxalines are a novel type of fluorine containing compounds designed and synthesized by our group, which exhibit excellent acaricidal and insecticidal activities. The reaction of o-phenylenediamine with ethyl trifluoropyruvate as a fluorine source, followed by chlorination leads to a common starting material for the aimed fluorine-containing chemicals. The reaction of the starting material with different types of nucleophiles such as amines gave a wide variety of 2-substituted-3-trifluoromethylquinoxalines. Some of the compounds synthesized showed excellent acaricidal activity against two-spotted spider mites (*Tetranychus urticae*), and some other compounds, in turn exhibited insecticidal activity.

Recently fluorine-containing compounds have gained importance as a source for useful agrochemicals. Of them, trifluoromethyl (CF_3)-containing compounds were major target in obtaining new molecules, and a large number of compounds possessing CF_3 moiety in the aryl ring were designed and commercialized as useful agrochemicals. On the other hand, limited number of CF_3-containing heterocyclic compounds have been so far designed and commercialized.

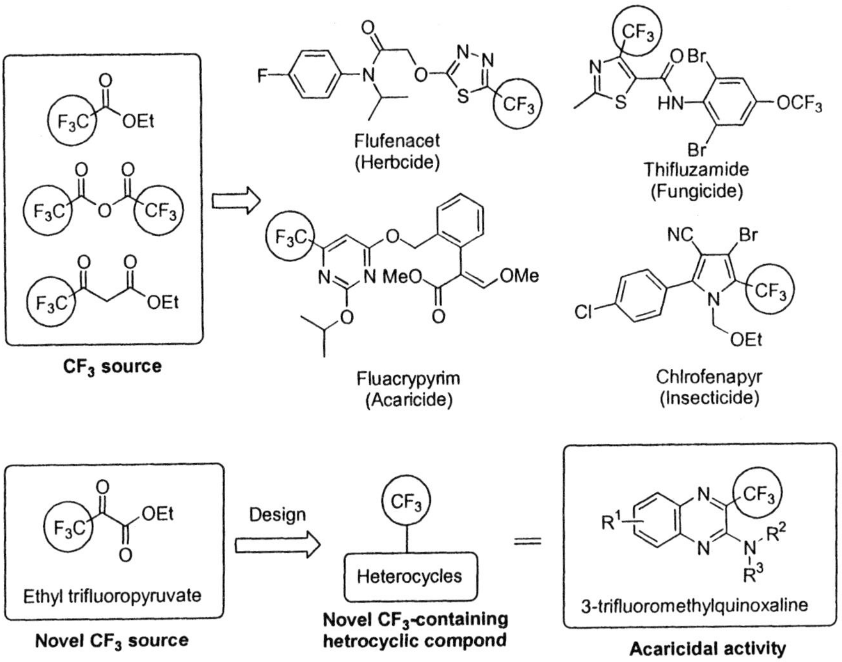

Figure 1. Design of novel CF$_3$-containinig heterocyclic compounds

As shown in Figure 1, flufenacet (herbicide) possessing trifluoromethyl moiety in the thiadiazole ring, thifluzamide (fungicide) having trifluoromethyl moiety in the thiazole ring, fluacrypyrim (acaricide) possessing trifluoromethyl moiety in the pyrimidine ring, and chlorfenapyr (insecticide) having trifluoromethyl moiety in the pyrrole ring are a few examples for the CF$_3$-containing heterocyclic agrochemicals (*1-5*). These compounds are usually prepared from trifluoroacetate, trifluoroacetic anhydride or trifluoroacetoacetate as a trifluoromethyl source. In order to create novel class of agrochemicals, which contain CF$_3$ moiety in the heterocyclic ring, we have designed and synthesized a series of 2-substituted-3-trifluoromethylquinoxalines (3-TFQ) by utilizing ethyl utilizing ethyl trifluoropyruvate. The pyruvate is considered to be a novel CF$_3$ source for heterocyclic ring. The compounds thus synthesized possess unique structure in the molecule among various CF$_3$-containing heterocyclic compounds, and proved to be acaricidal active.

Synthesis

The 2-substituted-3-TFQ were prepared according to the synthesis scheme shown in Figure 2. The first step is to react ethyl trifluoropyruvate **I** with a variety of *o*-phenylenediamines **II** in the presence of *p*-toluenesulfonic acid monohydrate (*p*-TsOH) in toluene to obtain 3-TFQ-2-one **III**, which in turn was treated with phosphorus oxychloride to give 2-chloro-3-TFQ **IV**. The compound **IV** was then reacted with different types of nucleophiles such as amines in the presence of potassium carbonate in *N*, *N*-dimethylformamide (DMF) to give the aimed product 2-substituted-3-TFQ **V**.

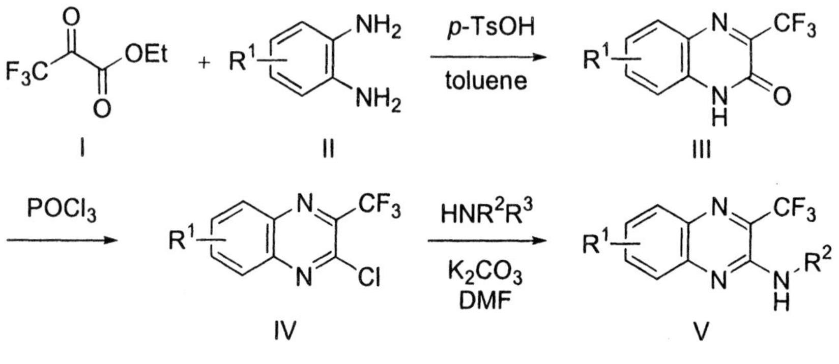

Figure 2. Scheme for Synthesis of 2-Substituted-3-TFQ (V)

Biological testing method

Acaricidal and insecticidal activities of the synthesized compounds were determined as follows. In the case of two-spotted spider mite (*T. urticae*), female adult mites were placed on the excised leaf of kidney bean seedling, and then, the test solution (test compounds were dissolved in acetone at 20 to 200 ppm) was sprayed onto the leaf. Mortality was determined 2 days after the treatment. In the case of western flower thrips (*F. occidentalis*), first instar larvae were released on released on the excised leaf of kidney bean seedling (15 larvae per a disk), which was then sprayed with the test solution. Mortality was determined 2 days after the treatment. In the case of silver leaf whitefly (*B. argentifolii*), adult whiteflies were released on the cabbage leaf and allowed to oviposit for 24 hours. Adults were removed from the leaf and the leaf was stood for 7 days. A leaf disk of 20 millimeters in diameter was excised from the leaf and sprayed with the test solution. Mortality was determined 6 days after the treatment. The acaricidal and insecticidal activities were indicated as activity indexes **3**, **2** and **1** which represent the mortality of 100%, 99% to 50% and less than 49%, respectively.

Structure Activity Relationships

Data for the acaricidal and insecticidal activities of different types of 3-TFQ and their related compounds were shown in Table I to VI. Table I shows the effect of R^2 substitution at 2-position of 3-TFQ on acaricidal activity. The compounds **1** to **7**, R^2 being methyl to n-heptyl and compounds **14** to **16**, R^2 being n-tetradecyl to n-hexadecyl showed very poor acaricidal activity. On the other hand, the compounds **8** to **13** having n-octyl to n-tridecyl group as R^2 showed acaricidal activity. Compound **9** to **11**, R^2 moiety being n-nonyl to n-undecyl showed excellent activity at 200 ppm.

Table II shows the effect of R substitution at 3-position of 3-TFQ on acaricidal activity. In this case, 2-position of 3-TFQ is n-decylamino and benzene moiety is unsubstituted. As shown in the table, phenyl substituted compound showed some activity, but the benzyl, hydrogen and chloro substituted compounds did not show any activity, or showed very poor activity. The results suggest that the CF_3 group at 3-position is essential for the activity.

Table I. Effect of R^2 Substitution at 2-Position of 3-TFQ on Acaricidal Activity

Compound	R^2	Acaricidal activity [a] (ppm) 200
1	CH_3	1
2	CH_2CH_3	1
3	$(CH_2)_2CH_3$	1
4	$(CH_2)_3CH_3$	1
5	$(CH_2)_4CH_3$	1
6	$(CH_2)_5CH_3$	1
7	$(CH_2)_6CH_3$	1
8	$(CH_2)_7CH_3$	2
9	$(CH_2)_8CH_3$	3
10	$(CH_2)_9CH_3$	3
11	$(CH_2)_{10}CH_3$	3
12	$(CH_2)_{11}CH_3$	2
13	$(CH_2)_{12}CH_3$	2
14	$(CH_2)_{13}CH_3$	1
15	$(CH_2)_{14}CH_3$	1
16	$(CH_2)_{15}CH_3$	1

[a] Activity index; 3: mortality 100%, 2: mortality 99~50%, 1: mortality 49~0%

Table II. Effect of R Substitution at 3-Position of 3-TFQ on Acaricidal Activity

Compound	R	Acaricidal activity [a] (ppm) 200
10	CF_3	3
17	Ph	2
18	CH_2Ph	1
19	H	1
20	Cl	1

[a] Activity index; 3: mortality 100%, 2: mortality 99~50%, 1: mortality 49~0%

Effect of R^1 Substitution of 3-TFQ on acaricidal activity was then determined. As shown in Table III, the disubstituted compound (**21**), 6,7-dimethyl substituted, showed enhanced activity compared to the unsubstituted compound (**10**). The 6,7-dichloro substituted compound **22** gave 100 % mortality at 100 ppm. On the other hand, of the monosubstituted compounds which were obtained as a mixture of two monosubstituted compounds, compound **23**, R^1 being 6 and 7-methyl, **24**, R^1 being 5 and 8-methyl, **25**, R^1 being 6 and 7-chloro, **26**, R^1 being 6 and 7-fluoro, being 6 and 7-chloro, **26**, R^1 being 6 and 7-fluoro, and **27**, R^1 being 6 and 7-trifluoromethyl showed equal activity to the disubstituted compounds.

Because of excellent acaricidal activity of **22** (6,7-dichloro substituted compound), the effect of substitution at R^2 position of 6,7-Cl_2-3-TFQ on acaricidal activity was investigated. Data for acaricidal and insecticidal activities are summarized in Table IV to VI. In Table IV, the data for acaricidal activity of substituted compounds on benzene ring (compounds **5** to **13**) was re-shown for comparison. The compounds **31** to **34** and **22** having *n*-hexyl to *n*-decyl as R^2, showed excellent activity even at 100 ppm. Their activity level was equal to the compounds possessing *n*-nonyl, *n*-decyl and *n*-undecyl as R^2, i.e., compounds **9**, **10** and **11**, which belong to the unsubstituted compounds on benzene ring. Suitable length of alkyl chain at the R^2 position for acaricidal activity is different between unsubstituted and dichlorosubstituted compounds on benzene ring, this probably due to the difference of total lipophilicity of the compounds.

Data for the effect of substitution of R^2 position of 6,7-Cl_2-3-TFQ on the insecticidal activity are shown in Table V. In contrast to the acaricidal activity, insecticidal activity (thrips) is higher with the compounds such as **31** to **33** having shorter alkyl chain length at R^2, R^2 being *n*-hexyl to *n*-octyl. As shown before, acaricidal activity was higher with a chain length of *n*-hexyl to *n*-decyl. The compound **31**, R^2 being *n*-hexyl, showed excellent activity against whitefly. Over all, compound **31** showed the most preferable acaricidal and insecticidal activities against thrips and whitefly.

Table III. Effect of R¹ Substitution of 3-TFQ on Acaricidal Activity

Compound	R^1	Acaricidal activity[a] (ppm) 200	100
10	H	3	1
21	6,7-Me$_2$	3	2
22	6,7-Cl$_2$	3	3
23[b]	6-Me and 7-Me	3	2
24[b]	5-Me and 8-Me	3	3
25[b]	6-Cl and 7-Cl	3	3
26[b]	6-F and 7-F	3	3
27[b]	6-CF$_3$ and 7-CF$_3$	2	2
28[b]	6-NO$_2$ and 7-NO$_2$	1	1
29[b]	6-PhCO and 7-PhCO	1	1

[a] Activity index; 3: mortality 100%, 2: mortality 99~50%, 1: mortality 49~0%
[b] Mixture of isomers

Table IV. Effect of Substitution at R² position of 6,7-Cl$_2$-3-TFQ on Acaricidal Activity (Comparison with Unsubstituted Compound on the benzene ring)

Compd.	R^2	Activity[a] (ppm) 200	100	Compd.	R^2	Activity[a] (ppm) 200
30	(CH$_2$)$_4$CH$_3$	3	1	5	(CH$_2$)$_4$CH$_3$	1
31	(CH$_2$)$_5$CH$_3$	3	3	6	(CH$_2$)$_5$CH$_3$	1
32	(CH$_2$)$_6$CH$_3$	3	3	7	(CH$_2$)$_6$CH$_3$	1
33	(CH$_2$)$_7$CH$_3$	3	3	8	(CH$_2$)$_7$CH$_3$	2
34	(CH$_2$)$_8$CH$_3$	3	3	9	(CH$_2$)$_8$CH$_3$	3
22	(CH$_2$)$_9$CH$_3$	3	3	10	(CH$_2$)$_9$CH$_3$	3
35	(CH$_2$)$_{10}$CH$_3$	3	2	11	(CH$_2$)$_{10}$CH$_3$	3
36	(CH$_2$)$_{11}$CH$_3$	2	1	12	(CH$_2$)$_{11}$CH$_3$	2
37	(CH$_2$)$_{12}$CH$_3$	1	1	13	(CH$_2$)$_{12}$CH$_3$	1

[a] Activity index; 3: mortality 100%, 2: mortality 99~50%, 1: mortality 49~0%

Table V. Effect of Substitution at R^2 Position of 6,7-Cl_2-3-TFQ on Insecticidal Activity

Compound	R^2	Insecticidal activity[a] (ppm)		Acaricidal activity[a] (ppm)	
		WFT^b 200	SLW^c 200	TSM^d 200	100
30	$(CH_2)_4CH_3$	1	1	1	1
31	$(CH_2)_5CH_3$	3	3	3	3
32	$(CH_2)_6CH_3$	3	2	3	3
33	$(CH_2)_7CH_3$	3	1	3	3
34	$(CH_2)_8CH_3$	2	1	3	3
22	$(CH_2)_9CH_3$	1	1	3	3
35	$(CH_2)_{10}CH_3$	1	1	3	2
36	$(CH_2)_{11}CH_3$	1	1	3	1
37	$(CH_2)_{12}CH_3$	1	1	3	1

[a] Activity index; 3: mortality 100%, 2: mortality 99~50%, 1: mortality 49~0%
[b] WET: western flower thrips (*F. occidentalis*)
[c] SLW: silver leaf whitefly (*B. argentifolii*)
[d] TSM: two-spotted spider mite (*T. urticae*)

Data for the insecticidal and acaricidal activities for the substituted compound of R^2 position with N, N-dialkylaminoalkyl, alkoxyalkyl, alkoxycarbonylalkyl or acyloxyalkyl are shown in Table VI. As already mentioned, the compound 31 possessing n-hexyl group as R^2 showed excellent acaricidal and insecticidal activities. Therefore, the compound having alkylamino, N, N-dialkylaminoalkyl, alkoxyalkyl, alkoxycarbonylalkyl or acyloxyalkyl as R^2, length of chain being similar to n-hexyl, were synthesized and their insecticidal and acaricidal activities were determined. As shown in the table, diethylaminopropylamino substituted compond, i.e., compound 38 showed excellent acaricidal activity at 20 ppm. Ethoxypropylamino substituted compound (41) also showed excellent insecticidal and acaricidal activities, activity level being equal to the n-hexylamino substituted compound 31.

Conclusion

In order to create a novel class of acaricide and insecticide, a series of 2-substituted-3-TFQ were synthesized by utilizing ethyl trifluoropyruvate as a novel source for CF_3 moiety, and acaricidal and insecticidal activity were determined.

Table VI. Effect of R^2 Position of 6,7-Cl_2-3-TFQ on Acaricidal and Insecticidal Activities

Compd.	R^2	Insecticidal activity[a] (ppm)		Acaricidal activity[a] (ppm)		
		WFT^b 200	SLW^c 200	TSM^d 200	100	20
31	⌇∕∖∕∖∕	3	3	3	3	1
38	⌇∕∖∕N(Et)₂	2	1	3	3	3
39	⌇∕∖∕NHMe	3	1	3	2	1
40	⌇∕N(Et)₂	3	2	3	2	1
41	⌇∕∖O∕∖	3	3	3	3	1
42	⌇∕∖O∕	3	2	3	1	1
43	⌇∕C(O)O∕∖	3	1	3	1	1
44	⌇∕∖OC(O)Me	3	1	3	1	1

[a] Activity index; 3: mortality 100%, 2: mortality 99~50%, 1: mortality 49~0%
[b] WET: western flower thrips (*F. occidentalis*)
[c] SLW: silver leaf whitefly (*B. argentifolii*)
[d] TSM: two-spotted spider mite (*T. urticae*)

Many of the compounds prepared showed excellent acaricidal activity against two-spotted spider mites (*T. urticae*), and some of them also showed good insecticidal activity against western flower thrips (*F. occidentalis*) and silver leaf whitefly (*B. argentifolii*). From the structure-activity relationship study, the CF_3 group at 3-position of quinoxaline structure proved to be decisive for the acaricidal activity against two-spotted spider mites. The 6,7-Cl_2 substitution on benzene ring of quinoxaline structure enhanced the acaricidal activity. The best acaricidal activity of substituted quinoxaline at 2-position was found with diethylaminopropylamino group. When the 2-position of quinoxaline was substituted with *n*-hexyl group or ethoxypropylamino group, these compounds showed excellent insecticidal activity for both insects tested as well as acaricidal activity.

References

1. Deege, R. BCPC-Weeds, **1991**, 87.
2. Bloomberg, J. R. Abst. WSSA, **1995**, 5.
3. O'Reilly, P. Brighton Crop Pro. Conf.-Pests and Diseases, **1992**, 427.
4. Kirstgen, R; Oberdorf, K; Schütz, F; Theobald, H; Harries, V. WO 9616047, 1996.
5. Addor, R. W.; Babcock, T. J.; Black, B. C.; Brown, D. G.; Diehl, R. E.; Furch, J. A.; Kameswarm, V.; Kamhi, V. M.; Kremer, K. A.; Kuhn, D. G.; Lovell, J. B.; Lowen, G. T.; Wright, D. P.: "Synthesis and Chemistry of Agrochemicals III", American Chemical Society, **1992**, pp 281.

Human Vector Control

Chapter 30

Emerging Vectorborne Diseases and Their Control

John D. Edman

Center for Vectorborne Diseases, University of California, Davis, CA 95616

ABSTRACT: Principally a combination of human population growth, environmental degradation, public health inadequacies and rapid global transportation has put the world at mounting risk from vectorborne and other infectious diseases. Health problems associated with the emergence of new pathogens and the re-emergence and spread of old ones have gained a new urgency due to the global threat of bioterrorism. A large percentage of emerging diseases are vectorborne and over one-third of the agents on the list of greatest concern from bioterrorists are vectorborne. Many of these diseases are caused by viruses for which there are no effective drugs for treatment or vaccines for prevention. Drug and insecticide resistance has further complicated our ability to respond effectively to these growing threats. Parasitic diseases like malaria, leishmmaniasis and African trypanosomiasis are resurging in many regions. Onchocerciasis and to a lesser extent filariasis control programs are reducing the impact of these diseases. This has largely been possible because of a long-lasting new filaricide (Ivermectin) and targeted application of insecticides to highly specialized development sites (i.e., riverine rapids). Chagas disease also has declined significantly in the southern part of its range through home improvement and indoor application of insecticides against domiciliary kissing bug vectors. Current outbreak of West Nile virus in North America and dengue worldwide are examples of emerging threat from mosquito-borne arboviruses, even in developed countries. Lyme disease, the most prevalent vectorborne disease in the U.S. with nearly 18,000 cases annually, was not recognized until the early 1980's and was followed by the emergence of another tick-borne disease, ehrlichiosis, in the 1990's. The ongoing West Nile epidemic highlights the fact that while mosquito control remains the first line of defense, the main adulticides and application strategies available for combating epidemics of mosquito-borne diseases today are the same as those used >25 years ago. Adulticides alone have not adequately protected populations from further transmission and disease spread. The demise of vector control programs and their qualified staffs during the 1980's caught many states unprepared. The U.S. is experiencing a vector control crisis at a time when disease threats and control needs are increasing. The recent Institute of Medicine report: "Microbial Threats to Health," offers remedies for reversing this imbalance. It recommends investment to increase the armamentarium for vector control. Even without new tools, many lives could be saved with current technology if political will and adequate resources were committed to reducing the burden of these diseases.

VECTORBORNE AND OTHER INFECTIOUS DISEASES ARE INCREASING

The 1st half of the 20th century saw reductions in many infectious diseases and a rise in life expectancy. Developed countries shift resources from infectious diseases to non-communicable diseases e.g., heart disease and cancer. Despite advanced technology and new vaccines and therapeutics, this trend was not sustained. Instead, infectious diseases increased to the point where today >25% of deaths and nearly 50% of premature deaths among those under 45 year of age are due to infectious diseases; 63% for children <4 years of age. Even in developed counties like the U.S., mortality from infectious diseases has risen from ~40 to ~70 per 100,000 during the last 20 years. Although infections such as TB, HIV and diarrhea account for ~65% of this mortality, a significant portion is caused by vectorborne diseases, especially malaria (~11%). Moreover, among epidemics with >10,000 cases during the 20 year span 1970-1990, five of the eleven outbreaks involved vectorborne diseases (i.e. dengue, Rift Valley fever, visceral leishmaniasis and Typhus) (Fig. 1).

Unexpected outbreaks during the 5-year period 1994-99, reveal a similar pattern in that >1/3 (12/33) were vectorborne (Fig. 2). These disease patterns gave rise to the terms "emerging" and "re-emerging" to denote new or previously known diseases that occur in places or in magnitudes that are unexpected based on recent experience.

CONTRIBUTING FACTORS

The new Institute of Medicine report "Microbial Threats to Health" (IOM 2003) identifies 13 major causes for this resurgence. Although interrelated, six of these particularly apply to vectorborne diseases: (1) Climate and Weather, (2) Changing Ecosystems, (3) Development and Land Use (4) Human Demographics and Behavior, (5) International Travel and Commerce, and (6) Breakdown of Public Health Measures. Global warming and the possible impact of global climate change on the incidence of vectorborne disease is controversial. Most agree that even modest climate change can cause significant redistribution of where and when diseases are endemic and epidemic because temperature, humidity and rainfall can influence vector density, survival rate and development time as well as parasite incubation time in the vector.

The single most important factor---contributing to most others, is mushrooming population. Now at 6.2 billion people and growing by 78 million annually, our planet may already have exceeded its sustainable carrying capacity. Population is expected to grow to ~9 billion in the next 50 years; >95% of this growth will occur in the poorest countries. Of the >35 diseases emerging since 1970, a substantial portion are related to manipulation of the earth's ecology. Dams and irrigation projects dramatically impacted populations of vectors with aquatic immatures. Examples are dams on the Volta River in West Africa leading to onchocerciasis downstream, and the Aswan in Egypt and the Mahaweli irrigation project in Sir Lanka leading to malaria invasion. Projects such as these are justified on bringing food, water and electricity to rapidly growing populations in these regions. Deforestation, mining and road construction, especially in tropical rainforests, have greatly impacted vectorborne diseases. Malaria epidemics have

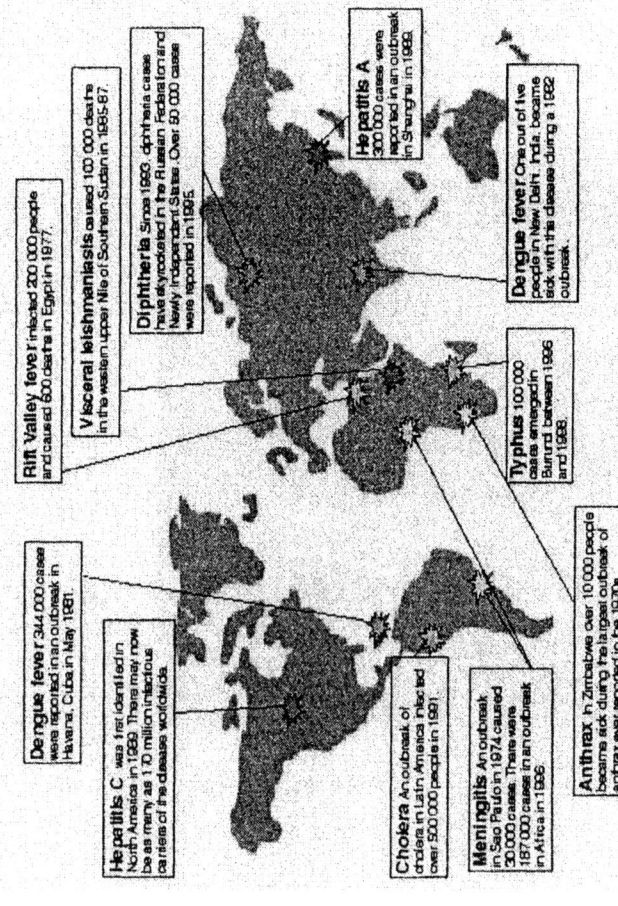

Figure 1. Twenty Year History of Large Disease Outbreaks (WHO website)

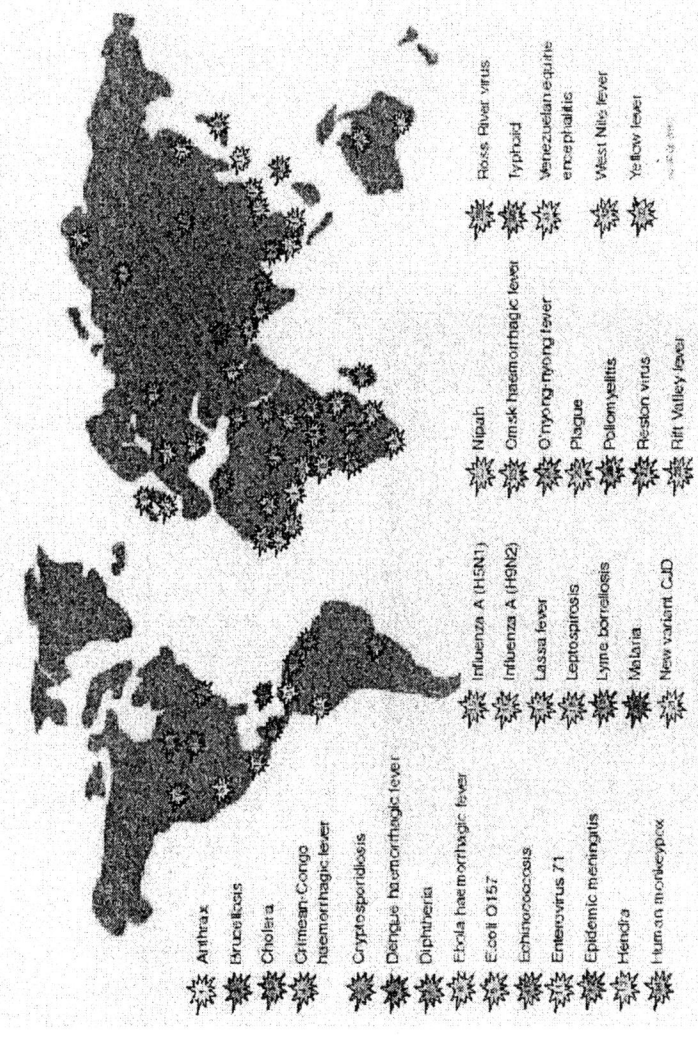

Figure 2. Disease epidemics during the 5 year period 1994-99 (WHO website)

accompanied these ecosystem changes, especially in Brazil and Indonesia where most uncut rain forests exist.

Rapid growth in the international airline industry in the last 50 years has changed the mobility of human populations It is now possible to be in a remote Asian jungle one day and dining in a Los Angeles restaurant the next---far less time than the incubation period of most pathogens. Recently SARS demonstrated this point emphatically It is sobering to contrast population growth with travel time to circumnavigate the earth (Fig. 3).

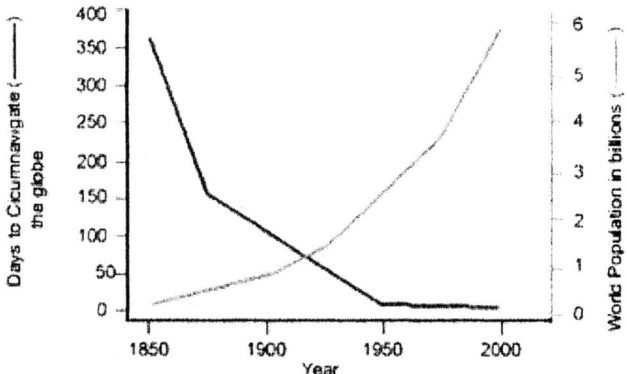

From: Murphy and Nathanson. Semin Virol 5, 87, 1994

Figure 3. Relationship between population growth and ease of travel

Rapid urbanization throughout much of the developing world has had a significant impact on vectorborne diseases like dengue and filariasis whose vectors thrive in urban conditions where water and waste management systems are marginal or lacking. Crowded, unsanitary conditions also create a higher risk for epidemics of flea- and rat-borne plague and louse-borne typhus. Table 1 is a summary of contributing factors to outbreaks of several vectorborne diseases (taken from Gubler, 1998). Malaria appears in all columns.

Table 1. Influences on Emergent/Resurgent Vectorborne Diseases (Gubler 1998)

Urbanization	Deforestation	Agricultural Practices
Dengue Fever	Loaiasis	Malaria
Malaria	Onchocerciasis	Japanese encephalitis
Yellow Fever	Malaria	St. Louis encephalitis
Chickungunya	Leishmaniasis	West Nile fever
Epidemic polyarthritis	Yellow Fever	Oropouche
West Nile fever	Kyasanur Forest disease	WE encephalitis
St. Louis encephalitis	La Crosse encephalitis	VE encephalitis
Lyme disease	EE encephalitis	
Ehrlichiosis	Lyme disease	
Plague		

CURRENT STATUS OF VECTORBORINE DISEASES

Vectorborne diseases fall into three categories: parasitic, microbial and arboviral. The more important diseases in each category are discusses below.

Parasitic Diseases.
Parasitic diseases receive the most attention and so most control efforts and control successes involve these diseases. Malaria, babesiosis, leishmaniasis and African trypanosomiasis best fit the definition of emerging diseases but it is also instructive to examine where and why some of the successes in disease elimination are occurring.

Onchocerciases (riverblindness). Control of this insidious disease is the biggest success story. WHO administered control began in West Africa in 1974 with a program involving seven countries, and expanded to include coffee growing regions in Latin America in 1975 with the Guatemala-Japan project. Programs initially relied on pesticides like abate and Bti to control stream and river dwelling larval stages of the black fly vector. Success led to expansion of both programs. Discovery of the filaricide ivermectin for treating infected people (the only reservoir for this parasite), led to a broadening of the control strategy to include mass drug treatment and prophylaxis. Drug donation by Merck provided for treatment of millions of people for the past several years and the impact has been dramatic. Case estimates that once stood at ~40 million have been reduced to ~18 million over the last 20 years. West Africans regions have returned to resettle and till the fertile lands along rivers where the disease had once driven them away. A new generation can now look forward to a life free of blindness, disfigurement and intense itching.

Lymphatic filariasis. Filaricide success in the onchocerciasis programs, precipitated an attack on the equally incideous lymphatic filariasis or elephantiasis. Transmitted mainly by house mosquitoes *(Culex pipiens)* that develop in polluted waters in urban areas, this disease has a wide tropical distribution. Control is again based on a combination of vector reduction and treating infected or at risk people with ivermectin. Mosquito control options are varied and include the use of bed nets and larvae control with pesticides or sanitation (=source reduction). This program is still in its early stages but optimism can be drawn from the experience in southern China where filariasis has been eliminated as a major infectious disease using this strategy.

Trypanosomiasis. American trypanosomiasis or Chagas disease is another parasitic disease where significant progress is occuring. This infection is transmitted by night-feeding kissing bugs that are permanent residents in thatch roofed or mud walled houses in mountainous regions from Mexico to Argentina. Human trypanosomes are zoonotic parasites with natural animal reservoirs that make long-term control more difficult. Moreover, diagnosis is often problematic and drug treatment more complicated and expensive. Thus, control has focused on eliminating the domesticated bugs by fumigating or spraying houses with residual insecticides, and by structural improvements that make houses inhospitable to bugs. Chagas has been eliminated or greatly reduced in many parts of its southernmost distribution in recent years.

In contrast, African trypanosomiasis has recently resurged in parts of its sub-Saharan range. Efforts are underway to transform novel research on tsetse fly traps into large, area-wide control efforts against sleeping sickness. Numerous wild animal reservoirs make eliminating the day-biting vector an attractive strategy for controlling this disease. The general plan is to widely deploy highly attractive and inexpensive tsetse fly traps and then to possibly followup with release of sterile male flies to eradicate residual populations. These strategies draw strength from the low reproductive capacity of this vector (8-10 offspring per female). Island demonstration projects have shown the potential of these methods but the long flight range of tsetse requires a regional approach. If successful, benefits to animal production will be greater than those to human health.

Leishmaniasis. This important human parasite has the widest distribution of the trypanosomal infections, affecting an estimated 10 million people. Its many forms and varied clinical picture make it difficult to characterize. Visceral forms mimic malaria symptoms (i.e. enlarged liver and spleen) and can be highly fatal. Large outbreaks have occurred in recent years, especially in northern and eastern Africa, central Asia and Brazil. Sand fly vectors are difficult to control and no serious vector control campaigns have been mounted as a way of managing this still emerging parasitic disease.

Malaria. No vectorborne disease is more studied or important than malaria and yet it continues to expand and re-emerge in many regions. Currently endemic in 92 countries, malaria dramatically resurged following the end of the WHO eradication effort in the late 1960's. Between 1.5 and 2 million deaths, mostly young children, and ~300-500 million cases are attributed annually to this mosquito-borne disease. The situation in Africa, where broad eradication was never attempted, is particularly acute. Malaria has recently invaded higher altitude areas, and returned to several large urban centers and southern regions where the disease had been absent for many years. Malaria causes severe economic damage because 3-4 days of work are lost with each episode of malaria and workers in many African countries have multiple episodes each year. Even in the U.S., millions of dollars were spent last year to stamp out a small malaria outbreak in Virginia where malaria had not occurred in >60 years. A 25 year effort to develop a malaria vaccine has taught us much about immunity to this parasite but has not resulted in a vaccine. A combination of insecticide treated bednets and anti-malarial drugs still offers hope for effectively managing this disease wherever the resources and political will exist. The WHO Roll Back Malaria Program is primarily based on these two strategies. Use of pyrethroid-impregnated bednets has sparked new concerns about resistance and resistance management but field studies in Africa have demonstrated significant disease reduction even when resistance is present.

Bacterial Diseases.

Vectorborne bacterial diseases include some of the most recently discovered ones like Lyme disease, ehrlichiosis and cat scratch fever but also some of the oldest known epidemic diseases like typhus and plague. Although treatable with antibiotics, early diagnosis and prompt administration is critical for success. Antibiotic resistance is a continuing concern. This group of mostly zoonotic pathogens also includes many of the vectorborne agents of special concern in bioterrorism because of their ability to persist in the environment and to also be transmitted through direct routes.

Plague. Plague, tularemia and typhus are on the list of bioagents of special concern from terrorists but plague's epidemic history along with its potential for airborne transmission once pneumonic symptoms develop, make it particularly worrisome. Recent plague outbreak in India reminded us that the risk of large natural epidemics still exists in many regions. Even in California, plague frequently causes the closing of campgrounds due to epizootic die offs among native rodents and consequent high risk to campers.

Lyme Disease. First recognized in 1975, Lyme is now the most common vectorborne disease in the USA, with >18,000 cases annually (Fig. 4). Despite novel insecticide application methods for controlling ticks on deer and rodent reservoirs, Lyme continues to increase in both the number of cases and the areas reporting cases. A new vaccine was briefly available but then withdrawn from the market due to unprofitability. Emergence of both Lyme disease and ehrlichiosis are attributable to reforestation and the return of forest-dependent wildlife such as mice and deer and the vector ticks that depend on them. Suburban development in tick-infested forests is also a contributing factor. Treatment of residential premises with acaricides during high-risk months has proven useful for homeowner protection but community-wide efforts to control Lyme disease are lacking.

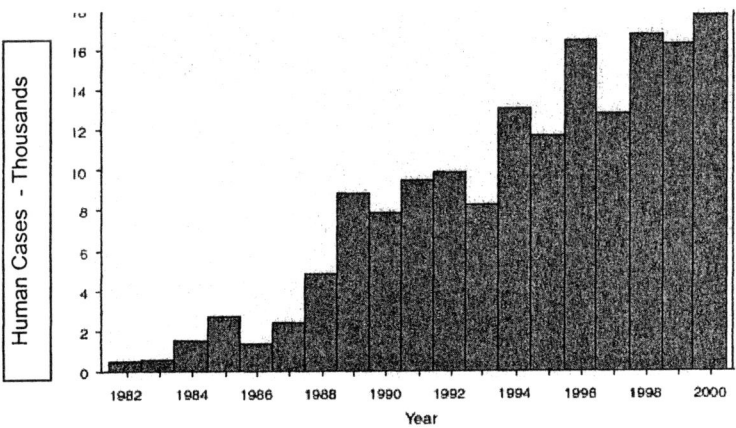

Figure 4. Lyme Diseases Cases in the United States 1982-2000.

Arboviral Diseases.

The arboviruses represent the largest and most diverse group of emerging infections. They include some of the most notable like dengue, West Nile, yellow fever and Rift Valley fever but also many that are relatively uncommon and not well understood. Effective vaccines are available for a few (Japanese encephalitis and yellow fever) but vaccines will not eliminate the zoonotic cycles of these viruses. Several arboviruses, especially the hemorrhagic fever viruses, are of significant concern as agents of bioterrorism due to their high mortality and a lack of effective treatment or prevention.

Dengue. Dengue causes an estimated 100 million cases each year and is the most widespread and rapidly emerging arboviral disease. The spread of multiple strains (up to 4) to new areas in Latin America, greatly increased the risk of more serious hemorrhagic manifestations correlated with sequential infection by different strains. There were major epidemic of dengue between the 17th and early 20th centuries but it was completely eliminated from the new world following the near eradication of its mosquito vector, *Aedes aegypti,* during the 1950's and 1960's in concert with PAHO yellow fever control efforts in the region. Similar efforts were not undertaken in Asia where dengue became the most important childhood disease following WW II. In the 1980's the picture changed as dengue returned to the new world (first in Cuba) and invaded the Pacific Islands. Each decade the number of cases has doubled to the point where nearly a million hospitalized cases a year are now estimated worldwide (Fig. 5). Because of the strong domestic habits of the main vector, community-based campaigns have been used in efforts to control dengue. Successes were unsustainable without government assistance. Highly efficient control is needed to eliminate disease because even low populations can maintain transmission. Work on vaccines is continuing and efforts to introduce new vector control strategies is underway as well but the worldwide epidemic continues to grow unabated.

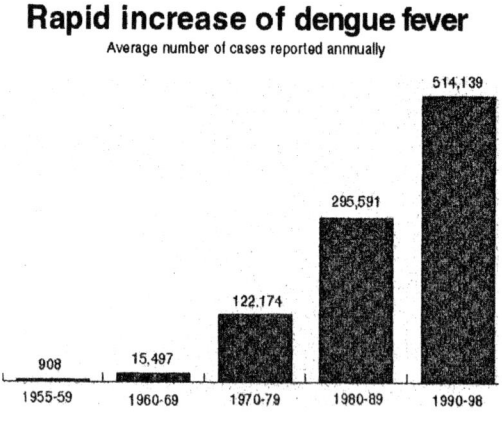

Figure 5. Re-emergence of Dengue in the Western Hemisphere.

West Nile. West Nile virus invaded the western hemisphere in 1999 and caught the U.S. unprepared. It should not have. It crystallized the fact that we live with unrealistic expectations. Microbial invasion is part of the landscape. WN exposed public health shortcoming and demonstrated our inability to effectively contain a major mosquito-borne epidemic. The same ULV fog trucks, spray planes and pesticides used >25 years ago were still our only emergency defense. Perhaps even more distressing is the fact that the effectiveness of these tools remains uncertain. Controversy over pesticide use resurfaced but for a change there were strong proponents because people were dying and others felt threatened. The question on most lips was how did an African virus find its way to New York City? Although there are several possibilities, the truth will never be known. Based on oligonucleotide fingerprints, the NY virus most closely matched an isolate from Israel. One of the many daily non-stop flights between Israel and NYC could easily have provided the entry for an infected mosquito.

The hot dry summer of 1999 promoted polluted water and produced an abundance of bird-feeding *Culex pipiens* mosquitoes which quickly amplified the virus. Crows and other corvids in particular have high viremias and many die. Dead crows are an important surveillance indicator of virus activity. Over the next 4 years the virus spread rapidly across the U.S. reaching the West Coast late in 2002. The number of human and horse cases rose precipitously in the summer of 2002 when large metropolitan areas in the Midwest and rural areas in the Great Plains were invaded (Table 2).

Table 2. Summary West Nile in the USA, 1999-2002*

YEAR	STATES	HORSES CASES	STATES	HUMANS CASES
1999	1	25	1	62
2000	7	60	3	21
2001	20	738	10	66
2002	40	14,717	40	4,156

* Fatality rates ~7% for hospitalized humans and ~30% for horses.

The summer of 2003 began with significant early activity in birds, horses and humans and the wet spring produced and abundance of mosquitoes in many regions. The U.S. is bracing for an even greater number of cases in 2003. Surveillance and control programs have been strengthened in many states as a direct result of this epidemic which is continuing to expand. West Nile is now a permanent part of the North American landscape but its long-term impact remains uncertain.

DISEASE CONTROL STRATEGIES

Most vectorborne infections are zoonotic, with wild animal reservoirs, which means that complete elimination of these diseases is unlikely. Only typhus, dengue and malaria rely on human reservoirs. Drugs for treating parasitic and bacterial diseases are generally available but not for arboviral diseases. No effective vaccines yet exist for any human parasitic disease and few vaccines against vectorborne bacterial diseases have been developed. Those that exist are not widely used. Vaccines against some arboviruses exist or are under development but the one most needed---for dengue, has been exceedingly challenging. Integrated vector control is the only real option for managing many vectorborne diseases. History tells use that vector control can be effective (Table 3).

Table 3. Successful Vector Control/Elimination Programs (Gubler 1998)

Yellow Fever	Cuba	1900-1901
Yellow Fever	Panama	1904
Yellow Fever	Brazil	1932
Anopheles gambiae	Brazil	1942
Anopheles gambiae	Egypt	1942
Epidemic Typhus	Italy	1942
Malaria	Sardinia	1946
Aedes aegypti/Dengue/YF	Americas	1947-1970's
Malaria	Americas	1955-75
Malaria	Global	1955-75
Yellow Fever	West Africa	1950-70
Onchocerciasis	West Africa	1974-present
Bancroftian Filariasis	South Pacific	1970's
Chagas Disease	South America	1991-present

Unfortunately the biggest success (i.e. the worldwide malaria eradication campaign) also turned into the biggest failure because the goal of eradication was not achieved. This along with the broadscale attack on pesticides during the environmental movement beginning in the 1960's gave vector control an undeserved black eye. If people were more aware of these successes and of the health and economic benefits that could be achieved, they may demand that current tools and technologies (such as impregnated bed nets, and biopesticides) be utilized in a more concerted effort to reduce the burden of these diseases. Much more could be achieved if resources and political will were present. It has been estimated that nearly half of all deaths from infectious disease are preventable if the knowledge and tools already available were only applied.

NEW AND BETTER VECTOR CONTROL TOOLS

The recent American Institute of Medicine report (IOM 2003) calls for a renewed effort to rebuild the public health infrastructure needed to conduct disease surveillance and oversee vector control programs in the United States. It also calls for training and for research that will provide:

- Improved Pesticides
- Novel Strategies to Prolong Pesticide Usage
- New Repellents
- New Biopesticides and Biocontrol Agents to Augment Chemical pesticides
- Novel Strategies to Interrupt Pathogen Transmission

The threat of bioterroism has energizes these discussions in ways that none could have predicted. Now that the threat is in our own communities and public awareness and concern has been heightened by WN and SARS, perhaps some fundamental changes will take place. New advocates and investments in vector control seem likely and vector control could be on the verge of experiencing its own re-emergence.

REFERENCES

1. Gubler, D.J., 1998, Resurgent Vector-Borne diseases as a global health problem., Emerging Infectious Diseases. 4: 442-450.
2. Githeko, A.K., S.W. Lindsay, U.E. Confalonieri and J.A. Patz, 2000, Climate change and vector-borne diseases: a regional analysis. Bull. World Health Organization. 78:1136-1147.
3. Institute of Medicine, 2003, Microbial Threats to Health, Emergence, Detection and Response., Smolinski, Hambureg & Lederberg (eds). The National Academies Press. 396 pp.
4. WHO and CDC websites

Chapter 31

Olyset Net, a Long Lasting Insecticidal Net for Malaria Control

Takaaki Itoh

Sumitomo Chemical Company, Ltd., 5–33, Kitahama 4-chome, Chuo-ku, Osaka 541-8550, Japan

Olyset net is incorporated with 2% permethrin during manufacturing of yarn. It has a long lasting effect on mosquitoes for more than 7 years. WHO Pesticide Evaluation Scheme evaluated the efficacy of Olyset net and highly recommended its use for malaria control. Roll Back Malaria initiated by WHO with its partners set the objective at 50% reduction of malaria burden by 2010 and is requesting a huge number of long lasting insecticidal nets for scaling up the campaign, especially in Sub Sahara. Since an affordable price for African people is essential for scaling up, collaborative work for local production of LLIN in Tanzania to reduce costs is in progress among partners of RBM.

More than one million people are killed by malaria each year, about 3,000 daily. It is estimated that over 700,000 children under 5 years of age die of malaria and at least 300 million people suffer from malaria each year.

The objective of Roll Back Malaria (RBM) initiated by WHO and its partners in 1998 was a 50% reduction of malaria burden by 2010 through the following actions.

1. Providing prompt access to effective treatments
2. Preventing and controlling malaria during pregnancy
3. Promoting the use of insecticide-treated mosquito nets as a means of prevention
4. Dealing effectively with malaria in emergency and epidemic situations.

To prevent malaria transmission, RBM originally promoted the campaign with mosquito nets impregnated with insecticides at an appropriate interval through community participation. However, RBM came to recognize that re-treatment was impossible in remote areas and re-treatment ratio after the 1st washing was very low even in urban areas. Thus, it changed the strategy from the system of re-treatment to the use of long lasting insecticidal nets (LLIN) which are effective throught the whole lifespan of the net itself.

Olyset net was developed to meet the above requirement. The net is incorporated with 2 % permethrin during the manufacturing process of yarn and characterized with the following points.

1. Long lasting effect for more than 7 years under normal use conditions
2. Good ventilation for sleepers with large mesh size.
3. No necessity for re-treatment after washing.
4. Long durability of net itself.

Characteristics of Olyset net

1) Large mesh size

The females of *Culex pipiens pallens* were allowed to pass through different mesh sizes of nets and observed their behaviors by which the mosquitoes passed through.

When the mosquitoes passed through the net with small mesh (0.8 x 0.8 cm), all of them landed on it first, then walked through a mesh. It is reasonable to assume that since the wing expanse of flying mosquitoes is larger than 0.8 cm, they cannot fly through the mesh without touching the net. When mosquitoes passed through the mid-size mesh (1.6 x 1.6 cm), a few of them landed on the net, however, none of them rested on the large (3.2 x 3.2 cm) mesh (Table I). These observations indicate that a net mesh size slightly smaller than the unfolded wings of a flying mosquito will cause the mosquito to rest on or contact the net when it attempts to pass through.

Table I. Observation of female *Culex pipiens pallens* passing through nets of different mesh sizes.

Mesh size (cm)	No. passing through net	No. landing on net	Landing % on net	Av resting time
0.8 x 0.8	10.7	10.7	100	23 sec
1.6 x 1.6	32.0	2.0	6.3	23 sec
3.2 x 3.2	20.0	0.0	0.0	-

(Itoh, T. et al.1986)

Thus, the mesh size of Olyset net was decided to be approximately 0.4 x 0.4 cm, while that of conventional nets is approximately 0.2 x 0.2 cm.

2) **Durability to washing**

Olyset net and polyethylene net treated with permethrin at 0.5g per m^2 were washed four times with soap and water,, then subjected to a 10 min contact bioassay using females of *Aedes aegypti* and *Anopheles maculatus*. The mortalities of *Ae. aegypti* and *An. maculatus* were 90.3 and 86.7%, respectively, for Olyset net, and 5 and 3.3%, respectively, for polyethylene net after the 4th washing (Table II),.

Table II Mortality (%) of mosquitoes after 24 hrs of momentary contact to the nets

Net	Mosquitoes	1st wash	2nd wash	4th wash
Olyset net	*Ae. aegypti*	100	93.8	90.3
	An. maculatus	100	95.8	86.7
Polyethylene net	*Ae. aegypti*	62.5	41.7	5.0
	An .maculatus	58.3	18.3	3.3

(Vythilingam et al.1996)

3) Regeneration of efficacy after washing

Unused Olyset net was washed with acetone to remove the active ingredient from the surface of thread. The washed net was kept at prescribed temperatures and submitted to 3 min contact bioassay using females of *Ae.aegypti*. When the washed nets were kept at 25 to 50°C for 5 hours, mortalities of mosquitoes were lower than that of unwashed net, whereas the nets kept at 60°C for 5 hours showed mortalities that were equivalent to that of an unwashed net (Table III). Thus, bleeding of the active ingredient from the inside of thread to the surface was accelerated by keeping net at 60 to 70°C for a certain hours after washing.

Table III Percentages of knock down and mortality of mosquitoes after 24 hrs of momentary contact to the nets

Temperature (°C)	Exposure Time (hr)	Knock down % after 20 min.	Mortality %
25	-	30	20
40	5	10	17
50	1	27	33
50	5	100	37
60	1	100	50
60	5	100	100
70	1	100	100
70	5	100	100
Unwashed	-	100	100

(Itoh and Okuno, 1996)

4) Impact of Olyset net on transmission of malaria in endemic areas

<Profile of field experiment>

Study area: Forest and mountain area, 350 km away from Phnom Penh

1. Olyset: 165 houses comprising 183 families with a total population of 860 inhabitants.
2. Control: 174 houses comprising 208 families with a total population of 1000 inhabitants, 1 km away from Olyset area

Species of Vector: *Anopheles dirus* and *An. minimus*
Malaria situation: 60 % of *Plasmodium falciparum*, 30 % of *P. vivax* and 10% of the mixed pathogens.

Olyset nets and conventional mosquito nets were distributed to Olyset and control areas, respectively, on July 10-13, 1994.. The mass blood survey was done on mid June 1994 for baseline data and repeated once every month in both areas during the use of nets for the comparison of slide positive rate (SPR). In each survey a sample of 100 slides (50 from children >5 years and 50 from children < 5 years of age) was collected.

Table IV Slide Positive Rate(%) by the mass blood survey [comments; how was the reduction % calculated? What did the reduction % in presample (baseline) data (Jun 12-14) mean?]

Date	Olyset area	Control area	Reduction%
June 12-14 (Pre-)	38.0	41.0	7.3
July 12-14	22.0	29.0	24.1
Aug.12-14	10.0	16.0	37.5
Sep.12-14	4.0	11.0	63.6
Oct.12-14	0	9.0	100
Nov.12-14	0	7.0	100
Dec.12-14	0	8.0	100

(Ministry of Health, Combodia, 1994)

The monthly slide positive rate from mass blood surveys clearly decreased after bed net use in both Olyset and control area (Table IV). However, the impact of Olyset use was apparently greater than that of conventional mosquito nets on the transmission of malaria.

5) Effective duration of Olyset under practical use conditions

Olyset nets which have been used daily in Tanzania for 7 years were randomly collected through WHO and analyzed for permethrin.. Results indicated that the average persistence of permethrin in the collected nets was 42.0% (Table V).

Table V Percentage of permethrin Remaining in the nets used daily for 7 years in Tanzania

[comments: I do not know what is permethrin % and remaining %. You may need to place a note under the table. I assume the remaining % is (the amount of permethrin extracted now/2%). Why are some of the values different?]

Net No.	Permethrin %	Remaining %
1	0.72	36.1
2	0.59	42.6
3	0.48	24.2
4	0.82	41.4
5	0.67	33.5
6	0.42	21.2
7	1.17	58.3
8	1.27	63.7
9	0.89	44.7
10	1.08	54.2
Average	0.81	42.0

(Itoh, T.,2002)

Nets 1, 6 and 8 in the above table were washed with soap and water, heated under various conditions and subjected to a 3 min contact bioassay using females of *Ae. aegypti* in comparison with unused net. [comments: Please add a sentence expressing What is the conclusion for this study] (Table VI)?

Table VI Percentage of knock down and mortality of mosquitoes after 24 hrs of momentary contact to the nets.

Sample No (Permethrin %)	Regeneration	KT_{50} (min)	Mortality (%)
No.1 (0.72%)	Room Temp.	>20	0
	70°C for 2 hr	7.2	65
	90°C for 2 hr	2-5	100
	Boiling for 30 min	2-5	100
No.6 (0.42%)	Room Temp.	>20	0
	70°C for 2 hr	>20	20
	90°C for 2 hr	2-7	90
	Boiling for 30 min	13.1	20
No.8 (1.27%)	Room Temp.	>20	5
	70°C for 2 hr	6.0	74
	90°C for 2 hr	2.0	100
	Boiling for 30 min	2.0	100
Unused Olyset	-	2.0	100
Blank net	-	>20	0

(Itoh,T.,2002)

The Olyset nets which have been used daily for 5 years are still expected to show insecticideal activity against mosquitoes by appropriate regeneration processes. [Comments: The 7 years old nets still show good insecticidal activity. Why do 5 years old nets need do regenetion to get insecticidal activity?]

Perspective as a partner of RBM

WHO Pesticide Evaluation Scheme has completed the evaluation of efficacy of Olyset nets and recommended their use for malaria control. Since RBM is requiring a large number of LLITN to cover the endemic area in the Sub Sahara of Africa at an affordable price, we are planning to transfer Olyset technology to African net manufacturers to accelerate local production of LLIN in Tanzania

Reference

1) Itoh, T. et al. (1986) Studies on wide mesh netting Impregnated with insecticides against *Culex* mosquitoes. J.Am.Mosq.Contol Associ. 2(4):503-506
2) Itoh, T. and T.Okuno (1996) Evaluation of the polyethylene net incorporated with permethrin during manufacture of thread on efficacy against *Aedes aegypti* (Linnaeus). Med. Entomol. Zool. 47(2):171-174

4) Ministry of Health, Combodia (1994) Final report on a field trial of Olyset net for the control of malaria transmitted by *Anopheles dirus* and *Anopheles minimus* in Rattanak Kiri Province, Cambodia 1994. Unpublished data.
5) Vythilingam, I. et al.(1996) Assessment of a new type of permethrin impregnated mosquito net. J.Biosci. 17(1) 63-70

Chapter 32

Process of Action of Dipteran-Specific Insecticidal Crystal Proteins from *Bacillus thuringiensis* subsp. *israelensis*

Hiroshi Sakai and Masashi Yamagiwa

Department of Bioscience and Biotechnology, Faculty of Engineering, Okayama University, Okayama 700–8530, Japan

Dipteran-specific insecticidal protein Cry4A is produced as a protoxin of 130 kDa in *Bacillus thuringiensis* subsp. *israelensis* (Bti). The 130-kDa protoxin was processed *in vitro* into a 60-kDa intermediate, which is subsequently converted into two protease-resistant fragments of 20 and 45 kDa through intramolecular cleavage. The 20-kDa and 45-kDa fragments, each of which alone was inactive, were associated to form a complex actively toxic against the larvae of mosquito *Culex pipiens*. Thus, it was strongly suggested that the complex composed of the two fragments is an active form of Cry4A. Immunohistochemical analyses revealed that Cry4A bound to the epithelial cells of mosquito larval midgut. Several Cry4A-specific binding proteins were detected by the ligand blotting experiments using digoxigenin-labeled Cry4A as a probe. Moreover, it was suggested that the mode of Cry4A binding to the target cell membrane was distinct from that of the Cry1 toxin family. Bti produces another dipteran-specific insecticidal protein Cry11A. Upon *in vitro* processing of the 70-kDa Cry11A protoxin, the 36-kDa and 32-kDa fragments were produced. The two fragments, each of which alone was inactive, were associated to form a complex actively toxic against the larvae of mosquito.

Introduction

Bacillus thuringiensis is a gram-positive soil bacterium that produces crystalline inclusions consisting of highly specific insecticidal proteins called δ-endotoxins during sporulation. These inclusions consist of insecticidal proteins called δ-endotoxins toxic to lepidopteran, dipteran, and coleopteran insect larvae (*1, 2*). The δ-endotoxins are biosythesized as protoxins in Bt cells and, upon ingestion by susceptible insect larvae, solubilized and proteolytically activated by gut proteases in the larval midguts (*3*). The activated toxin binds to a receptor in the microvilli of epithelial cells of the midgut (*4-6*), forming a pore in the cell membrane. Colloid-osmotic swelling and lysis of the cell result in the death of the larvae.

B. thuringiensis subsp. *israelensis* produces crystalline inclusions containing dipteran-specific δ-endotoxins, Cry4A, Cry4B and Cry11A, and non-specifically cytotoxic Cyt1A (*2, 7, 8*). The 130-kDa protoxins of the lepidopteran-specific δ-endotoxins are converted to protease-resistant active forms of 60- to 70- kDa through processing by the gut proteases (*2*). However, the protoxins of the dipteran-specific Cry4A and Cry11A are processed and activated differently.

Binding of activated toxins to a specific binding site(s) in the brush border membrane of midgut epithelial cells is thought to be a major determinant of the specificity of Cry insecticidal proteins (*9-11*). In the last few years, several toxin-binding proteins have been purified, and genes encoding them have been cloned. However, in the case of dipteran-specific δ-endotoxins, no receptor protein has been found so far (*12, 13*).

In the present study, the process of action of the dipteran-specific δ-endotoxin, Cry4A and Cry11A, was examined. We revealed the processing pathway of the Cry4A and Cry11A protoxins followed by production of the complexes actively toxic against the larvae of mosquito *Culex pipiens*. Membrane binding characteristics of an active form of Cry4A was examined. Moreover, Cry4A-specific binding proteins were detected in the target cell membrane from the mosquito larvae.

Materials and Methods

***In vitro* Processing of Cry Proteins.** The crystal was solubilized in 100 mM Na_2CO_3 (pH10.5)/10 mM DTT at 4°C followed by treatment with trypsin or the gut extracts of *C. pipiens* at 30°C. At appropriate times during the treatment, each sample was taken and immediately frozen at -20°C after adding PMSF to give a final concentration of 1 mM.

Bioassay of the Mosquito Larvicidal Activity of Cry Proteins. Bioassay of the Cry proteins was done essentially by the reported method (*14*). The mosquito larvicidal activities were assayed on 4th instar larvae of *C. pipiens*. One of each of the mosquito larva was transferred to 200 μl of distilled water in each well of a 96-well plate followed by incubation at 25°C for 8 hr. The Cry proteins, which had been adsorbed to latex beads if necessary, were added. The mortality was scored after 12-hr incubation at 25°C. The assay was performed more than three times, in each of which 48 larvae were used. The efficiency of adsorption of protein to the latex beads was almost 100% in a preliminary experiment.

Coprecipitation Experiments. A fusion polypeptide, in which a histidine hexamer was attached to the C-terminus, had been adsorbed to Ni-matrix. To this was added another fusion polypeptide, in which GST was attached to the N-terminus. After incubation at 4°C for 2hr, the mixture was centrifuged. The supernatant and the precipitate of Ni-matrix were analysed by SDS/14% PAGE followed by western blotting with anti-GST antibody.

Immunohistochemical Staining Experiments. Sections 8 μm thick were cut from the fixed larvae of *C. pipiens* on a cryostat, thaw-mounted onto APS-coated slides and air-dried. To the mosquito larval section was applied DIG-labeled Cry4A (DIG-Cry4A) in the *in vitro* experiments. In the *in vivo* experiments, the Cry4A crystal was given to the mosquito larvae before preparing the thin sections. The bound DIG-Cry4A was visualized with immunostaining using anti-DIG antibody, HRP-conjugated secondary antibody, and DAB.

Preparation of BBMV. The BBMV was prepared from whole larvae of *C. pipiens* essentially by the reported method (*15*).

Ligand Blotting. The proteins of BBMV were separated by SDS/14% PAGE, and transferred to a PVDF membrane. The ligand blotting experiment was performed essentially by the reported method (*16*). The DIG-labeled Cry4A bound to the membrane protein was visualized with anti-DIG antibody conjugated with HRP using ECL western blotting detection reagents (Amersham Pharmacia Biotech).

Insertion Assay. To examine the interaction of Cry proteins with the mosquito larval BBMV, the insertion assay was done. The DIG-labeled Cry protein was incubated with the BBMV for 2 hr at room temperature, and the mixture was centrifuged. Proteins in the pellet of the membrane fraction was analyzed by SDS/14% PAGE, and transferred to a nitrocellulose membrane. The DIG-labeled Cry protein bound to the BBMV was detected with the anti-DIG antibody conjugated with HRP using ECL western blotting detection reagents (Amersham Pharmacia Biotech).

Results and Discussion

In the *B. thuringiensis* subsp. *israelensis* cells, Cry4A is produced as a protoxin of 130kDa. Upon ingestion by susceptible insect larvae, the Cry4A protoxin is processed, through proteolytic cleavage with proteases in the digestive juice, producing an active form. That the processing of the Cry4A protoxin proceeds in two stages was found *in vitro* (*17*). Upon treatment with the gut extracts from the larvae of mosquito *C. pipiens*, the 130-kDa protoxin was processed into the 60-kDa intermediate through proteolytic cleavage. Then, further proteolysis of the 60-kDa intermediate generated the 20-kDa and the 45-kDa fragments **(Figure 1)**.

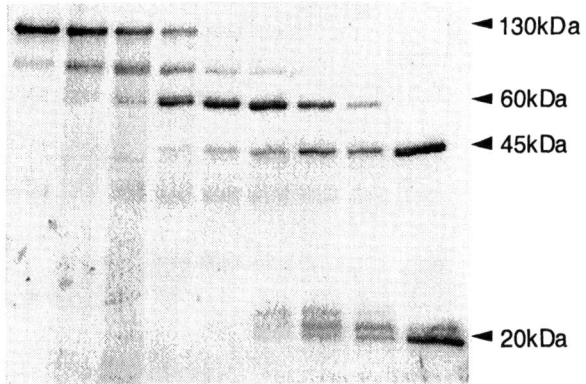

Figure 1. In vitro *processing of Cry4A was examined. Fifty micrograms of the Cry4A protoxin were treated with 5 mg of the gut extracts from larvae of* C. pipiens *at 30°C in a solution of 100 mL. At the indicated times during the treatment, each sample containing 1 mg of the Cry4A protein was taken and analysed with SDS/14% PAGE followed by CBB staining.*

When the crystal of Cry4A was ingested by the larvae of *C. pipiens*, two fragments of Cry4A were produced in 15 min, the sizes of which were quite similar to those of the Cry4A fragments produced *in vitro* (*17*). N-Terminal amino acid sequences of the Cry4A fragments were determined. The N-terminal amino acids of the 60-kDa, 20-kDa and 45-kDa fragments were Glycine58, Glycine58 and Glutamine236, respectively. Therefore, in the *in vitro* processing, the 60-kDa intermediate was produced through a proteolytic removal of the N-terminal 57 amino acid residues and the C-terminal tract of the 130-kDa protoxin of Cry4A. Then, the 20-kDa and 45-kDa fragments were produced through the intramolecular cleavage at Glutamine236 in the 60-kDa intermediate (**Figure 2**).

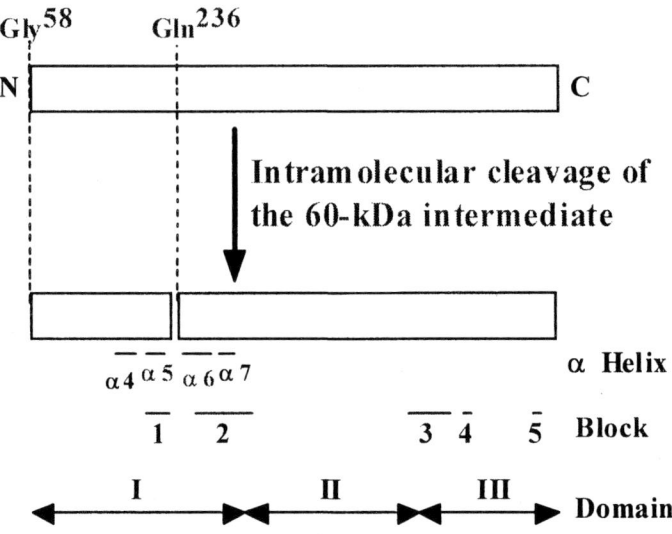

Figure 2. In vitro *processing products from Cry4A with the gut extracts of* C. pipiens *larvae.*

Fusion proteins were constructed by attaching GST or a histidine hexamer to the Cry4A fragments (**Figures 3 and 4**). GST-20 and GST-45 were fusion proteins in which GST was attached to the N-termini of the 20-kDa and 45-kDa fragments, respectively. 45His was a fusion protein in which a histidine hexamer was attached to the C-terminus of the 45-kDa fragment. Mutants of the fusion proteins were also constructed. GST-20Δα5 and GST-20Δ(α4+α5) were derivatives of GST-20 in which the putative α5, and both the putative α4 and α5 helices were deleted, respectively. 45HisΔα6 and 45HisΔ(α6+α7) were derivatives of 45His in which the putative α6, and both the putative α6 and α7 helices were deleted, respectively. Functional characteristics of the 20-kDa and 45-kDa fragments were examined with these fusion proteins.

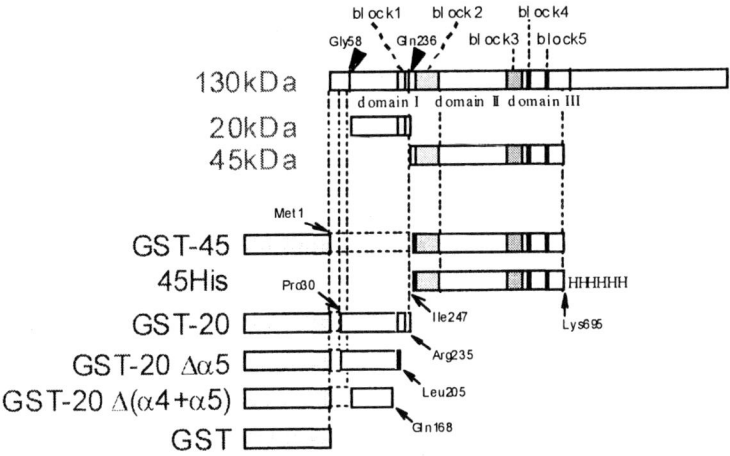

Figure 3. Maps of the fusion proteins of the Cry4A processed fragments and the mutants of the 20-kDa fragment. Putative domains and blocks are depicted.

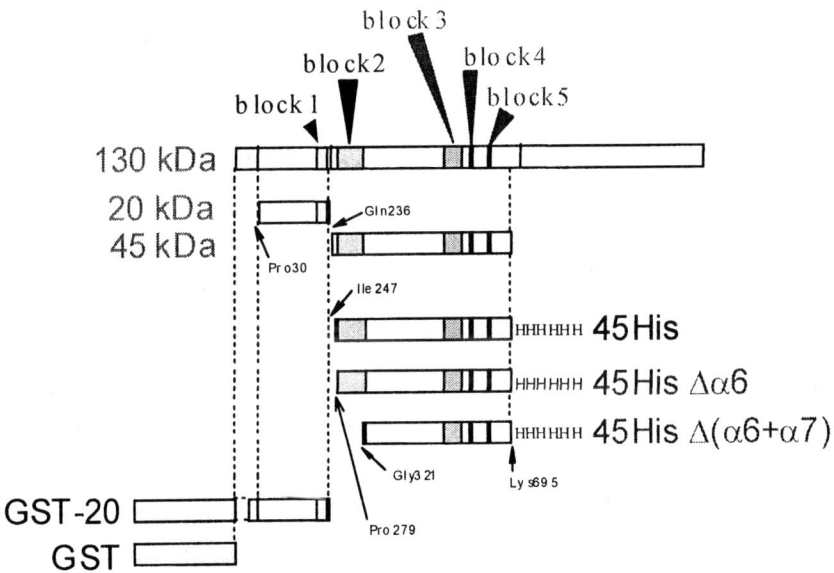

Figure 4. Maps of the fusion proteins of the Cry4A processed fragments and the mutants of the 45-kDa fragment. Putative domains and blocks are depicted.

Association of the 20-kDa and 45-kDa fragments was examined (data not shown). GST-20 was added to 45His that had been bound to the Ni-matrix. After centrifugation of the mixture, GST-20 in the supernatant and in the precipitate of 45His/Ni-matrix was analysed through gel electrophoresis followed by western blotting using anti-GST antibody. GST-20 was detected in the precipitate, not in the supernatant. Since GST itself was detected in the supernatant, but not in the precipitate, it was demonstrated that the 20-kDa and 45-kDa fragments actually associated to form a complex. Similar experiments were done with GST-20 and the mutants 45HisΔα6 and 45HisΔ(α6+α7). In the absence of the putative α6 helix, the association of the two fragments was not observed. In this case, no significant insecticidal activity was restored. The experimental results suggested that the putative α6 helix was involved in the association of the 20-kDa and 45-kDa fragments of Cry4A **(Figure 5)**.

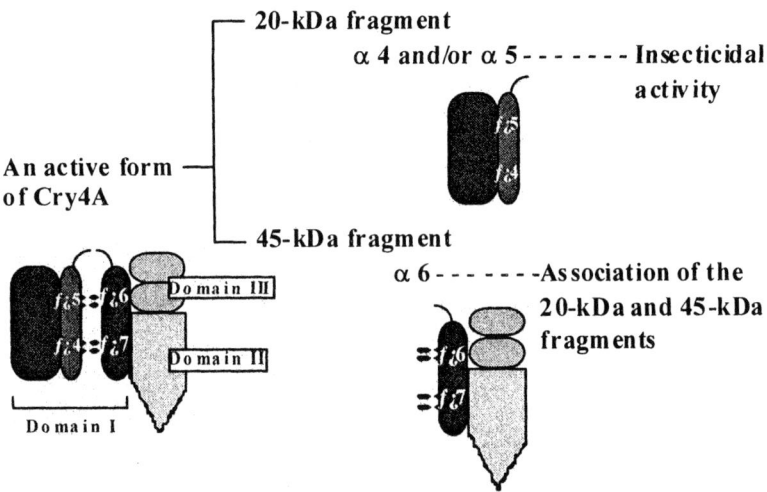

Figure 5. Schematic representation of an active form of Cry4A and its subunits. Putative α4 and/or α5 helices may be involved in determining the insecticidal activity. Putative α6 helix may be involved in the association of the 20-kDa and 45-kDa fragments.

Insecticidal activities of the Cry4A fragments were examined. Neither GST-20 nor GST-45 was toxic toward the larvae of *C. pipiens*. When GST-20 and GST-45 coexisted, the significant insecticidal activity was observed (*17*). The insecticidal activity was also examined with 45His and the mutant GST-20Δα5 or GST-20Δ(α4+α5) **(Figure 6)**. While the significant insecticidal activity was detected in the absence of the putative α5 helix, no significant activity was observed in the absence of both the putative α4 and α5 helices. According to the results of the present experiments and additional mutational analyses (data not shown), it was suggested that the putative α4 and/or α5 helices were involved in determining the insecticidal activity. Thus, we propose that an active form of Cry4A is a complex of the 20-kDa and 45-kDa fragments **(Figure 5)**.

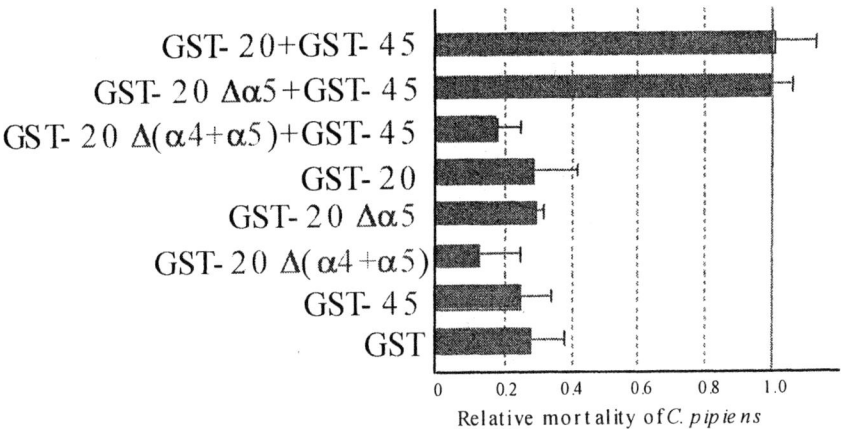

Figure 6. Insecticidal activity of the fusion proteins of the Cry11A fragments. The Cry4A fusion proteins were adsorbed to the latex beads, and given to the larvae of C. pipiens *to give a final concentration of 0.5 mg/mL.*

With immunohistochemical staining, we have examined binding of the processed Cry4A to epithelial cells of the larval midgut of *C. pipiens* (*18*). In the *in vivo* binding experiments, thin sections of the mosquito larvae were prepared after ingestion of the Cry4A crystal. And in the *in vitro* experiments, the thin sections were prepared with no toxin given to the larvae. Judging from the results of immunohistochemical staining with the thin sections from the larvae, the processed Cry4A bound to the epithelial cells of the anterior midgut and of the posterior midgut. The *in vitro* competitive binding experiments with DIG-labeled Cry4A and thin sections of the posterior midgut suggested that the binding to the epithelial cell was specific (data not shown).

In the ligand blotting experiments using processed and DIG-labeled Cry4A as a probe, some proteins that bound specifically to Cry4A were found in the BBMV from the larvae of *C. pipiens* (*18*). Sizes of the proteins ranged from 20 to 40kDa. However, analyses of the processed Cry4A binding to the BBMV from the mosquito larvae in the insertion assay suggested that, in addition to the receptor-mediated binding, Cry4A bound directly to the membrane in a manner independent of the receptor (*18*). Thus, we propose that Cry4A binds to the membrane through two modes of interaction. One is receptor-mediated binding, leading to the formation of functional channels. And the other is direct and receptor-independent binding to the membrane (**Figure 7**).

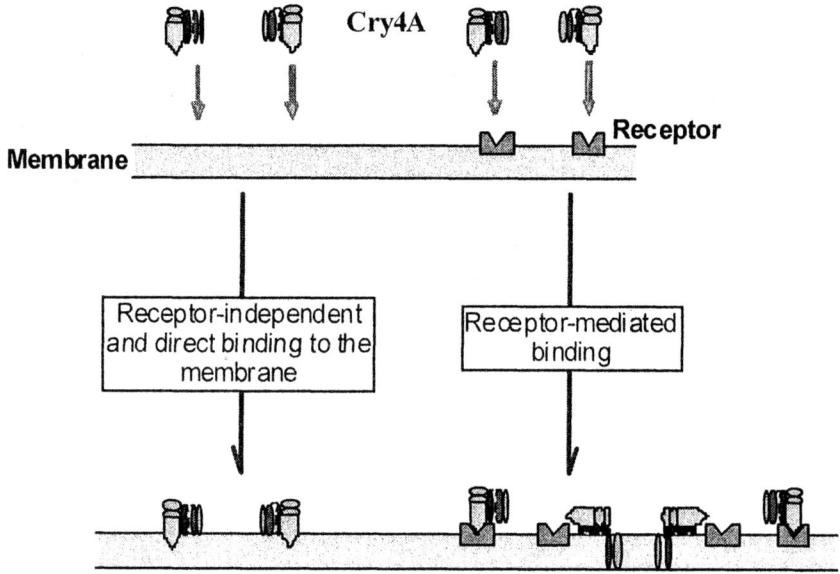

Figure 7. The predicted modes of membrane interaction of Cry4A.

Cry11A is another dipteran-specific insecticidal protein of 70kDa produced by Bti. The 70-kDa protoxin of Cry11A was processed into the 36-kDa and 32-kDa fragments upon *in vitro* treatment with trypsin. When the protoxin was treated *in vitro* with the gut extract from the larvae of *C. pipiens*, it was processed into the 34-kDa and 32-kDa fragments. The 36/34-kDa and 32-kDa fragments seemed to be associated since the two behaved together in the gel filtration chromatography. The aggregate of the 36/34-kDa and 32-kDa fragments was actively toxic toward the larvae of *C. pipiens*. The insecticidal activity was lower than that of the solubilized Cry11A crystal. Moreover, the processed Cry11A bound to the BBMV from the larvae of *C. pipiens* in a concentration-dependent manner (*19*).

We have determined amino acid sequences in the N-terminal regions of the Cry11A fragments. For the tryptic fragments, the N-terminal amino acid residues of the 36-kDa and 32-kDa fragments were Serine10 and Aspartate361, respectively **(Figure 8)**. For the fragments produced by the treatment with the larval midgut extracts, the N-terminal amino acid residues of the 34-kDa and 32-kDa fragments were Isoleucine28 and Aspartate361, respectively **(Figure 8)**. The results suggested that, upon *in vitro* processing, the 70-kDa protoxin of Cry11A was proteolytically cleaved at Serine10 or Isoleucine28, and at Aspartate361, producing the 36-kDa or 34-kDa, and 32-kDa fragments **(Figure 8)**.

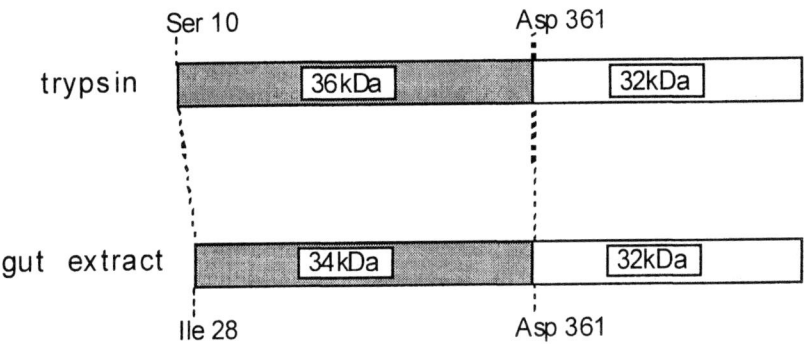

Figure 8. In vitro *processing of Cry11A with trypsin and the gut extracts from larvae of* C. pipiens.

Fusion proteins were constructed by attaching GST or a histidine hexamer to the tryptic fragments of Cry11A **(Figure 9)**. GST11A-36k and GST11A-32k were fusion proteins, in which GST was attached to the N-termini of the 36-kDa and 32-kDa fragments, respectively. 32k-His was a fusion protein, in which a histidine hexamer was attached to the C-terminus of the 32-kDa fragment. The fusion protein GST11A was actively toxic to the larvae of *C. pipiens* **(Figures 9 and 10)**. The mixture of GST11A-36k and GST11A-32k was as active as the fusion protein GST11A (data not shown). However, each of the two Cry11A fragments alone exhibited no significant insecticidal activity **(Figure 10)**. To examine the association of the 36-kDa and 32-kDa fragments,

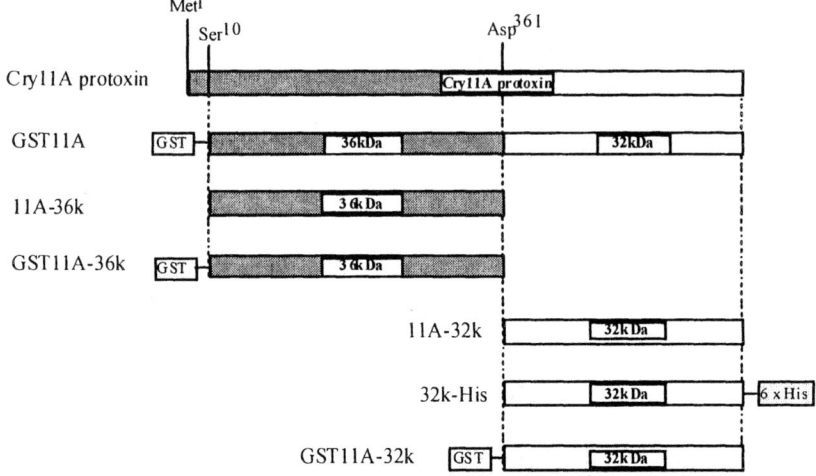

Figure 9. Maps of the fusion proteins of the Cry11A processed fragments

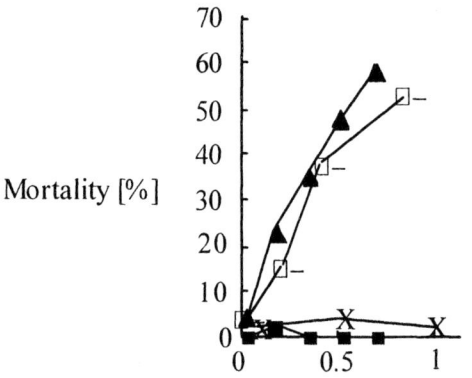

Figure 10. The insecticidal activity of the fusion proteins of the Cry11A processed fragments. ▲, GST11A; ■, GST11A-36k; X, GST11A-32k; *, GST11A-36k + GST11A-32k.

coprecipitation experiments were done with the fusion proteins of the Cry11A fragments. It was indicated that GST11A-36k and 32k-His were associated with each other (data not shown). These results indicated that the proteolytic cleavage at Aspartate361 in the middle of the 70-kDa protoxin of Cry11A has essentially little effect on the biological function of Cry11A, and that the aggregate of the 36-kDa and 32-kDa fragments of Cry11A is one of the possible active forms of Cry11A.

Abbreviations

Bt, *Bacillus thuringiensis*; DTT, dithiothreitol; PMSF, phenylmethylsulfonyl fluoride; GST, glutathion-S-transferase; PAGE, polyacrylamide gel electrophoresis; DIG, digoxigenin; HRP, horseradish peroxidase; DAB, diaminobenzidine tetrahydrochloride; BBMV, brush border membrane vesicle; Bti, *Bacillus thuringiensis* subsp. *israelensis*

Acknowledgements

We are grateful to the Dainihon Jochugiku Co., Ltd. for providing us with eggs of *C. pipiens*. This work was supported by Grant-in-Aid for Young Scientists (B) to M. Y, and by Grant-in-Aid for Scientific Research (C) to H. S. from the Ministry of Education, Culture, Sports, Science and Technology of the Japanese Government.

References

(1) Gill, S. S.; Cowles, E. A.; Pietrantonio, P. V. The mode of action of *Bacillus thuringiensis* endotoxins. *Annu. Rev. Entomol.* **1992** *37*, 615-636.
(2) Höfte, H.; Whiteley, H. R. Insecticidal crystal proteins of *Bacillus thuringiensis*. *Microbiol. Rev.* **1989** *53*, 242-255.
(3) Lilley, M.; Ruffell, R. N.; Somerville, H. J. Purification of the insecticidal toxin in crystals of *Bacillus thuringiensis*. *J. Gen. Microbiol.* **1980** *118*, 1-11.
(4) Hofmann, C.; Vanderbruggen, H.; Höfte, H.; Van Rie, J.; Jansens, S.; Van Mellaert, H. Specificity of *Bacillus thuringiensis* delta-endotoxins is correlated with the presence of high-affinity binding sites in the brush

border membrane of target insect midguts. *Proc. Natl. Acad. Sci. U. S. A.* **1988** *85*, 7844-7848.

(5) Hofmann, C.; Luthy, P.; Hutter, R.; Pliska, V. Binding of the δ-endotoxin from *Bacillus thuringiensis* to brush border membrane vesicles of the cabbage butterfly (*Pieris brassicae*). *Eur. J. Biochem.* **1988** *173*, 85-91.

(6) Van Rie, J.; Jansens, S.; Höfte, H.; Degheele, D.; Van Mellaert, H. Receptors on the brush border membrane of the insect midgut as determinants of the specificity of *Bacillus thuringiensis* δ-endotoxins. *Appl. Environ. Microbiol.* **1990** *56*, 1378-1385.

(7) Aronson, A. I.; Beckman, W.; Dunn, P. *Bacillus thuringiensis* and related insect pathogens *Microbiol. Rev.* **1986** *50*, 1-24.

(8) Insell, J. P.; Fitz-James, P. C. Composition and toxicity of the inclusion of *Bacillus thuringiensis* subsp. *israelensis*. *Appl. Environ. Microbiol.* **1985** *50*, 56-62.

(9) Knight, P. J.; Crickmore, N.; Ellar, D. J. The receptor for *Bacillus thuringiensis* Cry1A(c) delta-endotoxin in the brush border membrane of ther lepidopteran *Manduca sexta* is aminopeptidase N. *Mol. Microbiol.* **1994** *11*, 429-436.

(10) Sangadala, S.; Walters, F. S.; English, L. H.; Adang, M. J. A mixture of *Manduca sexta* aminopeptidase and phosphatase enhances *Bacillus thuringiensis* insecticidal Cry1A(c) toxin binding and $^{86}Rb^+$ - K^+ efflux *in vitro*. *J. Biol. Chem.* **1994** *269*, 10088-10092.

(11) Vadlamudi, R. K.; Ji, T. H.; Bulla Jr., L. A. A specific binding protein from *Manduca sexta* for the insecticidal toxin of *Bacillus thuringiensis* subsp. *berliner*. *J. Biol. Chem.* **1993** *268*, 12334-12340.

(12) Ravoahangimalala, O.; Charles, J.-F.; Schoeller-Raccaud, J. Immunological localization of *Bacillus thuringiensis* serovar *israelensis* toxins in midgut cells of intoxicated *Anopheles gambiae* larvae (Diptera: Culicidae). *Res. Microbiol.* **1993** *144*, 271-278.

(13) Ravoahangimalala, O.; Charles, J.-F. *In vitro* binding of *Bacillus thuringiensis* var. *israelensis* individual toxins to midgut cells of *Anopheles gambiae* larvae (Diptera: Culicidae). *FEBS Lett.* **1995** *362*, 111-115.

(14) Schnell, D. J.; Pfannenstiel, M. A.; Nickerson, K. W. Bioassay of solubilized *Bacillus thuringiensis* var. *israelensis* crystals by attachment to latex beads. *Science* **1984** *223*, 1191-1193.

(15) Silva-Filha, M. H.; Nielsen-Leroux, C.; Charles, J. F. Binding kinetics of *Bacillus sphaericus* binary toxin to midgut brush-border membranes of *Anopheles* and *Culex* sp. mosquito larvae. *Eur. J. Biochem.* **1997** *247*, 754-761.

(16) Krieger, I. V.; Revina, L. P.; Kostina, L. I.; Buzdin, A. A.; Zalunin, I. A.; Chestukhina, G. G.; Stepanov, V. M. Membrane proteins of *Aedes aegypti*

larvae bind toxins Cry4B and Cry11A of *Bacillus thuringiensis* ssp. *israelensis*. *Biochemistry (Mosc.)* **1999** *64*, 1163-1168.
(17) Yamagiwa, M.; Esaki, M.; Otake, K.; Inagaki, M.; Komano, T.; Amachi, T.; Sakai, H. Activation process of dipteran-specific insecticidal protein produced by *Bacillus thuringiensis* subsp. *israelensis*. *Appl. Environ. Microbiol.* **1999** *65*, 3464-3469.
(18) Yamagiwa, M., Kamauchi, S.; Ogawa, R.; Esaki, M.; Otake, K.; Amachi, T.; Komano, T.; Sakai, H. Binding properties of *Bacillus thuringiensis* Cry4A toxin to the apical microvilli of larval midgut of *Culex pipiens*. *Biosci. Biotechnol. Biochem.* **2001** *65*, 2419-2427.
(19) Yamagiwa, M.; Ogawa, R.; Yasuda, K.; Natsuyama, H.; Sen, K.; Sakai, H. Active form of dipteran-specific insecticidal protein Cry11A produced by *Bacillus thuringiensis* subsp. *israelensis*. *Biosci. Biotechnol. Biochem.* **2002** *66*, 516-522.

Chapter 33

Development of Vaccines for the Control of Blood-Feeding Arthropods: The Combined Use of Proteomic and Genomic Strategies

Stephen K. Wikel

Center for Microbial Pathogenesis, School of Medicine, University of Connecticut Health Center, 263 Farmington Avenue, MC3710, Farmington, CT 06030

Hematophagous ectoparasites and the infectious agents they transmit are of vast medical and veterinary public health importance. Control of both vectors and infectious agents is challenging due to insecticide/acaricide resistance, drug resistance, and the lack of vaccines or effective drugs for many vector-borne infectious agents. Interest in vaccine-based control of hematophagous arthropods and vector-blocking vaccines to prevent pathogen transmission continues to increase. Identification, isolation, and characterization of candidate immunogens by standard biochemical methods are highly effective approaches, but they are relatively slow and labor intensive processes. Combining genomic, especially expressed sequence tags, and proteomic approaches provides promising alternative methods for identification of vaccine candidates that complement the more traditional approaches. Increasingly, interest is being focused on pharmacologically active molecules essential for successful blood-feeding and/or pathogen transmission as candidate immunogens for novel vaccines.

INTRODUCTION

Blood feeding insects and ticks are of vast medical and veterinary importance due to their roles in transmitting infectious agents to humans, wildlife, and domestic animals, which cause significant morbidity and mortality. Economic losses due to direct effects of ectoparasitic arthropods and vector-borne diseases are tremendous. Arthropod-transmitted diseases resulted in more human morbidity and mortality than all other causes during the period from the seventeenth to the early twentieth centuries (*1*). We are faced with previously unrecognized, emerging, infectious diseases and diseases once though to be under control are resurging in both incidence and changing geographic distribution (*2, 1, 3*). Latin America and the Caribbean are now hyperendemic regions for dengue and dengue hemorrhagic fever (*4*). Since 1982, 15 previously unknown tick transmitted bacterial pathogens have been identified (*3*). Emerging vector-borne wildlife diseases pose risks to the health of domestic animals and may potentially be zoonotic diseases (*5*).

The key element in control of arthropod-borne infectious agents is vector suppression (*6, 7*). In light of current control problems, novel strategies are needed to achieve effective protection against vector-borne diseases. Anti-arthropod vaccines have the potential to reduce blood feeding and pathogen transmission. Enhanced interest in this novel approach is due in part to drug resistance of pathogens (*8*) and arthropod resistance to insecticides and acaricides (*9, 10*). Advantages of anti-arthropod vaccines include: safety, cost, ease of administration, and target species specificity. Another factor contributing to interest in immunological-based control is the successful development of the anti-*Boophilus microplus* vaccine, which is based on induction of immunity to a tick gut antigens (*11*, 12, *13*). Research is being conducted to identify vaccine candidate immunogens for a number of ectoparasitic arthropods (*14, 12, 13, 15, 16*). The power of genomics and proteomics is contributing greatly to ongoing efforts to identify candidate immunogens and develop anti-ectoparasitic arthropod vaccines.

Ectoparasitic arthropods and host immune defenses

Hematophagous and tissue residing arthropods induce host innate and specific acquired immune responses that can reduce feeding, impair development, and kill the arthropod. Host immune responses that are damaging to an ectoparasitic arthropod are not universal phenomena. Acquired immunity to infestation depends on both the species of arthropod and host (*17*). Tick-host immune interactions are the most extensively studied (*18, 19, 20, 21, 22*). Host-arthropod immune interactions have been the subject of reviews for following groups: sucking lice, fleas, and bugs (*23*); flies, mosquitoes, and myiasis producing flies (*24*); scabies mites (*25*); and, chiggers and mange mites (*26*).

Blood feeding arthropod modulation of host immune defenses is important for both successful acquisition of the blood meal and transmission of infectious agents (*27*). Modulation of host immune defenses has been reviewed for black flies (*28*), sand flies (*29*), and ticks (*30, 21, 31*).

Arthropod saliva contains pharmacologically active molecules

Both hematophagous and tissue residing arthropods are confronted with a number of host associated threats to their existence: hemostasis, pain/itch responses, and immunity.

Pharmacologically active molecules in saliva of blood-feeding arthropods and released by tissue dwelling species are essential for their survival (*32, 33*). Likewise, blood-feeders and myiasis larvae have multiple ways to counteract host coagulation pathways, platelet aggregation, and vasoconstriction (*32, 34, 35, 28, 36, 29, 37*).

Rather than examining different groups of vector arthropods, this section will focus on ticks to provide an example of the complexity of the mixture of activities in the saliva of a blood feeding arthropod. Tick saliva contains a complex mixture of pharmacologically active molecules that facilitate both blood feeding and pathogen transmission (*38, 27*). Tick salivary gland pharmacology has been reviewed (*24, 17, 38, 39*). Ticks evolved countermeasures to host hemostasis, immune defenses, and pain/itch responses. Indicative of their importance, redundant mechanisms exist within individual tick species to counteract host defenses (*39*), and different tick species may utilize different strategies for modulating or counteracting the same host defensive pathway (*38*).

Pain and/or itch are important host responses that alert an individual to the presence of a tick, which in turn results in grooming behavior to remove the tick. Bradykinin is a peripheral mediator of the sensation of itch (*40*), and pain (*41*). *Ixodes scapularis* saliva contains a metallo dipeptidyl carboxypeptidase capable of degrading bradykinin (*42*). Bradykinin also has roles in inflammation and increases vascular permeability (*43*). Histamine is one of several mediators that transmit the sensation of itch from peripheral sensory nerve endings (*40*). Histamine is an important mediator of innate and acquired immune responses (*44*). Furthermore, both histamine and serotonin inhibit tick-feeding (*45*). Histamine-binding proteins are produced by the salivary glands of many tick species. Histamine-blocking activity was initially observed in the salivary gland homogenates of partially fed adult *Rhipicephalus sanguineus* (*46*). *Rhipicephalus appendiculatus* has three closely related histamine–binding proteins with two occurring in females and one in males (47). These proteins are unique lipocalins in that they possess one high affinity and a second binding site with low affinity for histamine (48). The second binding site of the *Dermacentor reticulatus* histamine-binding lipocalin is now known to bind serotonin, while the first binding site binds histamine (49). This serotonin binding site has higher affinity for serotonin than membrane receptors on target cells (49). A dual-binding lipocalin possibly reflects the need to modulate combinations of mediators produced by diverse host species (49).

Additional histamine-binding proteins were putatively identified by characterization of cDNAs prepared from mRNAs of salivary glands of *I. scapularis* (50) and *Amblyomma americanum* (51). As genomic and proteomic studies of tick salivary glands progress additional molecules that modulate these important mediators will certainly be described.

Blood coagulation results from activation of the extrinsic, tissue factor, and intrinsic pathways, which converge to activate factor X (Xa) to form the common pathway of coagulation (52). Factor Xa and activated factor V (Va) convert prothrombin to thrombin, which in turn covers fibrinogen into fibrin, which is the basis of all blood clots (52, 53). The intrinsic pathway is approximately 50 times more efficient at activating factor X than the extrinsic pathway (52). Pharmacologists are focusing considerable attention on factor X as a key target for development of novel drugs to control coagulation (53).

Blood feeding arthropods are expert pharmacologists, having evolved ways to circumvent or modulate host defenses (38). Ticks predominantly block host coagulation by targeting factor Xa and thrombin (54). Inhibitors of factor Xa occur in the saliva or salivary glands of the following tick species: *Ornithodoros moubata* (55), *R. appendiculatus* (56), *Hyalomma turicatum* (57), *Ornithodoros savignyi* (58), *Hyalomma dromedarii* (59), and *I. scapularis* (60). *Ixodes scapularis* saliva contains an inhibitor of activation of factor X by utilizing both factors X and Xa as scaffolds for inhibition of activated factor VII/tissue factor of the extrinsic coagulation pathway (*54*).

Blocking the action of thrombin prevents conversion of fibrinogen to fibrin (*52*). Thrombin inhibitors are produced by the salivary glands of *Ixodes holocyclus* (*61*), *Ixodes ricinus* (*62*), *A. americanum* (*63*), and *O. savignyi* (*64*). *Dermacentor andersoni* salivary glands contain inhibitors of factor V, an integral part of the complex with Xa that converts prothrombin to thrombin, and factor VII, a component of the extrinsic pathway (*65*). *Ornithodoros moubata* produces an inhibitor of the intrinsic pathway component, factor IX (*66*).

Platelet aggregation can be inhibited by apyrase, which hydrolyses both ATP and ADP to AMP and inorganic phosphate (*38*). Salivary apyrases are present in *O. moubata* (*67*), *O. savignyi* (*68*), *I. scapularis* (*69*), and many other blood feeding arthropods (*38*). *Ornithodoros moubata* can inhibit platelet aggregation with a disintegrin-like molecule (*70*) and an inhibitor of collagen-mediated platelet aggregation (*71*). *Dermacentor variabilis* salivary glands inhibit platelet aggregation with an antagonist of fibrinogen and vibronectin receptors (*72*). Metalloproteases degrade extracellular matrix (*73*). Snake venom metalloproteases prevent platelet aggregation by altering platelet binding to fibrin (*74*). Genes encoding putative metalloproteases are present in the salivary glands of *I. scapularis* (*50*). Tick saliva metalloproteases would provide another means of preventing platelet aggregation. In addition, the disruption of extracellular matrix likely contributes to maintaining the feeding lesion around tick mouthparts.

In addition to inhibiting platelet aggregation, prostaglandin E_2 is also a vasodilator. Prostaglandin E_2 occurs in the salivary glands of *Boophilus microplus* (*75, 76*), *I. scapularis* (*69*), and *A. americnum* (*77*). Vasodilation increases blood flow to the bite site, which is beneficial to engorging ticks. Saliva of *A. americanum* contains phospholipase A2, which possesses hemolytic activity (*78*), likely valuable in maintaining the feeding site and for uptake of blood.

Rapid advances are being made in identification of tick salivary gland molecules that modify inflammation, as well as innate and specific acquired immune responses. Many of those molecules likely possess additional activities. *Amblyomma americanum* calreticulin was the first tick salivary gland gene cloned and expressed (*79*). Calreticulin is a highly conserved molecule with a wide range of biological functions: acting as a molecular chaperone, regulation of integrin-mediated adhesion, and interfering with the action of complement component C1q (*80, 81*). Antibodies to *A. americanum* calreticulin revealed that it is secreted in saliva of both *A. americanum* and *D. variabilis* females (*79*). *Boophilus microplus* calreticulin is expressed, during each developmental phase, in all tissues (*82*). Repeatedly infested cattle do not develop antibodies to *B. microplus* calreticulin (*82*). However, some humans bitten by *I. scapularis* develop cross-reactive serum antibodies to *A. americanum* calreticulin (*83*).

Tick salivary glands produce proteins that bind mediators of host immune defenses. Salivary gland extracts of *Amblyomma variegatum*, *Ixodes hexagonus*, and *R. appendiculatus* each contain IgG-binding proteins, which have different molecular weights for each tick species (*84*). Two additional IgG-binding proteins were observed in the salivary glands of male, but not female, *R. appendiculatus* after six days of feeding. Host immunoglobulins in the blood meal can pass from the digestive tract into the hemolymph (*85*). Saliva of feeding female *R. appendiculatus* contains host IgG and IgG-binding proteins occurs in both the hemolymph and salivary glands (*86*). Potentially, IgG-binding proteins represent a way for a tick to cope with host immunoglobulin in the blood meal that crosses into the hemolymph.

Ixodes scapularis saliva contains an IL-2 binding protein that complexes with IL-2 in the fluid phase rather than by acting directly upon the cytokine producing cell (*87*). This IL-2 binding protein can potentially influence any IL-2 responsive cell population, and it has similar affinities for human and mouse IL-2

(87). As stated earlier, histamine-binding proteins occur in many tick species, and they potentially have anti-itch, anti-inflammatory, and immunomodulatory properties *(40, 44, 88)*. Histamine is reported to alter immune regulation by changing the Th1/Th2 polarizing capacity of immature dendritic cells; modifying cytokine production by monocytes; and, stimulating IL-10 production *(88)*. Blocking the action of histamine could have many advantages for the feeding tick.

Blocking alternative pathway of complement activation is not unexpected, since that pathway was linked to resistance to tick feeding *(89)*. *Ixodes scapularis* saliva contains inhibitors of both the alternative pathway and anaphylatoxins *(90, 91)*. Tick host range has been linked to the ability to impair host alternative pathway activity *(92)*. An anti-complement protein was purified from *I. scapularis* saliva, cloned and expressed *(93)*. This inhibitor accelerates the uncoupling of factor Bb from the alternative pathway C3 convertase and inhibits C3b binding *(93)*. At concentrations inhibitory to the alternative pathway, the classical pathway was not affected. By inhibiting the alternative pathway, this protein could reduce the release of mediators from mast cells and other leukocytes in the vicinity of the bite site; impair attraction of inflammatory cells; and, reduce the multiple biological activities of factors generated downstream by the effector pathway.

An intriguing finding is expression of the pro-inflammatory cytokine, macrophage migration inhibitory factor (MIF) in the salivary glands and midgut of *A. americanum* *(94)*. This pleotropic lymphocyte/macrophage cytokine is reported to: possess oxidoreductase activity; inhibit lysis by NK cells; act as a neurohumoral mediator; prevent random migration of macrophages; and, inhibit delayed type hypersensitivity and generation of antigen specific B and T-lymphocyte responses *in vivo* *(95, 96)*. Specific roles of this molecule in tick saliva and midgut are undefined.

Several tick salivary molecules have been identified that modulate innate and acquired immune defenses, although the modes of action of most of these proteins remain undefined. Saliva of *D. andersoni* contains a 36 kdal (p36) inhibitor of mitogen, concanavalin A (Con A), stimulated T-lymphocyte proliferation (*97*). Antibodies to native p36 cross-reacted with polypeptides of 33 and 101 kdal in *D. variabilis* salivary gland extract. An *I. ricinus* salivary gland protein (Iris) inhibits Con A driven *in vitro* proliferation of mouse splenocytes (*98*). Using human peripheral blood mononuclear cells stimulated with the T-cell mitogen phytohemagglutinin, Iris reduced the number of cells producing IFN-γ, but did not alter the number producing IL-10. Similar cells stimulated with multiple activators produced reduced amounts of IFN-γ, IL-6, IL-8, and TNF-α in the presence of Iris (*98*). In general, Iris inhibits pro-inflammatory cytokines.

Several *I. scapularis* antigens were identified by screening a cDNA library prepared from engorged nymphs with antisera from infested laboratory hosts (*99, 100, 101*). These proteins are referred to as Salps, and they have diverse characteristics. Salp 15 inhibits activation of CD4 T-cells by repressing T-cell receptor engagement initiated calcium signals (*101*). Salp 25D is a glutiothione peroxidase homologue, which is a phylogenetically conserved antioxidant protein (*100*). Induced during feeding, Salp 16 represents 0.2% of the saliva protein and its activity is unknown (*99*). Salp 20 is a homologue of the previously described anti-complement protein of *I. scapularis* (*100*).

Saliva of ticks and other blood feeding arthropods contain complex mixtures of pharmacologically active molecules that counteract host defenses in diverse ways that are becoming clearer as genomic and proteomic tools are being used along with more traditional biochemical purification strategies (*102*). New insights will rapidly merge from analyses of salivary gene transcripts and proteins. In addition, the importance and modes of action of many already identified molecules in tick saliva need to be defined. The complexity and sophistication of the pharmacological tools of ticks will likely be a source of amazement.

Pathogen transmission linked to saliva molecules

Hematophagous insect and tick saliva potentates pathogen transmission and establishment within the host (*19, 20, 21, 30, 31, 103, 27*). Targeting molecules essential for pathogen transmission, rather than immunogens of each individual pathogen transmitted by a blood-feeding arthropod, could be an effective strategy to protect humans and other animal species from these important diseases. Hence, the idea of vaccines directed against relevant molecules in vector saliva.

Infection with *Leishmania major* is enhanced by saliva of both Old World, *Phlebotomus papatasi*, and New World, *Lutzomyia longipalpis*, sand flies (*104, 105*). The saliva of both species contains molecules that modulate host immune defenses relevant to this parasite. The immunomodulator maxadilan is produced by *L. longipalpis* (*31*). In the absence of maxadilan, saliva of *P. papatasi* reduces macrophage expression of an inducible nitric oxide synthase gene and production of nitric oxide (*106*), and *P. papatasi* salivary gland lysate enhances lymph node cell IL-4 while reducing IFN-γ, IL-12, and nitric oxide synthase (*107*).

Arbovirus transmission is enhanced by mosquito saliva. Cache Valley virus introduced into feeding sites of *Aedes aegypti*, *Aedes triseriatus*, or *Culex pipiens* enhances infection (*108*). Likewise, saliva of *A. triseriatus* potentiates vesicular stomatitis virus transmission (*109*). Tick saliva facilitates transmission of Thogoto virus (*110*), tick-borne encephalitis virus (*111*), vesicular stomatitis virus (*112*), *Theileria parva* (*113*), and *Francisella tularensis* (*114*).

Insights are emerging into the mechanisms by which vector saliva might potentiate pathogen transmission. Infestation with

Rhipicephalus appendiculatus larvae or salivary gland extract reduces interferon induced MX1-mediated A2G mouse resistance to Thogoto virus (*115*). *Ixodes ricinus* salivary gland extract reduces *in vitro* macrophage production of nitric oxide and superoxide, which in turn reduce mouse peritoneal macrophage killing of *Borrelia afzelii* spirochetes (*116*). Salivary gland extract of partially fed *Dermacentor reticulates* inhibits L cell production of interferon α/β, resulting in diminished anti-viral defenses (*117*). *Ixodes ricinus* salivary gland extract enhancement of *F. tularensis* proliferation after injection into mouse skin is associated with reduction of interleukin (IL)-12, tumor necrosis factor (TNF) α, and interferon (IFN)-γ mRNA in skin (*114*). Both TNF-α and IFN-γ are associated with protective immunity in tularemia (*118*).

Passive administration of host cytokines reduced by tick feeding resulted in protection against transmission of *Borrelia burgdorferi* by infected *I. scapularis* nymphs, providing evidence for the role of tick induced immunomodulation in transmission of an infectious agent (*119*). Furthermore, enhancement of spirochete burdens in mouse tissues is linked to the specific species of spirochete and the salivary gland extract of its relevant tick vector (*120*). A lysate of *I. scapularis* salivary glands enhanced the target tissue burden of *Borrelia burgdorferi*, but not the Portuguese strain of *Borrelia lusitaniae*. However, *I. ricinus* salivary gland lysate affected enhanced the tissue burden of *B. lusitaniae*, but not that of *B. burgdorferi*.

Pathogen transmission can be blocked by immunity to the arthropod vector

Exposure of hosts to bites of pathogen-free, blood-feeding arthropods can prevent or reduce transmission of disease causing agents by subsequent bites from the same species of infected arthropod. Importantly, vector blocking responses are not necessarily associated with host responses that prevent or reduce arthropod feeding. Rabbits infested with pathogen-free *D. andersoni* were significantly resistant to *D. andersoni* transmission of *Francisella tularensis* than tick naive rabbits infested with infected ticks (*121*). Four repeated infestations of BALB/c mice with pathogen-free *I. scapularis* nymphs stimulated resistance to transmission of *B. burgdorferi* by a subsequent infestation with infected nymphs (*122*). Acquired resistance to *I. scapularis* did not occur. Repeated infestations of guinea pigs with pathogen-free *I. scapularis* nymphs induced both acquired resistance to tick feeding and transmission of *B. burgdorferi* (*123*).

A similar phenomenon has been reported for a rapidly feeding insect. Exposure of mice to bites of uninfected sand flies, *P. papatasi*, resulted in protection against fly transmitted *L. major* (*124*). Protection was attributed to an intense delayed type hypersensitivity response at the bite site with production of IFN-γ where parasites were deposited along with fly saliva. The protective response has been linked to a 15 kdal salivary protein of *P. papatasi* (*125*).

ANTI-ECTOPARASITE VACCINES

Anti-ectoparasite vaccine research emerged with the report of the successful induction of guinea pig resistance to larvae of *Dermacentor variabilis* by immunization with an extract of whole tick larvae (*126*). Many investigators subsequently used whole arthropod extracts, homogenates of specific tissues, and, physiologically relevant molecules of an ectoparasitic arthropod. Anti-ectoparasite vaccines have been the subject of reviews (*127, 128, 129, 130, 131, 14, 12, 13, 15, 16*).

Ticks

Ticks have been the predominant focus of anti-ectoparasite vaccine research, since the report by Trager (*126*). During the 64 years since that publication, many attempts have been made to induce anti-tick immunity with whole tick extracts, salivary gland extracts, and most recently with defined immunogens. Commercial development of the anti-*B. microplus* vaccines TickGARD™ Plus in Australia and GAVAC™ Plus in Latin America establish the feasibility and utility of anti-tick vaccines in the veterinary market (*12, 22*).

Variable levels of anti-tick immunity were reported for immunization with whole tick extracts of *Ixodes holocyclus* (*132*), *Amblyomma americanum* (*133*), *Rhipicephalus appendiculatus* (*134*), and *Amblyomma hebraeum* and *Amblyomma marmoreum* (*135*). Likewise, immunity induced to challenge infestations was variable for animals immunized with salivary gland extracts of *B. microplus* (*136*), *D. andersoni* (*137*), and *Hyalomma anatolicum anatolicum* (*138*). Vaccination with whole tick extracts, likely containing hundreds of different molecules, will induce responses to immunodominant molecules, which might not be relevant to anti-tick/vector-blocking immunity. Molecules which might be in low abundance but are physiologically important may be ideal candidates for induction of host protective immune responses.

Another approach was fractionation and characterization of molecules in tissues that would be available targets for host humoral and cell mediated immune responses. Due to its intimate contact with the blood meal, tick digestive tract was a logical target. Tick gut cells would be encounter host antibodies, lymphocytes, monocytes and granulocytes in a blood meal obtained from an immunized animal. Damage to the midgut could disrupt feeding, molting, reproduction/ development, and transmission of infectious agents. Trager (*139*) was the first to use tick midgut as an immunogen. Female *D. andersoni* that fed on guinea pigs immunized with a mixture of female *D. andersoni* midgut combined with reproductive tract produced fewer ova and none of those ova yielded viable larvae (*140*). Immunization with isolated brush border of *A. americanum* digestive tract cells induced resistance to that species (*141*). Immunization of sheep with vitellin derived from B. microplus eggs partially protected sheep from a *B. microplus* infestation and reduced tick oviposition (*142*). Tic encoded serine protease inhibitors have been proposed as candidate immunogens for anti-tick vaccine development (*143*).

Vaccination of Hereford cattle with a 100,000 x g membrane pellet of *B. microplus* midgut induced an antibody response that reduced *B. microplus* egg production by 91% (*144, 145*). Other Australian researchers were pursuing a line of investigation that resulted in a gut antigen for an anti-*B. microplus* vaccine. Immunization of cattle with an extract of whole female *B. microplus* induced partial immunity (*146*), and ticks that fed on those immunized cattle had damaged digestive tracts (*147*). From an estimated 50,000 partially fed ticks, approximately 100 micrograms of a midgut membrane molecule were isolated, which provided to be a highly effective immunogen for induction of immunity to *B. microplus* (*148*). The immunogen purification strategy involved detergent solubilization of a membrane extract, lectin affinity chromatography, preparative isoelectric focusing, and HPLC gel filtration chromatography (*149*). The isolated glycoprotein immunogen, Bm86, has a molecular weight of 89 kdal with an isoelectric point of 5.1 to 5.6. Cattle vaccinated with microgram quantities of Bm86 developed significant anti-tick immunity, which was expressed as reduced engorgement weights, impaired ova production, and death of feeding ticks. Antibodies to Bm86 bound to the surface of tick gut cells and inhibited their endocytotic activity (*149*). Since this molecule was not introduced into the host during feeding, it was designated a "concealed" antigen (*150*).

Commercialization of the Bm86 vaccine has been reviewed (*11, 12, 151, 15*). Anti-tick immunity was enhanced by combined administration of Bm86 and a mucin-like membrane glycoprotein derived from partially fed female *B. microplus* (*152*). Immunity induced with Bm86 provided cross-protection against infestation with *Boophilus decoloratus, H. anatolicum anatolicm*, and *Hyalomma dromedarii* (*153*). However, immunization with Bm86 had no effect on infestation with *Amblyomma variegatum* or *R. appendiculatus* (*153*). Another protection inducing immunogen of *B. microplus*, Bm91, is an 86 kdal glycoprotein with an isoelectric point of 4.8 to 5.2, and this molecule is present in both digestive tract and salivary gland (*154*). Multiple potential protection inducing immunogens are likely to be found in an individual tick species. Clearly, some protection inducing epitopes are shared within a genus and across genera.

Insects

Initial anti-insect vaccine studies often yielded contradictory results (*131*). Crude antigen extracts were used and variable responses to immunization would be expected (*14*). This situation is similar to early efforts to induce anti-tick immunity with whole tick or tissue extracts (*127*). Fortunately, there is continued interest in vaccines against hematophagous insects and myiasis larvae, and promising results are being obtained (*155, 24, 156, 157, 158, 159, 160*).

The first report of vaccination against an insect was administration to humans of a whole extract of the dog flea, *Ctenocephalides canis*, which was reported to provide "protection" against flea bites (*161*). The immunization regimen was similar to that used to desensitize immediate hypersensitivity responses to an allergen. Repeated immunizations were administered with increasing doses of antigen. The regimen essentially achieved the goal of reduced reactivity to flea bites. Immunization of cats with gut membranes of the cat flea, *Ctenocephalides felis felis*, was not effective in protecting against flea bites (*162*). However, allergens of the cat flea were identified in efforts to define the immunologic basis of canine flea bite hypersensitivity (*163, 164*).

Due to host specificity and importance, sucking lice are logical candidates for vaccine based control. Immunization with a whole louse sonicate of the mouse louse, *Polyplax serrata*, induced resistance to infestation, which was expressed as a 62% reduction in weight of the louse burden (*165*). An extract of *Polyplax spinulosa* induced rat resistance to infestation (*166*). Subsequently, a 31 kdal protein immunogen was detected in midgut epithelia, partially digested gut contents, and feces of *P. spinulosa* (*167*).

Rabbits immunized with midgut of the human body louse, *Pediculus humanus humanus*, developed immunity that resulted in reduced blood meal size, inhibited development, decreased ova production, and increased mortality (*168*). Many lice that obtained a blood meal from immunized rabbits were red, indicating that their midguts were disrupted. Protection inducing immunogens were detected both on the microvilli of midgut cells (*169*) and in louse feces (*170*). Feces of *P. humanus humanus* were used to isolate antigen for immunization of rabbits (*171*). Engorgement weights were reduced 29% for females that fed on immunized rabbits. Likewise, consumption of blood from an immunized rabbit resulted in a significant reduction in the mean number of ova produced. However, immunization did not reduce louse survival. Cross-reactive immunogens occur among *P. humanus humanus*, the cattle louse, *Haematopinus africanus*, and the goat louse, *Linognathus stenopsis* (*170*).

Anti-biting fly vaccines could complement the use of insecticides in integrated control strategies. Immunization of rabbits with different tissues of the stable fly, *Stomoxys calcitrans*, resulted in immune responses that damaged feeding flies (*172*). Immunization of rabbits with flight muscle induced responses that resulted in significant fly mortality. Interestingly, flies obtaining a blood meal from those rabbits displayed leg and wing paralysis and difficulty in probing. Vaccination with whole fly, abdomen, or gut extract stimulated antibody responses that recognized eight antigens from gut and 12 each from whole fly and abdomen extracts (*173*). Immunization with gut antigen produced the highest fly mortality and the lowest percentage of viable eggs (15.5%). Mortality of tsetse flies, *Glossina morsitans*, was significant when flies obtained a blood meal from rabbits immunized with cuticle and adhering hypodermal cells and wing buds of *S. calcitrans* (*172*).

Horn flies, *Haematobia irritans irritans*, and buffalo flies, *Haematobia irritans exigua*, are serious pests of cattle. The close associations of these flies with their hosts make them ideal targets for vaccine based control (*174, 175, 157*). Cattle exposed to the bites of horn flies produce antibodies to salivary gland antigens (*174*). Buffalo flies infesting cattle with high antibody titers to fly immunogens did not die more frequently than flies fed on control animals (*176*). There was no apparent effect when buffalo flies were fed bovine blood *in vitro*, containing anti-peritrophin antibodies (*157*).

Effective anti-mosquito/vector-blocking vaccines could have an enormous impact on improving global public health. Mosquito molecular biology is well studied when compared to other blood feeding arthropods. Mosquito EST and genome sequencing projects provide additional resources for combined genomic and proteomic strategies for identifying relevant candidate immunogens for blocking blood-feeding and transmission/establishment of infectious agents.

The first report of immunization against mosquitoes described administration of extracts of whole female *Anopheles quadrimaculatus* to rabbits, which resulted in an immune response that did not impact feeding during a subsequent challenge (*177*). Several studies have reported attempts to immunize with whole mosquito extracts that induced variable levels of anti-mosquito immunity. Rabbits were vaccinated with one of the following homogenates of *Anopheles stephensi*: low speed supernatant of whole mosquitoes, low speed pellet of whole mosquitoes, or midgut (*178*). Mosquitoes that obtained blood meals from midgut immunized hosts had the highest level of mortality. Immunization with an homogenate of whole, sugar fed, *Aedes aegypti* followed by exposure to *Ae. aegypti* and *Culex tarsalis* reduced fecundity but did not effect viability of the mosquitoes. Mortality of mosquitoes was modestly increased for *Ae. Aegypti* obtaining a blood meal from mice immunized with whole mosquito or midgut extracts (*179*).

Female *A. aegypti* were used as a source of immunogens 24 hours after blood-feeding to prepare extracts of head/thorax, midgut, and the abdomen minus midgut tissues (*180*). Mosquitoes which blood fed on immunized or control rabbits did not differ in regard to mortality. However, fecundity was reduced for mosquitoes in each treatment group. A similar experiment was performed with *Anopheles tessellatus* (*181*). Reduced egg production occurred for *A. tessellatus* but not for *Culex quinquefasciatus*, which fed on *A. tessellatus* immunized rabbits. Only *Cu. quinquefasciatus* had increased mortality. The same experimental design was used to investigate mosquitoes feeding on *A. tessellatus* immunized mice, resulting in a reduction of *A. tessellatus* egg production up to 29% (*182*). Subsequently, anti-midgut antibodies inhibited peritrophic membrane formation by *An. tessellatus* (*183*). Longevity of *Anopheles stephensi* fed on midgut vaccinated mice was significantly decreased (*184*).

The concept of a vector-blocking vaccine has focused on the use of anti-mosquito midgut antibodies. *Aedes aegypti* were significantly less susceptible to infection with Murray Valley encephalitis and Ross River viruses after engorging on blood containing antibodies to both mosquito midgut and virus (*185*). Fewer *Plasmodium berghei* oocysts were detected in *Anopheles farauti* fed on midgut immunized mice (*186*). Similar findings were reported when *A. stephensi* ingested anti-midgut antibodies combined with *P. berghei* (*187*) and when *A. tessellatus* ingested *Plasmodium vivax* infected erythrocytes in the presence of anti-midgut antibodies (*188*). Antibodies to mosquito midgut blocked development of both *Plasmodium falciparum* and *P. vivax* in multiple species of *Anopheles* mosquitoes and reduced mosquito fecundity and survival (*189*).

Myiasis is the invasion of tissues or open body cavities by fly, dipteran, larvae. Myiasis can be caused by fly larvae of many different species. Vaccine based control of myiasis is very appealing. Fly larvae can reside in direct contact with elements of the host immune system for months. Not unexpectedly, natural infections do not generally induce inflammatory responses (*190*) or host immunity to re-infestation (*191*). The most extensively studied causes of myiasis are the warble flies, *Hypoderma bovis* and *Hypoderma lineatum*, and the sheep blowfly, *Lucilia cuprina* (*24*).

The feasibility of vaccination against *H. bovis* and *H. lineatum* was established when vaccinated bovines were capable of killing fly larvae *in vivo* (*192*). Cell mediated immune responses to the enzymes, hypodermin A, B, and C, of *H. lineatum* were stimulated in vaccinated cattle, resulting in development of fewer larvae (*193*). Hypodermin A has been used to induce high levels of antibodies that provided protection (*194*). Sequence and gene expression have been determined for hypodermins A, B, and C in larvae of *H. lineatum* (*155*). In addition, the three-dimensional structure of hypodermin C has been described (*195*).

An important consideration regarding vaccination against hypodermosis is the ability of hypodermins to modulate host immunity. Hypodermins A and B degrade complement component C3 (*196*) and hypodermin A also degrades IgG (*197*). Hypodermin A impairs *in vitro* proliferation of bovine T-cells, which is likely due to its ability to reduce IL-2 production by T-lymphocytes (*198*).

Lucilia cuprina myiasis can rapidly kill a sheep (*24*). Infestations occur most commonly in young sheep, suggesting that acquired resistance develops with age (*24*). Significant progress has been made in developing an anti-*L. cuprina* vaccine (*199*). Peritrophic membrane molecules, peritrophins, induced sheep antibodies that inhibited development of *L. cuprina* larvae when tested both *in vitro* and *in vivo*, however, significant larval death was not evident *in vivo* (*200*). Both cDNA and amino acid sequences were determined for peritrophin 48 (*201*). First instar larvae of *L. cuprina* were fractionated by preparative isoelectric focusing and a fraction with a pH range of 5.9 to 6.7 contained proteins that induced sheep antibodies that when ingested reduced *in vitro* growth of larvae by 84±7% (*156*). Protective antibodies bound to larval peritrophic membrane, cuticle, and less well to microvilli and basement membranes of digestive tract epithelial cells (*156*). The glycoprotein, peritrophin-95 was isolated from larval peritrophic matrix and found to induce an immune response in sheep, which inhibited larval growth (*202*). Growth inhibitory activity was found in antibodies to both the peritropin-95 polypeptide and oligosaccharide (*202*). Peritrophin-55, a mucin-like protein, induced immunity that inhibited larval growth by 51 to 66%, when larvae fed on sera of vaccinated sheep (*160*). Vaccination was used to control Old World screwworm fly, *Chrysoma bezziana* (*158*). Immune responses induced by larval extracts, peritrophic membrane, and cardia reduced the growth of *C. bezziana* larvae.

Immunization induced protection must take into account the ability of *L. cuprina* to counteract host immunity. Excretory/secretory products of larvae reduce antibody responses (*203*) and secreted larval enzymes degrade C3 (*204*) and IgG (*205*).

GENOMICS AND PROTEOMICS

Powerful methods have emerged that are of great value for identification of the genes expressed in tissues of blood feeding arthropods. An expressed sequence tag (EST) is a partial nucleic acid sequence obtained from randomly selected clones from a cDNA library generated from the tissue of interest (*206*). Central to the success of these analyses is the availability for comparison of gene sequences, translated amino acid sequences, and motifs in the public databases. Many blood feeding arthropods have not been the focus of extensive study. Initially, many gene sequences might not have matches in the databases. An important point is that a database match does not necessarily mean that two genes encode proteins with similar biological activities. In order to effectively realize the power of these approaches, high throughput functional genomic strategies are needed (*207*).

The majority of EST studies on blood feeding arthropods have focused on salivary glands, and the topic has been recently reviewed (*102*). Vector insect salivary gland EST projects were reported for *Glossina morsitans morsitans* (*208*), *Anopheles gambiae* (*54*), and *Ae. aegypti* (*209*). The first tick EST studies focused on larvae of *B. microplus* (*210*) and larvae and adults of *A. americanum* (*211*). Tick salivary gland EST projects have been reported for *I. ricinus* (*212*) and *I. scapularis* (*213*). Differential display has been used to compare genes expressed in the salivary glands of *A. americanum* and *D. andersoni* males (*51*).

FUTURE CHALLENGES AND AVENUES OF RESEARCH

Vaccination to limit hematophagous arthropods has gone from a possibility to a reality. Previous concerns about the feasibility of anti-arthropod vaccines were largely due to variability of protective responses stimulated by whole arthropod or whole tissue extracts. Characterized immunogens have been more effective. Advances in delivery systems, such as DNA immunization, and adjuvant strategies will further facilitate the development of these important vaccines. Knowing the spectrum of the pharmacologically active molecules in blood-feeding arthropod saliva will result in identifying candidate immunogens. Only a few activities of saliva have been linked to specific molecules. Not only do these molecules represent potential targets of anti-arthropod/ vector-blocking vaccines, many likely have utility as potential drugs.

Genomics and proteomics will likely provide the platform for significant advances. These powerful tools should go hand-in-hand with biochemical purification schemes and high throughput screening strategies for activities of interest. Analysis of expressed sequence tags (ESTs) prepared from mRNA derived from the salivary glands of blood-feeding arthropods should be vigorously pursued. Genomic strategies must be linked to studies of gene expression at the protein level (*214*).

The rapid advances in immunobiology, vector biology, genomics, proteomics, and vaccine technology make it very likely that anti-arthropod and vector-blocking vaccines will become important tools for control of ectoparasitic arthropods and vector-borne infectious agents.

ACKNOWLEDGMENTS

This work was supported in part by Grant AI46676 from the National Institute of Allergy and Infectious Diseases, National Institutes of Health; Cooperative Agreement number U50/CCU119575 from the Centers for Disease Control and Prevention; and, Contract Number DAMD 17-03-1-0075 from the U.S. Army Medical Research and Materiel Command to S.K.W.

REFERENCES

1. Gubler, D.J.: *Emerg. Infect. Dis.* **1998**, *4*, 442-450.
2. Gratz, N.G.: *Ann. Rev. Entomol.* **1999**, *44*, 51-75.
3. Parola, P.; Raoult, D.: *Clin. Infect. Dis.* **2001**, *32*, 897-928.
4. Isturiz, R.E.; Gubler, D.J.; Brea del Castillo, J.: *Infect. Dis. Clin. N. Amer.* **2000**, *14*, 121-140.
5. Daszak, P.; Cunningham, A.A.; Hyatt, A.D.: *Science* **2000**, *287*, 443-449.
6. Laird, M.: *Experientia* **1985**, *41*, 446-456.

7. Curtis, C.F.; Davies, C.R.: *Med. Vet. Entomol.* **2001**, *15*, 231-235.
8. Molyneux, D.H.: *Int. J. Parasitol.* **1998**, *28*, 927-934.
9. Hemingway, J.; Ranson, H.: *Ann. Rev. Entomol.* **2000**, *45*, 371-391.
10. Mitchell, M.: *Trop. Anim. Prod.* **1996**, *28*, 53S-58S.
11. Tellam, R.L.; Smith, D.; Kemp, D.H.; Willadsen, P. (1992) Vaccination Against Ticks, in *Animal Parasite Control Utilizing Biotechnology* (Yong, W.K., Ed.), pp. 303-331, CRC Press, Boca Raton, Florida.
12. Willadsen, P.: *Vet. Parasitol.* **1997**, *71*, 209-222.
13. Willadsen, P.: *Vet. Parasitol.* **2001**, *101*, 353-367.
14. Wikel, S.K.; Bergman, D.K.; Ramachandra, R.N. (1996) Immunological-based Control of Blood-feeding Arthropods, in *The Immunology of Host-Ectoparasitic Arthropod Relationships* (Wikel, S.K., Ed.), pp. 290-315, CAB International, Wallingford, United Kingdom.
15. Lee, R.; Opdebeeck, J.P.: *Infect. Dis. Clin. N. Amer.* **1999**, *13*, 209-226.
16. Pruett, J.H.: *Int. J. Parasitol.* **1999**, *29*, 25-32.
17. Ribeiro, J.M.C.: *Exptl. Appl. Acarol.* **1989**, *7*, 15-20.
18. Brossard, M.; Wikel, S.K.: *Med. Vet. Entomol.* **1997**, *11*, 270-276.
19. Wikel, S.K.: *Ann. Rev. Entomol.* **1996**, *41*, 1-22.
20. Wikel, S.K.: *BioScience* **1999**, *49*, 311-320.
21. Wikel, S.K.: *Int. J. Parasitol.* **1999**, *29*, 851-859.
22. Willadsen, P.; Jongejan, F.: *Parasitol. Today* **1999**, *15*, 258-262.
23. Jones, C.J. (1996) Immune Responses to Felas, Bugs, and Sucking Lice, in *The Immunology of Host-Ectoparasitic Arthropod Relationships* (Wikel, S.K., Ed.), pp. 150-174, CAB International, Wallingford, United Kingdom.
24. Sandeman, R.M. (1996) Immune Responses to Mosquitoes and Flies, in *The Immunology of Host-Ectoparasitic Arthropod*

Relationships (Wikel, S.K., Ed.), pp. 175-203, CAB International, Wallingford, United Kingdom.
25. Arlian, L.G. (1996) Immunology of Scabies, in *The Immunology of Host-Ectoparasitic Arthropod Relationships* (Wikel, S.K., Ed.), pp. 232-258, CAB International, Wallingford, United Kingdom.
26. Wrenn, W.J. (1996) Immune Responses to Mange Mites and Chiggers, in *The Immunology of Host-Ectoparasitic Arthropod Relationships* (Wikel, S.K., Ed.), pp. 259-289, CAB International, Wallingford, United Kingdom.
27. Schoeler, G.B.; Wikel, S.K.: *Ann. Trop. Med. Parasitol.* **2001**, *95*, 755-771.
28. Cupp, E.W.; Cupp, M.S.: *J. Med. Entomol.* **1997**, *34*, 87-94.
29. Kamhawi, S.: *Microbes Infect.* **2000**, *2*, 1765-1773.
30. Wikel, S.K.; Bergman, D.K.: *Parasitol. Today* **1997**, *13*, 383-389.
31. Gillespie, R.D.; Mbow, M.L.; Titus, R.G.: *Parasite Immunol.* **2000**, *22*, 319-331.
32. Ribeiro, J.M.C.: *Infect. Agents Dis.* **1995**, *4*, 143-152.
33. Moire, N.; Nicolas-Gaulard, I.; LeVern, Y.; Boulard, C.: *Parasite Immunol.* **1997**, *19*, 21-27.
34. Champagne, D.E., and Valenzuela, J.G. (1996) Pharmacology of Haemathophagous Arthropod Saliva, in *The Immunology of Host-Ectoparasitic Arthropod Relationships* (Wikel, S.K., Ed.), pp. 85-106, CAB International, Wallingford, United Kingdom.
35. Stark, K.R., and James, A.A. (1996) The Salivary Glands of Disease Vectors, in *The Biology of Disease Vectors* (Beaty, B.J. and Marquardt, W.C., Eds.), pp. 333-348, University Press of Colorado, Niwot.
36. Jones, D.: *Parasitology* **1998**, *116*, S73-81.
37. Basanova, A.V.; Baskova, I.P.; Zavalova, L.L.: *Biochemistry* (Moscow) **2002**, *67*, 167-176.
38. Ribeiro, J.M.C.: *Infect. Agents Dis.* **1995**, *4*, 143-152.
39. Ribeiro, J.M.C.: *Parasitol. Today* **1995**, *11*, 91-93.
40. Alexander, J.O'D.: *Parasitol. Today* **1986**, *2*, 345-351.

41. Clark, W.G.: *Handbook Exptl. Pharmacol.* **1979**, *25*, 311-356.
42. Ribeiro, J.M.C.; Mather, T.N.: *Exptl. Parasitol.* **1998**, *89*, 213-221.
43. Proud, D.; Kaplan, A.P.: *Ann. Rev. Immunol.* **1988**, *6*, 49-83.
44. Falus, A. *Histamine and Inflammation*; R.G. Landes Company: Austin, TX, 1994.
45. Paine, S.H.; Kemp, D.H.; Allen, J.R.: *Parasitolology* **1983**, *86*, 419-428.
46. Chinery, W.A.; Ayitey-Smith, E.: *Nature* **1977**, *265*, 366-367.
47. Paesen, G.C.; Adams, P.L.; Harlos, K.; Nuttall, P.A.; Stuart, D.I.: *Mol. Cell* **1999**, *3*, 661-671.
48. Paesen, G.C.; Adams, P.L.; Nuttall, P.A.; Stuart, D.I.: *Biochem. Biophys. Acta* **2000**, *1482*, 92-101.
49. Sangamnatdej, S.; Paesen, G.C.; Slovak, M.; Nuttall, P.A.: *Insect Mole. Bio.* **2002**, *11*, 79-86.
50. Valenzuela, J.G.; Francischetti, I.M.B.; Pham, V.M.; Garfield, M.K.; Mather, T.N.; Ribeiro, J.M.C.: *J. Exp. Biol.* **2002**, *205*, 2843-2864.
51. Bior, A.D.; Essenberg, R.C.; Sauer, J.R.: *Insect Biochem. Mole. Biol.* **2002**, *32*, 645-655.
52. Mann, K.G.: *Thrombosis Haemosta.* **1999**, *82*, 165-174.
53. Rai, R.; Sprengeler, P.A.; Elrod, K.C.; Young, W.B.: *Curr. Med. Chem.* **2001**, *8*, 101-119.
54. Francischetti, I.M.B.; Valenzuela, J.G.; Pham, V.M.; Garfield, M.K.; Ribeiro, J.M.C.: *J. Exp. Biol.* **2002**, *205*, 2429-2451.
55. Waxman, L.; Smith, D.E.; Arcuri, K.E.; Vlasuk, G.P.: *Science* **1990**, *248*, 593-596.
56. Limo, M.K.; Voigt, W.P.; Tumbo-Oeri, A.G.; Njogu, R.M.; Ole-MoiYoi, O.K.: *Exp. Parasitol.* **1991**, *72*, 418-429.
57. Joubert, A.M.; Crause, J.C.; Gaspar, A.R.; Clarke, F.C.; Spickett, A.M.; Neitz, A.W.: *Exptl. Appl. Acarol.* **1995**, *19*, 79-92.
58. Joubert, A.M.; Louw, A.I.; Joubert, F.; Neitz, A.W.: *Exptl. Appl. Acarol.* **1998**, *22*, 603-619.

59. Ibrahim, M.A.; Ghazy, A.H.; Maharem, T.M.; Khalil, M.I.: *Comp. Biochem. Physiol. B, Biochem. Mol. Biol.* **2001**, *130*, 501-512.
60. Narasimhan, S.; Koski, R.A.; Beaulieu, B.; Anderson, J.F.; Ramamoorthi, N.; Kantor, F.; Cappello, M.; Fikrig, E.: *Insect Mole. Biol.* **2002**, *11*, 641-650.
61. Anastopoulos, P.; Thurn, M.J.; Broady, K.W.: *Aust. Vet. J.* **1991**, *68*, 366-367.
62. Hoffmann, A.; Walsmann, P.; Riesener, G.; Paintz, M.; Markwardt, F.: *Pharmazie* **1991**, *46*, 209-212.
63. Zhu, K.; Bowman, A.S.; Brigham, D.L.; Essenberg, R.C.; Dillwith, J.W.; Sauer, J.R.: *Exp. Parasitol.* **1997**, *87*, 30-38.
64. Nienaber, J.; Gaspar, A.R.M.; Neitz, A.W.H.: *Exp. Parasitol.* **1999**, *93*, 82-91.
65. Gordon, J.R.; Allen, J.R.: *J. Parasitol.* **1991**, *77*, 167-170.
66. Hellmann, K.; Hawkins, R.I.: *Thromb. Diat. Haemorrhag.* **1967**, *18*, 617-625.
67. Ribeiro, J.M.C.; Endris, T.M.; Endris, R.: *Comp. Biochem. Physiol.* **1991**, *100A*, 109-112.
68. Mans, B.J.; Gaspar, A.R.M.D.; Louw, A.I.; Neitz, A.W.H.: *Comp. Biochem. Physiol. B* **1998**, *120*, 617-624.
69. Ribeiro, J.M.C.; Makoul, G.T.; Levine, J.; Robinson, D.R.; Spielman, A.: *J. Exp. Med.* **1985**, *161*, 332-344.
70. Karczewski, J.; Endris, R.; Connolly, T.M.: *J. Biol. Chem.* **1994**, *269*, 6702-6708.
71. Waxman, L.; Connolly, T.M.: *J. Biol. Chem.* **1993**, *268*, 5445-5449.
72. Wang, X.; Coons, L.B.; Taylor, D.B.; Stevens, S.E.,Jr.; Gartner, T.K.: *J. Biol. Chem.* **1996**, *271*, 17785-17790.
73. Ravanti, L.; Kahari, V.M.: *Int. J. Mol. Med.* **2000**, *6*, 391-407.
74. Swenson, S.; Bush, L.R.; Markland, F.S.: *Arch. Biochem. Biophys.* **2000**, *384*, 227-237.
75. Dickinson, R.G.; O'Hagan, J.E.; Shotz, M.; Binnington, K.C.; Hegarty, M.P.: *Aust. J. Exp. Biol. Med. Sci.* **1976**, *54*, 475-486.

76. Higgs, G.A.; Vane, J.R.; Hart, R.J.; Porter, C.; Wilson, R.G.: *Bull. Entomol. Res.* **1976**, *66*, 665-670.
77. Ribeiro, J.M.C.; Evans, P.M.; McSwain, J.L.; Sauer, J.: *Exp. Parasitol.* **1992**, *74*, 112-116.
78. Zhu, K.; Dillwith, J.W.; Bowman, A.S.; Sauer, J.R.: *J. Med. Entomol.* **1997**, *34*, 160-166.
79. Jaworski, D.C.; Simmen, F.A.; Lamoreaux, W.; Coons, L.B.; Muller, M.T.; Needham, G.R.: *J. Insect Physiol.* **1995**, *41*, 369-375.
80. Coppolino, M.G.; Dedhar, S.: *Int. J. Biochem. Cell Biol.* **1998**, *30*, 553-558.
81. Kovacs, H.; Campbell, I.D.; Strong, P.; Johnson, S.; Ward, F.J.; Reid, K.B.M.; Eggleton, P.: *Biochemistry* **1998**, *37*, 17865-17874.
82. Ferreira, C.A.S.; Eza, I.D.S.,Jr.; daSilva, S.S.; Haag, K.L.; Valenzuela, J.G.; Masuda, A.: *Exp. Parasitol.* **2002**, *101*, 25-34.
83. Sanders, M.L.; Glass, G.E.; Nadelman, R.B.; Wormser, G.P.; Scott, A.L.; Raha, S.; Ritchie, B.C.; Jaworski, D.C.; Schwartz, B.S.: *Amer. J. Epidemiol.* **1999**, *149*, 777-784.
84. Wang, H.; Nuttall, P.A.: *Parasitology* **1995**, *111*, 161-165.
85. Sauer, J.R.; McSwain, J.L.; Essenberg, R.C.: *Int. J. Parasitol.* **1994**, *24*, 33-52.
86. Wang, H.; Nuttall, P.A.: *Parasitology* **1994**, *109*, 525-530.
87. Gillespie, R.D.; Dolan, M.C.; Piesman, J.; Titus, R.G.: *J. Parasitol.* **2001**, *77*, 167-170.
88. Jutel, M.; Watanabe, T.; Akdis, M.; Blaser, K.; Akdis, C.A.: *Curr. Opin. Immunol.* **2002**, *14*, 735-740.
89. Wikel, S.K.: *Amer. J. Trop. Med. Hyg.* **1979**, *28*, 586-590.
90. Ribeiro, J.M.C.; Spielman, A.: *Exp. Parasitol.* **1986**, *62*, 292-297.
91. Ribeiro, J.M.C.: *Exp. Parasitol.* **1987**, *64*, 347-353.
92. Lawrie, C.H.; Randolph, S.E.; Nuttall, P.A.: *Exp. Parasitol.* **1999**, *93*, 207-214.
93. Valenzuela, J.G.; Charlab, R.; Mather, T.N.; Ribeiro, J.M.C.: *J. Biol. Chem.* **2000**, *275*, 18717-18723.

94. Jaworski, D.C.; Jasinskas, A.; Metz, C.N.; Bucala, R.; Barbour, A.G.: *Insect Mole. Biol.* **2001**, *10*, 323-331.
95. Petrovsky, N.; Bucala, R.: *Ann. N. Y. Acad. Sci.* **2000**, *917*, 665-671.
96. Lue, H.; Kleeman, R.; Calandra, T.; Roger, T.; Bernhagen, J.: *Microbes Infec.* **2002**, *4*, 449-460.
97. Bergman, D.K.; Palmer, M.J.; Caimano, M.J.; Radolf, J.D.; Wikel, S.K.: *J. Parasitol.* **2000**, *86*, 516-525.
98. Leboulle, G.; Crippa, M.; Decrem, Y.; Mejri, N.; Brossard, M.; Bollen, A.; Godfried, E.: *J. Biol. Chem.* **2002**, *277*, 10083-10089.
99. Das, S.; Marcantonio, N.; DePonte, K.; Telford, S.R.III; Anderson, J.F.; Kantor, F.S.; Fikrig, E.: *Amer. J. Trop. Med. Hyg.* **2000**, *62*, 99-105.
100. Das, S.; Banerjee, G.; DePonte, K.; Marcantonio, N.; Kantor, F.S.; Fikrig, E.: *J. Infect. Dis.* **2001**, *184*, 1056-1064.
101. Anguita, J.; Ramamoorthi, N.; Hovius, J.W.R.; Das, S.; Thomas, V.; Persinski, R.; Conze, D.; Askenase, P.W.; Rincon, M.; Kantor, F.S.; Fikrig, E.: *Immunity* **2002**, *16*, 849-859.
102. Ribeiro, J.M.C.; Francischetti, I.M.B.: *Annu. Rev. Entomol.* **2003**, *48*, 73-88.
103. Wikel, S.K.; Alarcon-Chaidez, F. J.: *Vet. Parasitol.* **2001**, *101*, 275-287.
104. Titus, R. G.; Ribeiro, J.M.C.: *Science* **1988**, *239*, 1306-1308.
105. Belkaid, Y.; Kamhawi, S.; Modi, G.; Valenzuela, J.; Noben-Trauth, N.; Rowton, E.; Ribeiro, J.; Sacks, D.L.: *J. Exp. Med.* **1998**, *188*, 1941-1953.
106. Waitumbi, J.; Warburg, A.: *Infect. Immun.* **1998**, *66*, 1534-1537.
107. Mbow, M.L.; Bleyenberg, J.A.; Hall, L.R.; Titus, R.G.: *J. Immunol.* **1998**, *161*, 5571-5577.
108. Edwards, J.F.; Higgs, S.; Beaty, B.J.: *J. Med. Entomol.* **1998**, *35*, 261-265.

109. Limesand, K.H.; Higgs, S.; Pearson, L.D.; Beaty, B.J.: *Parasite Immunol.* **2000**, *22*, 461-467.
110. Jones, L.D.; Hodgson, E.; Nuttall, P.A.: *J. Gen. Virol.* **1989**, *70*, 1895-1898.
111. Labuda, M.; Jones, L.D.; Williams, T.; Nuttall, P.A.: *Med. Vet. Entomol.* **1993**, *7*, 193-196.
112. Hajnicka, V.; Fuchsberger, N.; Slovak, M.; Kocakova, P.; Labuda, M.; Nuttall, P.A.: *Parasitology* **1998**, *116*, 533-538.
113. Shaw, M.K.; Tilney, L.G.; McKeever, D.J.: *Infect. Immun.* **1993**, *61*, 1486-1495.
114. Krocova, Z.; Macela, A.; Hernychova, L.; Kroca, M.; Pechova, J.; Kopecky, J.: *J. Parasitol.* **2003**, *89*, 14-20.
115. Dessens, J.T.; Nuttall, P.A.: *J. Virol.* **1998**, *72*, 8362-8364.
116. Kuthejlova, M.; Kopecky, J.; Stepanova, G.; Macela, A.: *Infect. Immunol.* **2001**, *69*, 575-578.
117. Hajnicka, V.; Kocakova, P.; Slovak, M.; Labuda, M.; Fuchsberger, N.; Nuttall, P.A.: *Parasite Immunol.* **2000**, *22*, 201-206.
118. Sjostedt, A.; North, R.J.; Conlon, J.W.: *Microbiology*, **1996**, *142*, 1369-1374.
119. Zeidner, N.; Dreitz, M.; Belasso, D.; Fish, D.: *J. Infect. Dis.* **1996**, *173*, 187-195.
120. Zeidner, N.S.; Schneider, B.S.; Nuncio, M.S.; Gern, L.; Piesman, J.: *J. Parasitol.* **2002**, *88*, 1276-1278.
121. Bell, J.F.; Stewart, S.J.; Wikel, S.K.: *Amer. J. Trop. Med. Hyg.* **1979**, *28*, 876-880.
122. Wikel, S.K.; Ramachandra, R.N.; Bergman, D.K.; Burkot, T.R.; Piesman, J.: *Infect. Immun.* **1997**, *65*, 335-338.
123. Nazario, S.; Das, S.; de Silva, A.M.; Deponte, K.; Marcantonio, N.; Anderson, J.F.; Fish, D.; Fikrig, E.; Kantor, F.S.: *Amer. J. Trop. Med. Hyg.* **1998**, *58*, 780-785.
124. Kamhawi, S.; Belkaid, Y.; Modi, G.; Rowton, E.; Sacks, D.: *Science* **2000**, *290*, 1351-1354.

125. Valenzuela, J.G.; Belkaid, Y.; Garfield, M.K.; Mendez, S.; Kamhawi, S.; Rowton, E.D.; Sacks, D.L.; Ribeiro, J.M.C.: *J. Exptl. Med.* **2001**, *194*, 331-342.
126. Trager, W.: *J. Parasitol.* **1939**, *25*, 57-81.
127. Wikel, S.K.: *Ann. Rev. Entomol.* **1982**, *27*, 21-48.
128. Opdebeeck, J.P.: *Vet. Parasitol.* **1994**, *54*, 205-222.
129. Kay, B.H.; Kemp, D.H.: *Amer. J. Trop. Med. Hyg.* **1994**, *50*, 87-96.
130. Barriga, O.O.: *Vet. Parasitol.* **1994**, *55*, 29-55.
131. Jacobs-Lorena, M.; Lemos, F.J.A.: *Parasitol. Today* **1995**, *11*, 144-147.
132. Bagnall, B.G. (1975) Cutaneous Immunity to the Tick *Ixodes Holocyclus*. Unpublished PhD Thesis, University of Sydney.
133. McGowan, M.J.; Barker, R.W.; Homer, J.T.; McNew, R.W.; Holscher, K.H.: *J. Med. Entomol.* **1981**, *18*, 328-332.
134. Mongi, A.O.; Shapiro, S.Z.; Doyle, J.J.; Cunningham, M.P.: *Insect. Sci. Appl.* **1986**, *7*, 471-477.
135. Tembo, S.D.; Rechav, Y.: *J. Med. Entomol.* **1992**, *29*, 757-760.
136. Brossard, M.: *Acta Tropica* **1976**, *33*, 15-36.
137. Wikel, S.K.: *Amer. J. Trop. Med. Hyg.* **1981**, *30*, 284-288.
138. Banerjee, D.P.; Momin, R.R.; Samantaray, S.: *Int. J. Parasitol.* **1990**, *20*, 969-972.
139. Trager, W.: *J. Parasitol.* **1939**, *25*, 137-139.
140. Allen, J.R.; Humphreys, S.J.: *Nature* **1979**, *280*, 491-493.
141. Wikel, S.K.: *Vet. Parasitol.* **1988**, *29*, 235-264.
142. Tellam, R.L.; Kemp, D.; Riding, G.; Briscoe, S.; Smith, D.; Sharp, P.; Irving, D.; Willadsen, P.: *Vet. Parasitol.* **2002**, *103*, 141-156.
143. Mulenga, A.; Sugino, M.; Nakajima, M.; Sugimoto, C.; Onuma, M.: *J. Vet. Med. Sci.* **2001**, *63*, 1063-1069.
144. Opdebeeck, J.P.; Wong, J.Y.M.; Jackson, L.A.; Dobson, C.: *Parasite Immunol.* **1988**, *10*, 405-410.

145. Jackson, L.A.; Opdebeeck, J.P.: *Parasite Immunol.* **1990**, *12*, 141-151.
146. Johnston, L.A.Y.; Kemp, D.H.; Pearson, R.D.: *Int. J. Parasitol.* **1986**, *16*, 27-34.
147. Agbede, R.I.S.; Kemp, D.H.: *Int. J. Parasitol.* **1986**, *16*, 35-41.
148. Willadsen, P.; McKenna, R.V.; Riding, G.A.: *Int. J. Parasitol.* **1988**, *18*, 183-189.
149. Willadsen, P.; Riding, G.A.; McKenna, R.V.; Kemp, D.H.; Tellam, R.L.; Nielsen, J.N.; Lahnstein, J.; Cobon, G.S.; Gough, J.M.: *J. Immunol.* **1989**, *143*, 1346-1351.
150. Willadsen, P.; Kemp, D.H.: *Parasitol. Today* **1988**, *4*, 196-198.
151. Willadsen, P.; Bird, P.; Cobon, G.S.; Hungerford, J.: *Parasitology* **1995**, *110*, 43-50.
152. McKenna, R.V.; Riding, G.A.; Jarmey, J.M.; Pearson, R.D.; Willadsen, P.: *Parasite Immunol.* **1998**, *20*, 325-336.
153. de Vos, S.; Zeinstra, L.; Taoufik, O.; Willadsen, P.; Jongejan, F.: *Exptl. Appl. Acarol.* **2001**, *25*, 245-261.
154. Riding, G.A.; Jarmey, J.; McKenna, R.V.; Pearson, R.; Cobon, G.S.; Willadsen, P.: *J. Immunol.* **1994**, *153*, 5158-5166.
155. Moire, N.; Bigot, Y.; Periquet, G.; Boulard, C.: *Mol. Biochem. Parasitol.* **1994**, *66*, 233-240.
156. Tellam, R.L.; Eisemann, C.H.: *Int. J. Parasitol.* **1998**, *28*, 439-450.
157. Wijffels, G.; Hughes, S.; Gough, J.; Allen, J.; Don, A.; Marshall, K.; Kay, B.; Kemp, D.: *Int. J. Parasitol.* **1999**, *29*, 1363-1377.
158. Sukarsih; Partoutomo, S.; Satria, E.; Wijffels, G.; Riding, G.; Eisemann, C.; Willadsen, P.: *Parasite Immunol.* **2000**, *22*, 545-552.
159. Colditz, I.G.; Watson, D.L.; Eisemann, C.H.; Tella, R.L.: *Vet. Parasitol.* **2002**, *104*, 345-350.

160. Tellam, R.L.; Vuocolo, T.; Eisemann, C.; Briscoe, S.; Riding, G.; Elvin, C.; Pearson, R.: *Insect Biochem. Mol. Biol.* **2003**, *33*, 239-252.
161. Cherney, L.S.; Wheeler, C.M.; Reed, A.C.: *Amer. J. Trop. Med.* **1939**, *19*, 327-332.
162. Opdebeeck, J.P.; Slacek, B.: *Int. J. Parasitol.* **1993**, *23*, 1063-1067.
163. Lee, S.E.; Jackson, L.A.; Opdebeeck, J.P.: *Parasite Immunol.* **1997**, *19*, 13-19.
164. Lee, S.E.; Johnstone, I.P.; Lee, R.P.; Opdebeeck, J.P.: *Vet. Immunol. Immunopathol.* **1999**, *69*, 229-237.
165. Ratzlaff, R.E.; Wikel, S.K.: *J. Med. Entomol.* **1990**, *27*, 1002-1007.
166. Volf, P.; Grubhoffer, L.: *Vet. Parasitol.* **1991**, *38*, 225-234.
167. Volf, P.: *Int. J. Parasitol.* **1994**, *24*, 1005-1010.
168. Ben-Yakir, D.; Mumcuoglu, K.Y.; Manor, O.; Ochanda, J.; Galun, R.: *Med. Vet. Entomol.* **1994**, *8*, 114-118.
169. Mumcuoglu, K.Y.; Rahamim, E.; Ben-Yakir, D.; Ochanda, J.O.; Galun, R.: *J. Med. Entomol.* **1996**, *33*, 74-77.
170. Mumcuoglu, K.Y.; Ben-Yakir, D.; Gunzberg, S.; Ochanda, J.O.; Galun, R.: *Med. Vet. Entomol.* **1996**, *10*, 105-107.
171. Mumcuoglu, K.Y.; Ben-Yakir, D.; Ochanda, J.O.; Miller, J.; Galun, R.: *Med. Vet. Entomol.* **1997**, *11*, 315-318.
172. Schlein, Y.; Lewis, C.T.: *Physiol. Entomol.* **1976**, *1*, 55-59.
173. Webster, K.A.; Rankin, M.; Goddard, N.; Tarry, D.W.; Coles, G.C.: *Vet. Parasitol.* **1992**, *44*, 143-150.
174. Baron, R.W.; Lysyk, T.J.: *J. Med. Entomol.* **1995**, *32*, 630-635.
175. East, I.J.; Allingham, P.G.; Bunch, R.J.; Matheson, J.: *Med. Vet. Entomol.* **1995**, *9*, 120-126.
176. Kerlin, R.L.; Allingham, P.G.: *Vet. Parasitol.* **1992**, *43*, 115-129.
177. Dubin, I.N.; Reese, J.D.; Seamans, L.A.: *J. Immunol.* **1948**, *58*, 293-297.

178. Alger, N.E.; Cabrera, E.J.: *J. Econ. Entomol.* **1972**, *65*, 165-168.
179. Hatfield, P.R.: *Med. Vet. Entomol.* **1988**, *2*, 331-338.
180. Ramasamy, M.S.; Ramasamy, R.; Kay, B.H.; Kidson, C.: *Med. Vet. Entomol.* **1988**, *2*, 87-93.
181. Ramasamy, M.S.; Srikrishnaraj, K.A.; Wijekoone, S.; Jesuthasan, L.S.B.; Ramasamy, R.: *J. Med. Entomol.* **1992**, *29*, 934-938.
182. Srikrishnaraj, K.A.; Ramasamy, R.; Ramasamy, M.S.: *Med. Vet. Entomol.* **1993**, *7*, 66-68.
183. Ramasamy, M.S.; Raschid, L.; Srikrishnaraj, K.A.; Ramasamy, R.: *J. Med. Entomol.* **1996**, *33*, 162-164.
184. Almeida, A.P.; Billingsley, P.F.: *Int. J. Parasitol.* **1998**, *28*, 1721-1731.
185. Ramasamy, M.S.; Sands, M.; Kay, B.H.; Fanning, I.D.; Lawrence, G.W.; Ramasamy, R.: *Med. Vet. Entomol.* **1990**, *4*, 49-55.
186. Ramasamy, M.S.; Ramasamy, R.: *Med. Vet. Entomol.* **1990**, *4*, 161-166.
187. Lal, A.A.; Schriefer, M.E.; Sacci, J.B.; Goldman, I.F.; Louis-Wileman, V.; Collins, W.E.; Azad, A.F.: *Infect. Immun.* **1994**, *62*, 306-318.
188. Srikrishnaraj, K.A.; Ramasamy, R.; Ramasamy, M.S.: *Med. Vet. Entomol.* **1995**, *9*, 353-357.
189. Lal, A.A.; Patterson, P.S.; Sacci, J.B.; Vaughan, J.A.; Paul, C.; Collins, W.E.; Wirtz, R.A.; Azad, A.F.: *Proc. Nat. Acad. Sci., U.S.A.* **2001**, *98*, 5228-5233.
190. Chabaudie, N.; Boulard, C.: *Vet. Immunol. Immunopathol.* **1992**, *31*, 167-177.
191. Sandeman, R.M.; Chandler, R.A.; Collins, B.J.; O'Meara, T.J.: *Int. J. Parasitol.* **1992**, *22*, 1175-1177.
192. Khan, M.A.; Connell, R.; Darcel, C.leQ.: *Canadian J. Comp. Med.* **1960**, *24*, 177-180.
193. Baron, R.W.; Colwell, D.D.: *Vet. Parasitol.* **1991**, *38*, 185-198.

194. Pruett, J.H.; Barrett, C.C.; Fisher, W.F.: *Southwestern Entomol.* **1987**, *12*, 79-88.
195. Broutin, I.; Arnoux, B.; Riche, C.; Lecroisey, A.; Keil, B.; Pascard, C.; Ducruix, A.: *Acta Crystalograph.* **1996**, *D52*, 380-392.
196. Boulard, C.: *Vet. Immunol. Immunopathol.* **1989**, *20*, 387-398.
197. Pruett, J.H.: *J. Parasitol.* **1993**, *79*, 829-833.
198. Nicolas-Gaulard, I.; Moire, N.; Boulard, C.: *Immunology* **1995**, *84*, 160-165.
199. Tellam, R.L.; Bowles, V.M.: *Int. J. Parasitol.* **1997**, *27*, 261-273.
200. East, I.J.; Fitzgerald, C.J.; Pearson, R.D.; Donaldson, R.A.; Vuocolo, T.; Cadogan, L.C.; Tellam, R.C.; Eisemann, C.H.: *Int. J. Parasitol.* **1993**, *23*, 221-229.
201. Schorderet, S.; Pearson, R.D.; Vuocolo, T.; Eisemann, C.; Riding, G.A.; Tellam, R.: *Insect Biochem. Mol. Biol.* **1998**, *28*, 99-111.
202. Tellam, R.L.; Eisemann, C.H.; Vuocolo, T.; Casu, R.; Jarmey, J.; Bowles, V.; Pearson, R.: *Int. J. Parasitol.* **2001**, *31*, 798-809.
203. Kerlin, R.L.; East, I.J.: *Parasite Immunol.* **1992**, *14*, 595-604.
204. O'Meara, T.J.; Nesa, M.; Raadsma, H.W.; Saville, D.G.; Sandeman, R.M.: *Res. Vet. Sci.* **1992**, *52*, 205-210.
205. Sandeman, R.M.; Chandler, R.A.; Seaton, D.S.: *Int. J. Parasitol.* **1995**, *25*, 621-628.
206. Adams, M.D.; Kelley, J.M.; Gocayne, J.D.; Dubnick, M.; Polymcropoulos, M.H.; Xiao, H.; Merril, C.R.; Wu, A.; Olde, B.; Morens, D.M.; Kerlavage, A.R.; McConnell, D.J.; Venter, J.C.: *Science* **1991**, *252*, 1651-1656.
207. Valenzuela, J.G.: *Insect Biochem. Mol. Biol.* **2002**, *32*, 1199-1209.
208. Li, S.; Kwon, J.; Aksoy, S.: *Insect Mol. Biol.* **2001**, *10*, 69-76.

209. Valenzuela, J.G.; Pham, V.M.; Garfield, M.K.; Francischetti, I.M.B.; Ribeiro, J.M.C.: *Insect Biochem. Mol. Biol.* **2002**, *32*, 1101-1122.
210. Crampton, A.L.; Miller, C.; Baxter, G.D.; Barker, S.C.: *Exptl. Appl. Acarol.* **1998**, *22*, 177-186.
211. Hill, C.A.; Gutierrez, J.A.: *Microb. Comp. Genom.* **2000**, *5*, 89-101.
212. Leboulle, G.; Rochez, C.; Louahed, J.; Rutti, B.; Brossard, M.; Bollen, A.; Godfroid, E.: *Amer. J. Trop. Med. Hyg.* **2002**, *66*, 225-233.
213. Valenzuela, J.G.; Francischetti, I.M.B.; Pham, V.M.; Garfield, M.K.; Mather, T.N.; Ribeiro, J.M.C.: *J. Exp. Biol.* **2002**, *205*, 2843-2864.
214. Naaby-Hansen, S.; Waterfield, M.D.; Cramer, R.: *Trends Pharmacol. Sci.* **2001**, *22*, 276-384.

Chapter 34

Control and Resistance Management of Human Pediculosis

Si Hyeock Lee[1], Kyong Sup Yoon[2], Jian-Rong Gao[2], Young-Joon Ahn[1], and J. Marshall Clark[2]

[1]School of Agricultural Biotechnology, Seoul National University, Seoul 151-742, Republic of Korea
[2]Department of Veterinary and Animal Science, University of Massachusetts, Amherst, MA 01003

Head louse resistance to commonly used pediculicides is wide spread worldwide. Resistance to permethrin and pyrethrum is 40- to 70-fold based on tolerance to knockdown response in US populations of head louse. Resistance to malathion is likewise widespread but at lower levels (2.0- to 3.3-fold). Currently, there appears to be little or no resistance to abamectin or lindane. Permethrin-resistant head lice are significantly more tolerant to knockdown and are cross-resistant to DDT, indicating a *kdr*-type of resistance mechanism. Sequence analysis of the full-length cDNA fragments of voltage-sensitive sodium channel α-subunit gene from permethrin-resistant head louse populations has identified three mutations (M815I, T917I, and L920F). All these mutations appear to exist *en bloc* as a haplotype and have functional significances in resistance. We have developed DNA-diagnostic protocols, including serial invasive signal amplification reaction (SISAR), for the detection of the conserved point mutations resulting in knockdown resistance to the pyrethrins, the pyrethroids, and DDT. A sound resistance management strategy based on efficient DNA-based genotyping techniques will greatly expand the lifespan of the valuable and effective pediculicides, such as permethrin, and ensure safe control of pediculosis.

Introduction

Pediculosis caused by *Pediculus capitis*, the human head louse, is the most prevalent parasitic infestation of humans worldwide, especially among school children of 3-12 years old (*1*). More than 6-12 million people in the United States (US) are estimated to have pediculosis. Symptoms of infestations are relatively mild, but the social, mental and economic consequences are substantial. Pediculosis is not widely tolerated in the US and is repeatedly treated by a variety of over-the-counter pediculicides, which are exclusively limited to those containing pyrethrin or permethrin as active ingredients. Because these pediculicides share a common mechanism of action on the voltage-sensitive sodium channel in the nervous system of insects (*2*), their repetitive and continued use has imposed a high pressure for the selection of resistant louse populations.

To date, head louse resistance to permethrin has been reported from France (*3*), Czech Republic (*4*), Israel (*5*), Britain (*6*), Argentina (*7,8*), and the US (*9-12*). Head louse populations from the US that were resistant to the lethal effect of permethrin also exhibited knockdown resistance (*kdr*) to the same material in behavioral knockdown bioassay (*10*). Through molecular cloning and sequencing analysis, the T929I and L932F mutations (T917I and L920F in louse sequence numbering, respectively) in a voltage-sensitive sodium channel α-subunit gene was identified to be associated with permethrin resistance via nerve insensitivity mechanism. One of the mutations, T929I, has already been functionally validated to be responsible for knockdown-type resistance in the diamondback moth, *Plutella xylostella* (*13*), indicating strongly that the same mutation in the human head louse plays a critical role in knockdown resistance to permethrin. Nevertheless, it is not known how widely permethrin-resistant head lice are distributed in the field nor is known the relative frequency of the *kdr* alleles within permethrin-resistant populations. Resistance to permethrin in the US head louse populations is of great concern because there currently are no effective alternative treatments available as replacements, which have the same low level of mammalian toxicity as do the pyrethroids. Therefore, there is an urgent need for developing monitoring and resistance management strategies and discovery of new alternative pediculicides.

In this article, the level of susceptibility to various insecticides in different human head louse populations in the US was investigated to assess the distribution of resistance. We also reported on the molecular mechanisms of head louse resistance to pyrethroid mediated by sodium channel mutations. Also discussed were the genotyping techniques for resistance monitoring and the attempts for the discovery of new alternative pediculicides from plants, which will be critical elements in constructing a sound pediculosis management system.

Extent and Mechanisms of Resistance to Pediculicides

Insecticide-susceptible strains of body louse from Israel (IS-BL) and head lice from Panama (PA-HL) and Ecuador (EC-HL) have been previously described (10,14). Another permethrin-susceptible strain of head louse was collected from Seoul, Korea (KR-HL). Permethrin-resistant strains of head louse were obtained from Western Massachusetts (MA-HL), Plantation, Florida (FL-HL), Homestead, FL (SF-HL), San Bernardino County, California (SC-HL) and four locations in Texas (TCC-HL, TMF-HL, TMS-HL and TSA-HL). A permethrin- and malathion-resistant strain (BR-HL) of head louse was provided by G. Coles (University of Bristol).

Resistance levels to permethrin in the FL-HL and MA-HL strains were 41- and 69-times higher, respectively, when compared by knockdown ratios and 3.8- and 4.3-times higher, respectively, when compared using mortality ratios, than those of the susceptible PA-HL strain (10). The higher resistance ratios determined by knockdown suggest that nerve insensitivity, most likely mediated by mutations in the sodium channel gene (kdr), is a principal resistance mechanism. Lack of synergism by piperonyl butoxide (PBO) and cross-resistance to DDT in the FL-HL strain substantiate this suggestion. The MA-HL (10) and SC-HL (14) strains, however, were PBO-synergized, possibly indicating the involvement of oxidative resistance. Recently, permethrin resistance was determined in southern California (SC-HL) (14) and south central Texas (TCC-HL, TMF-HL, TMS-HL, TSA-HL)(15) (Table 1), suggesting that resistance is widespread but not yet uniform. Cross-resistance to pyrethrum and PBO-synergized pyrethrum in the SF-HL and SC-HL strains was also apparent, indicating that the effective longevity of pediculicidal shampoos is now in jeopardy (14) (LT_{50} values, Tables 1 and 2).

Malathion resistance was not detected in the MA-HL and FL-HL populations (10) but levels from 2.0- to 3.3-fold were determined in the SF-HL and SC-HL populations (14) (Table 1). Malathion resistance in the SF-HL strain was completely synergized by DEF, indicating a role of malathion carboxylesterase, but had no effect on the SC-HL population (14) (Table 2 and 1). The SC-HL population, however, responded as a susceptible population when treated with malaoxon (14) (Table 1) and PBO actually protected this strain from malathion (data not shown), indicative of a reduced desulfuration activity (decreased formation of malaoxon) as a resistance mechanism. Similar findings using the malathion-resistant BR-HL population from the UK (14) validate the importance of reduced desulfuration. The lack of substantial cross-resistance to malathion in the permethrin-resistant US populations suggests that malathion may be an effective alternative treatment. However, presence of dual permethrin- and malathion-resistant populations in the UK (16) and our biochemical data (2.3 to 3.4-fold increase in esteratic activity in the dual resistant BR-HL strain, data not shown) indicate that its effectiveness will be short-lived unless augmented with alternative pediculicides and resistance management.

Table 1. Resistance Ratio (RR^a) of Head Lice from FL, CA, and TX

Treatment	Head Louse Strain				
	SF-HL	SC-HL	TSA-HL	TCC-HL	TMF-HL
1%Permethrin	8.5_a	5.3	4.5	4.6	5.5
0.33%Pyrethrum	2.8	2.6	-	-	-
0.5%Malathion	2	3.3	-	-	-
0.5%Malaoxon	0.6	0.8	-	-	-
10%Abamectin	1.7	1.8	-	-	-
1%Lindane	0.8	0.8	-	-	-

^{a}RR=LT_{50} of resistant population/LT_{50} of susceptible population (EC-HL)

Table 2. Synerglistic Effect of PBO and DEF on Pyrethrum-and Malathion-resistant Head Lice (SF-HL and SC-HL)

Louse Population and Treatment	N	LT_{50}(95%CL)	Slope ± SE	SR^b
0.33%pyrethrum + 4%PBO				
SF-HL	30	1505 (1464-1548)	3.0 ± 2.2	1
SC-HL^a	30	1296 (1257-1338)	6.6 ± 1.4	1.1
0.5%malathion + 0.1%DEF				
SF-HL^a	23	91 (59-126)	4.2 ± 0.6	2.6
SC-HL	27	399 (377-418)	17.0 ± 2.3	1

^{a}Logit regression is significantly different from unsynergized populations

Low levels of resistance to abamectin (1.7- to 1.8-fold) and lindane (0.8-fold) were observed in the tested US head lice (14) (Table 1). Because dexamethasone pretreatment protects lice from abamectin toxicity (data not shown), a 3"-O-demethylation of abamectin may function as a possible resistance mechanism (17).

Molecular Mechanisms of Head Louse Resistance to Pyrethroids and Development of Monitoring Tools for Resistance Management

Molecular analysis of kdr in permethrin-resistant head lice.

A 397-bp cDNA fragment of *para*-orthologous sodium channel α-subunit gene was obtained using a PCR cloning strategy that spans the two mutation sites most associated with knockdown resistance (M918T and L1014F). A 568-bp genomic DNA fragment corresponding to the cDNA fragment was obtained using the same PCR protocol (*10*). The sequences from the susceptible PA-HL, EC-HL, and KR-HL populations were identical over this region and are highly similar to other susceptible insects. Two mutations, T917I and L920F, occur only in the resistant FL-HL, SF-HL, MA-HL, TX-HL, SC-HL, and BR-HL lice, and are associated with resistance (Fig. 1).

```
               798                              827
EC-HL    TLCIVVNTLFMALDHHDMDKDMDRALKSGNY
IS-BL    -----------------I-------------
FL-HL    -----------------I-------------

               902                              932
EC-HL    LISIMGRTVGALGNLTFVLCIIIFIFAVMGM
PA-HL    -------------------------------
KR-HL    -------------------------------
FL-HL    ---------------I--F------------
SF-HL    ---------------I--F------------
MA-HL    ---------------I--F------------
TX-HL    ---------------I--F------------
SC-HL    ---------------I--F------------
BR-HL    ---------------I--F------------
```

Figure 1. Deduced amino acid sequences of the partial sodium channel α–subunit genes from several populations of head louse. The M815I, T917I and L920F mutations are shown in bold letter.

One additional novel mutation (M815I) in the IIS1-2 extracellular loop from permethrin-resistant lice has been identified (*18*). Absolute conservation of the Met815 residue at corresponding positions from all known susceptible insect populations implies that this mutation functions in resistance. Sequence analyses of cloned cDNA fragments and genomic DNA fragments from individual lice, both containing the three mutation sites, confirmed that all the mutations exist *en bloc* as a haplotype. A comparison of sodium channel gene sequences between head and body lice revealed only 26 polymorphisms, of which only one resulted in a conservative substitution (E11D) between the two species. This finding indicates that both body and head lice are con-specific, and justifies the use of body lice as a surrogate for head lice in biochemical and molecular biology studies. Conserved point mutations resulting in *kdr* resistance to pyrethroids are suitable for detection by various DNA-diagnostic protocols for monitoring, resistance management, and population genetic studies.

Functional analysis of the T917I and L920F mutations

We have inserted the T917I mutation (*10*) into the *Vssc1* sodium channel α-subunit gene from the house fly by site-directed mutagenesis and co-expressed it with the house fly Vsscβ subunit using the *Xenopus* oocyte expression system. We have obtained tetrodotoxin-sensitive currents using a two-electrode voltage clamp electrophysological recording system with both the wild type (*Wt*) and mutated (*TI*) forms of the expressed Vssc1 sodium channel (Fig. 2). The T917I mutation did not by itself substantially modify the voltage-dependent activation and inactivation kinetics of the expressed channel. However, the effect of increasing concentrations of cismethrin, a T-syndrome pyrethroid, on channel deactivation processes (sodium tail current) was greatly diminished. These findings confirms that the T917I mutation is a key element in *kdr*-type resistance mechanism.

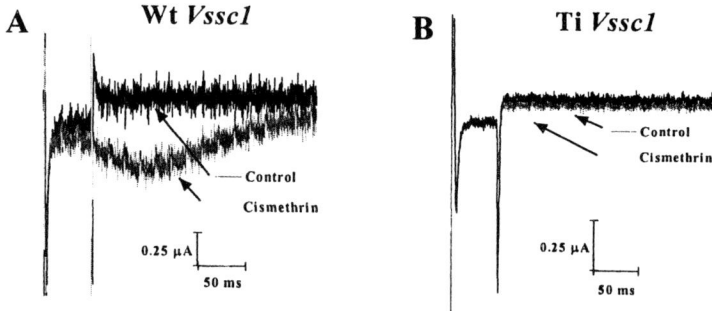

Figure 2. Effects of the T917I mutation on the response of Vssc1 to 100 μM cismethrin. Sodium channels were measured from oocytes expressing wt Vssc1 (A) or T917I (B). Sodium currents were measured and kinetic analysis performed as described previously (19).

Linkage analysis of the T917I and L920F mutations to the resistant phenotype

To determine a correlation between survivorship (due to permethrin resistance) and the frequency of *kdr*-type mutations (T917I, L920), genomic DNA was extracted from previously bioassayed lice from California, Florida, and Texas (reported above). Individual lice were determined to be phenotypically susceptible or resistant using the LT_{95} value of the permethrin-susceptible strains (PA-HL, EC-HL). Resistant or susceptible homozygous or heterozygous genotypes were determined from nucleotide sequences. The T917I and L920F mutations were tightly linked in the 424 louse samples examined. The logit mortality versus log survival time regression lines of susceptible- and resistant-homozygotes and heterozygotes displayed that the resistance trait is completely recessive (*15*) (Fig 3). The increasing presence of the T917I and L920F mutations correlated well with increased survival times following exposure to permethrin (non-parametric rank correlation test for independence was rejected, $r^2=0.94$, $P<0.05$) and indicates that *kdr* is causing permethrin resistance in US louse populations (*15*) (Fig. 4).

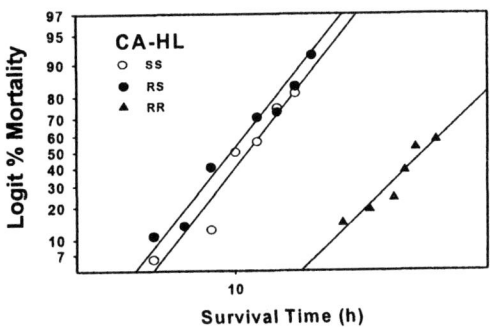

Figure 3. Mortality-survival time regression lines for susceptible, resistant homozygotes and heterozygotes exposed to 1% permethrin-impregnated filter paper. SS: Susceptible homozygote. RS: Heterozygote. RR: Resistant homozygote

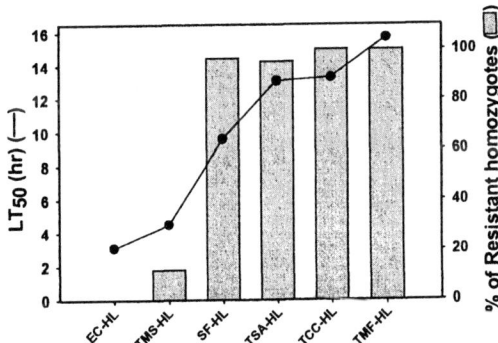

Figure 4. Relationship between phenotypic resistant level (LT_{50}) and percentage of resistant homozygotes (T917I and L920F) in different head louse populations from Ecuador, Florida and Texas

Detection of kdr mutations for monitoring of resistance.
We have developed a serial invasive signal amplification reaction (SISAR) that detects both the T917I and L920F mutations with the INVADAR DNA ASSAY (Third Wave Agbio, Madison, WI) using louse genomic DNA and the fluorescence resonance energy transfer (FRET) detection format (*20*)(Table 3). Using the L920 (C) to F920 (T) mutation for example, a C and T primary targets probes were able to distinguish individual susceptible (SS)- and resistant (RR)- homozygotes and heterozygotes (RS).

Table 3. Genotype Comparison of SISAR versus Sequencing Results Using the L920F Mutation from Head Lice Collected in Mathis, TX (TMA-HL)

	Head Louse Genotype			
	SS	RS	RR	Total
Individuals[a]	10	13	3	26
Average SISAR Ratio (±SE)	0.13 (0.02)	1.32 (0.15)	2.51 (0.12)	-
Correctly Predicted	10	12	3	25
% Predidted	100	92	100	96

[a] Individual genotype determined by direct sequencing.

Discovery of New Alternative Pediculicides

Plant essential oils may be a good alternative source for the discovery of novel insect control agents with low mammalian toxicity because they constitute a rich source of bioactive chemicals and are commonly used as fragrances and flavoring agents for food additives (*22*). As a traditional Chinese medicine, clove (*Eugenia caryophyllata* Thunberg), has long been considered to have medicinal properties such as a stimulant against digestive disorders and diarrhea. We investigated the insecticidal activity of the major components of clove bud and leaf oils (acetyleugenol, β–caryophyllene, eugenol, α–humulene, and methyl salicylate) and congeners of eugenol (isoeugenol and methyleugenol) against *P. capitis* females and eggs using direct contact application and fumigation methods, and compared with those of the widely used *d*–phenothrin and pyrethrum.

Table 4. Adulticidal and Ovicidal Activities of Test Compounds from Clove Oil against Head Louse as Determined by Filter Paper Diffusion Bioassay (Data from 23)

Compound	Adulticidal Activity[a]		Ovicidal Activity[b]	
	Dose, mg/cm^2	LT_{50}, min	Dose, mg/cm^2	Hatchability,% (mean ± SE)
Eugenol[c]	0.125	41.32	0.125	32 ± 1.7
	0.50	26.04	0.50	12 ± 1.7
Methyl salicylate[c]	0.25	35.86	0.25	0 ± 0.0
	0.50	28.26	0.50	0 ± 0.0
Acetyleugenol[c]	0.50	>300	5.0	77 ± 1.7
Isoeugenol	0.50	>300	5.0	67 ± 1.7
Methyleugenol	0.50	>300	5.0	68 ± 4.4
α–humulene[c]	0.50	>300	5.0	48 ± 1.7
β–caryophyllene[c]	0.50	>300	5.0	42 ± 1.7
d–phenothrin	0.25	25.22	5.0	83 ± 4.4
Pyrethrum	0.25	27.93	5.0	75 ± 0.0
Untreated	-	-	-	80 ± 2.9

[a] Exposed for 5 h.
[b] Exposed for 24 h
[c] *E. caryophyllata* bud and leaf oil compounds identified in this study.

In a filter paper diffusion bioassay, eugenol was most toxic against female *P. capitis* on the basis of LT_{50} values (Table 4), followed by methyl salicylate. Acetyleugenol, β–caryophyllene, α–humulene, isoeugenol, and methyleugenol were not effective. Potency of eugenol was comparable to d–phenothrin and pyrethrum. Against *P. capitis* eggs, methyl salicylate and eugenol were highly effective at the doses of 0.25-0.5 mg/cm^2 whereas little or no activity was observed with the other test compounds as well as with d–phenothrin and pyrethrum even at much higher dose (5 mg/cm^2). In fumigation tests with female *P. capitis* at 0.25 mg/cm^2, eugenol and methyl salicylate were more effective in closed cups than in open ones, indicating that the effect of the compounds was largely due to action in the vapor phase. Neither d–phenothrin nor pyrethrum exhibited fumigant toxicity. Based on these findings, the *Eugenia* bud and leaf essential oils, particularly eugenol and methyl salicylate, merit further study as potential *P. capitis* control agents or lead compounds.

Conclusion

Our results substantiate that permethrin resistance in US is widespread but not uniform. Permethrin resistance is highly correlated with the presence of the T917I and L920F mutations, which can be detected by a variety of DNA-based diagnostic techniques (21). These types of DNA-based diagnostic protocol will greatly facilitate the determination of resistant allele frequencies in head louse populations worldwide. Large-scale monitoring of permethrin resistance is possible utilizing these molecular techniques and would provide critical information necessary for the development of an effective resistance management program. It also appears possible that plant-derived novel pediculicides can be discovered and employed for the control of resistant louse populations.

Acknowledgements

This work was supported by the NIH/NIAID (PHS 1R01 AI 45062-01A1) and a grant-in-aid from Medicis Pharmaceutical Corporation, Phoenix, Arizona. S. H. Lee was supported in part by Brain Korea 21 Program. The authors sincerely thank Lice Source Services Inc. (Plantation, Florida), and FEST members for their assistance in louse collecions and shipments. The authors also wish to thank Dr. D. M. Soderlund for providing with the Vsscl clone.

References

1. Gratz, N. G.: World Health Organization, Switzerland (1997).
2. Soderland, D.M.; Clark, J.M.; Sheet, L.P.; Mullin, L.S.; Piccirillo, V.J.; Sargent, D.; Stevens, J.T.; Weiner, M.L.: *Toxicology* **2003**, *171*, 3.
3. Coz, J.; Combescot-Lang, C.; Verdier, V.: *Bull. Soc. Fr. Parasitol.* **1993**, *2*, 245.
4. Rupes, V.; Moravee, J.; Chmela, J.; Ledvinka, J.; Zelencová, J.: *Centr. Eur. J. Publ. Health.* **1995**, *3*, 30.
5. Mumcuoglu, K.Y.; Hemingway, J.; Miller, J.; Ioffe-Uspensky, I.; Klaus, S.; Ben-Ishai, F.; Galun, R.: *Med. Vet. Entomol.* **1995**, *9*, 427.
6. Burgess, I.F.; Brown, C.M.; Peock, S.; Kaufman, J.: *Br. Med. J.* **1995**, *311*, 752.
7. Picollo, M.I.; Vassena, C.; Casadío, A.; Mássimo, J.; Zerba, E.N.: *J. Med. Entomol.* **1998**, *35*, 814.

8. Picollo, M.I.; Vassena, C.V.; Cueto, G.A.M.; Vernetti, M.; Zerba, E.N.: *J. Med. Entomol.* **2000**, *37*, 721.
9. Pollack, R.J.; Kiszewski, A.; Armstrong, P.; Hahn, C.; Wolfe, N.; Rahman, H.A.; Laserson, K.; Telford III, S.R. ; Spielman, A.: *Adolesc. Med.* **1999**, *153*, 969.
10. Lee, S.H.; Yoon, K.S.; Williamson, M.S.; Goodson, S.J.; Takano-Lee, M.; Edman, J.D.; Devonshire, A.L.; Clark, J.M.: *Pestic. Biochem. Physiol.* **2000**, *66*, 130.
11. Meinking, T.L.; Entzel, P.; Villar, M.E.; Vicaria, M.; Lemard, G.; Porcelain, S.L.: *Arch. Dermatol.* **2001**, *137*, 287.
12. Meinking, T.L.; Serrano, L.; Hard, B.; Entzel, P.; Lemard, G.; Rivera, E.; Villar, M.E.: *Arch. Dermatol.* **2002**, *138*, 220.
13. Vias, H.; Williamson, M.S.; Usherwood, P.N.: *Pest Manag. Sci.* **2001**, *57*, 877.
14. Yoon, K. S.; Gao, J.-R.; Lee, S. H.; Clark, J. M.; Brown, L.; Taplin, D.: *Arch. Dermatol.* **2003**, (in press)
15. Gao, J. R.; Yoon, K. S.; Lee, S. H.; Takano-Lee, M.; Edman, J. D.; Meinking, T. L.; Taplin, D.; Clark, J. M.: *Pestic. Biochem. Physiol.* **2003**, (in press)
16. Downs, A. M. R.; Stafford, K. A.; Harvey, I.; Coles, G. C.: *Br. J. Dermatol.* **1999**, *141*, 508-511.
17. Yoon, K. S.; Clark, J. M.: *Pestic. Biochem. Physiol.* **2002**, *73*, 74-86.
18. Lee, S. H.; Gao, J. R.; Yoon, K. S.; Mumcuoglu, K. Y.; Taplin, D.; Edman, J. D.; Takano-Lee; Clark, J. M.: *Pestic. Biochem. Physiol.* **2003**, *75*, 79-91.
19. Lee, S. H.; Soderlund, D. M.: *Insect Biochem. Mol. Biol.* **2000**, *31*, 19-29.
20. Lyamichev, V. I.; Kaiser, M. K.; Lyamicheva, N. E.; Vologodskii, A. V.; Hall, J. G.; Ma, W.-P.; Allawai, H. T.; Neri, B. P.: *Biochemistry.* **2000**, *39*, 9523-9532.
21. Clark, J. M.; Lee, S. H.; Kim, H. J.; Yoon, K. S.; Zhang, A.: *Pest Manag. Sci.* **2001**, *57*, 1-7.
22. Isman, M. B.: *Pestic. Outlook* **1999**, *68*, 72
23. Yang, Y.-C.; Lee, S. H.; Lee, W. J.; Choi, D. H.; Ahn, Y. J.: *J. Agri. Food Chem.* **2003**, (in press)

Indexes

Author Index

Abe, Hiroshi, 239
Ahn, Young-Joon, 383
Ando, Tetsu, 226
Araya, Hiroshi, 63
Baerson, Scott R., 151
Bargar, Thomas M., 119
Beck, Michael, 74
Bräse, Stefan, 74
Cao, Song, 273
Chen, Audrey W., 295
Chua, Nam-Hai, 40
Chuman, Hiroshi, 142
Clark, J. Marshall, 383
Dayan, Franck E., 151
de Meijere, Armin, 74
Dick, Michael R., 119
Duke, Stephen O., 151
Edman, John D., 314
Es-Sayed, Mazen, 74
Fukase, Koichi, 87
Fukase, Yoshiyuki, 87
Fukushima, Yoshitaka, 304
Funke, Christian, 74
Gao, Jian-Rong, 383
Garvin, Gail M., 119
Gutteridge, Steven, 132
Hammond, Bruce G., 28
Hartnell, Gary F., 28
Hirose, Sakiko, 18
Hirose, Taro, 256
Huang, Qingchun, 273
Ikegami, Hiroshi, 256
Imai, Tetsuya, 304
Inui, Hideyuki, 40
Ishii, Naoki, 304
Islam, Md. Tofazzal, 202

Ito, Atsushi, 142
Itoh, Takaaki, 326
Izumi, Keiichi, 256
Kagan, Isabelle A., 151
Kamimura, Manabu, 191
Kather, Kristian, 74
Kawahigashi, Hiroyuki, 18
Kawaide, Hiroshi, 239
Kikuchi, Mami, 142
Kimura, Yasuo, 216
Kiyota, Hiromasa, 246
Kodama, Susumu, 40
Komiya, Mayumi, 239
Koyanagi, Fumie, 239
Kumazawa, Satoru, 142
Kusano, Miyako, 216
Kusumoto, Shoichi, 87
Kuwahara, S., 246
Lee, Si Hyeock, 383
Li, Zhong, 273
Limbach, Michael, 74
Lormann, Matthias E. P., 74
Lorsbach, Beth A., 99
Manker, Denise C., 283
Matsuda, Kazuhiko, 172
Matsumoto, Hiroshi, 161
Matsuo, Noritada, 256
Matsuo, Sanshiro, 256
Minakuchi, Chieka, 191
Miyagawa, Hisashi, 191
Nagai, Idumi, 183
Nagatomi, Toshio, 256
Nakagawa, Yoshiaki, 191
Nakahara, Satoshi, 216
Natsume, Masahiro, 239
Ohkawa, Hideo, 18, 40

Ohkawa, Yasunobu, 18
Ohta, Hiroto, 183
Oritani, T., 246
Ozoe, Yoshihisa, 183
Pasteris, Robert J., 109
Paulitz, Christian, 74
Pember, Steve O., 132
Qian, Xuhong, 273
Ridley, William P., 28
Ruiz, James M., 99
Saito, Shigeru, 256
Sakai, Hiroshi, 334
Sakamoto, Noriyasu, 256
Sasaki, Hideaki, 40
Sattelle, David B., 172
Sears, Mark K., 48
Siddall, Thomas L., 119
Song, Gonghua, 273
Strang, Harry, 2
Sudo, Keiichi, 142
Suzuki, Masaya, 256

Tahara, Satoshi, 202
Tanabe, Yoo, 267
Tao, Yong, 132
Tashiro, Nobuya, 239
Tsushima, Kazunori, 256
Turner, James A., 119
Umeda, Kimitoshi, 256
Umesako, Naomi, 87
Umetsu, Noriharu Ken, 304
Usui, Makio, 304
Utsumi, Toshihiko, 183
Walker, Mike, 132
Wikel, Stephen K., 348
Wroblowsky, Heiner, 74
Wu, LiHong, 132
Yamagiwa, Masashi, 334
Yoon, Kyong Sup, 383
Zhang, San-Qi, 87
Zimmermann, Viktor, 74

Subject Index

A

Abscisic acid (ABA) and analogs. See Plant hormones
Acaricidal activity. See 2-Substituted-3-trifluoromethylquinoxalines (3-TFQ)
Acetylaspyrone, nematicidal activity toward *Pratylenchus penetrans*, 220t
Action. See Mode of action; Site of action
Activists, exaggerating pesticide risks, 11–12
Activity, libraries screening, 106, 107t
Aedes aegypti, dengue, 322, 324t
Affinity separation. See Synthesis based on affinity separation (SAS)
Affordability, food in United States, 8, 9f
Agonist actions
  imidacloprid, 177f, 179f
  See also Ecdysone agonists, non-steroidal
Agrastatins, Serenade®, 291–292
Agricultural chemicals, world market, 3
Agricultural practices, vectorborne diseases, 318t
Agrochemical discovery
  leads, 120
  novel and actionable chemistry, 100
  See also Discovery
Agrochemical molecules, biological hypotheses, 100
Agro-research, combinatorial chemistry, 75
Aldol reactions
  features, 268
  lactone analogs of dihydrojasmone and *cis*-jasmone, 270–271
  synthesis of *R*-mintlactone and *R*-menthofuran, 271–272
Alkaloids, nematicidal activity toward *Pratylenchus penetrans*, 220t
Allelopathy
  activity in mushroom, 64–65
  allelochemicals, 63–64
  assay for mushroom litters, 67f
  estimation of activity of mushrooms, 66–67
  possible release routes of allelochemicals from common plants, 65f
  possible release routes of allelochemicals from mushrooms, 66f
  safety of allelochemicals, 69
  See also Mushrooms
Amaranthus gangeticus, regulation of developmental transition by, 210–211
Amblyomma americanum
  calreticulin, 354
  See also Ticks
American workers, high technology agriculture, 10
Ames, Dr. Bruce, cancer and protective role of fruits and vegetables, 13
Amino acid sequence
  Bombyx mori clone, B96Bom receptor, 185f
  nicotinic acetylcholine receptor (nAChR) subunits, 180t
  sodium channel α-subunit genes of head louse, 387f
  See also Cytochrome P450 14α-demethylase (CYP51)

399

Amino pyridazinones, synthesis, 279
2,5-Anhydro-D-glucitol (AhG)
   inhibiting lettuce root growth, 152
   schematic of mode of action, 153*f*
Anopheles gambiae,
   control/elimination program, 324*t*
Anti-biting fly, vaccination against, 363–364
Anti-ectoparasite vaccines
   anti-biting fly vaccines, 363–364
   anti-mosquito/vector-blocking, 364
   anti-tick immunity, 360
   commercialization of Bm86 vaccine, 361–362
   guinea pig resistance, 359
   horn flies, 364
   hypodermosis, 366
   immunization against mosquitoes, 364–365
   insects, 362–367
   Lucilia cuprina myiasis, 366–367
   myiasis, 366
   protection against flea bites, 362–363
   rabbit immunization, 363
   sucking lice, 363
   tick digestive tract, 360–361
   ticks, 360–362
   vaccination of Hereford cattle, 361
   vector-blocking vaccine concept, 365–366
Aphanomyces cochlioides
   complex mixtures affecting zoospore motility, 209
   differentiation of zoospores, 211
   motility inhibitory and cell lytic substances against zoospores, 208–209
   repellent activity of estrogenic compounds, 207–208
   stimuli triggering characteristic behavior and morphological changes, 204*t*
   zoospore bioassay, 203
   See also Oomycete phytopathogens

Arabidopsis transgenic plants
   absorption of ^{14}C-labeled 17β-estradiol and nonylphenol, 46*f*
   expressing green fluorescence protein (GFP) gene, 44–46
   expression plasmids, 45*f*
   recombinant estrogen receptor (ER) gene, 44–46
Arabidopsis thaliana, gene expression profile libraries, 158–159
Arboviral diseases
   dengue, 322
   West Nile virus, 323
   See also Vectorborne diseases
Arbovirus transmission, mosquito saliva, 357
Arthropods
   advantages of anti-arthropod vaccines, 349–350
   blocking pathogen transmission by immunity to arthropod vector, 358–359
   ectoparasitic, and host immune defenses, 350
   expressed sequence tag (EST), 367–368
   future challenges, 368–369
   genomics and proteomics, 367–368
   pyridalyl and beneficial, 265*t*
   research avenues, 368–369
   saliva containing pharmacologically active molecules, 350–357
   transmitting diseases, 349
   vector suppression for infectious agents, 349
   See also Ticks
Aryl hydrocarbon receptor (AhR), transgenic tobacco plants, 42–43
Asiatic corn borer, insect-growth regulators, 281
Asparagine synthetase, potential herbicide target site, 154–155
Aspergillus sp.

compounds from unidentified, 221–222
nematicidal activity of compounds toward *Pratylenchus penetrans*, 222t
See also Nematicidal compounds
Aspyrone, nematicidal activity toward *Pratylenchus penetrans*, 220t, 222t, 224t
Attribute-driven approach, lead discovery, 120, 121f
Azoles
fungicides, 143–144
See also Cytochrome P450 14α-demethylase (CYP51)
N-Azoyl phenoxypyrimidine, herbicide lead optimization, 114–115

B

B96Bom from silkworm
actions of formamidine insecticide on *Bombyx. mori* tyramine receptor, 187–189
affinity of biogenic amines for B96Bom receptor, 186–187
B. mori silkworm, 183, 184
functional properties of B96Bom receptor in cell line, 185–186
molecular cloning of, 184–185
Bacillus thuringiensis
δ-endotoxins, 335
green pesticides, 287–288
subspecies *israelensis*, 335
See also Dipteran-specific insecticidal crystal proteins
Bacillus thuringiensis corn pollen
dispersal, 54
exposure factors, 53
probability of monarch larvae exposure, 56–57
See also Monarch butterfly
Bacterial diseases

Lyme disease, 321
plague, 321
See also Vectorborne diseases
Barbituric acid
interaction with artificial receptor, 91–97
See also Synthesis based on affinity separation (SAS)
Benzotriazoles
Buchwald–Hartwig reaction on triazene-linked phenyl bromides, 78f
inverse Buchwald–Hartwig reaction, 78, 79f
method development, 79
Pythia results for, library, 77f
Pythia search engine, 76
Sanger reagent, 76f
solid-phase synthesis, 75–79
switching to Buchwald–Hartwig reaction, 77
synthesis of triazene-linked anilines, 78f
See also Combinatorial chemistry
Benzoylphenylureas, insect-growth regulator, 274
Bioassays
lettuce seedling growth, 217–218
nematicidal activity, 217
rice seedling growth, 218
Biochemical approach, fungal lead, 137–138
Biofungicide, Serenade®, 291–293
Biogenic amines, 184
Biological hypotheses, agrochemical molecules, 100
Biological properties, pyridalyl, 263–265
Biosynthesis, pheromones, 235–236
Biotech crops
body weights, kidney/body weight, and liver/body weight in rats, 36t
composition evaluation, 33–34
development, 29

materials and methods, 30–33
poultry performance evaluation, 34–35
rat safety assurance evaluation, 35–37
Roundup Ready®, 29
safety evaluation, 29
study plan for compositional analysis of maize, 30–31
study plan for nutritional performance of maize, 31–32
study plan for rat safety assurance, 32t, 33
Blood coagulation, ticks, 352
Blood feeding insects
host immune defenses, 350
transmitting infectious agents, 349
Bombyx mori. *See* B96Bom from silkworm
Boophilus microplus
calreticulin, 354
See also Ticks
Borrelia burgdorferi, tissue burden, 358
Botrytis cinerea, cytochrome P450 14α-demethylase (CYP51), 147–149
Bradykin, mediator of itch/pain, 351
Buchwald–Hartwig reaction
benzotriazole synthesis, 77–78
inverse, 78, 79f
Butterfly. *See* Monarch butterfly

C

Calreticulin, tick saliva, 354
Cancer, fruit and vegetable consumption, 13
Carotenoid biosynthesis, herbicides interfering, 162
Carrot cells
cell growth condition, 164
effect of metabolite on growth, 168, 169f
effect of pyrazole herbicides and metabolite on HPPD activity from, 168, 170f
See also Pyrazole herbicides
Cats, vaccination against fleas, 362–363
Cattle
anti-tick vaccination, 361
vaccine control against pests, 364
Center for Disease Control and Prevention (CDC), average life span, 14
Ceramide synthase biosynthesis, potential target site, 152–154
Chagas disease
control/elimination program, 324t
parasitic disease, 319–320
Chemistry to gene (C2G), target identification approach, 133–134, 138
Chemotaxis, specificity, 206
Chicken
amino acid sequences, 180t
interactions of α subunits with neonicotinoids, 176–177
interactions of loop F with neonicotinoids, 178
Chilo suppressalis
expression and functional analysis of ecdysone receptor (EcR) and ultraspiracle (USP) proteins, 193–194
relationship between molting hormonal activity and receptor-binding activity of ecdysone agonists, 196f
relationship between molting insecticidal activity and receptor-binding activity of ecdysone agonists, 196f
relationship of receptor-binding activities between insect orders, 196–198
relationships among receptor-binding, molting-hormonal and

insecticidal activities against, 194–196
structure-activity relationships (SAR), 193
See also Ecdysone agonists, nonsteroidal
Chlordimeform
actions on *Bombyx mori* tyramine receptor, 187–189
dose-response curve in inhibiting tyramine binding to B96Bom, 188*f*
Chlorfenapyr, trifluoromethyl containing insecticide, 305
1,4-Cineole, activity of asparagine synthetase, 154–155
Cinmethylin, proposed mode of action, 155
Civetone
practical synthesis of Z-, 268, 269
short step synthesis of E/Z-, 268–270
Claisen condensations
features of reactions, 268
formal synthesis of R-muscone, 270
short step synthesis of E/Z-civetone, 268–270
Ti- or Zr-, 267–268
CLASS expert system, candidate library screening, 104
Cloning. See B96Bom from silkworm
Clove oil, compounds against head louse, 391*t*
Coagulation, ticks, 352
Combinatorial chemistry
agro-research, 75
benzotriazole synthesis, 75–79
designing libraries for herbicide lead discovery, 121–122
discovery processes, 110
geometrically defined peptidomimetics, 80–84
hit validation and optimization, 84–85

libraries, 75
Pythia search engine, 76, 77*f*, 84
role in biologically driven discovery research, 120
traditional vs. attribute-driven approach, 120, 121*f*
See also Benzotriazoles; Parallel synthesis technologies; Peptidomimetics, geometrically defined
Compositional analysis, maize event NK603, 30–31, 33–34
Computer aided library design. See Ring-fused 2-pyridinone esters
Concanamycin A/B
analysis method, 241
Streptomyces scabies, 240
structure, 240*f*
See also *Streptomyces* species
Concanamycins A/B, 242
Condensations. See Claisen condensations
Conformation analysis, metconazole, 147, 148*f*
Conjugated diene system, type I pheromones, 228–229
Consumers, pesticide benefits, 10, 11*t*
Contamination, endocrine disruptors (EDs), 41–42
Corn pollen. See Monarch butterfly
Crop Disease Management (CDM). See Nicotinic acid
Crop protection
industry consolidation, 4, 5*f*
market overview, 3
See also Biotech crops; Pesticides; Site of action
Crown ether
ammonium ion interaction, 88–91
See also Synthesis based on affinity separation (SAS)
Crystal proteins. See Dipteran-specific insecticidal crystal proteins
Culex pipiens
lymphatic filariasis, 319

West Nile virus, 323
See also Dipteran-specific insecticidal crystal proteins; Malaria
Cyclopropylideneacetates. See Peptidomimetics, geometrically defined
CYP1A1 and CYP2B6 genes
  germination tests of transgenic rice plants, 20, 22–23
  metabolism of herbicides in transgenic rice plants expressing, 23, 24f
  See also Transgenic rice plants
Cystospores, *Aphanomyces cochlioides* zoospores, 211
Cytochrome P450 14α-demethylase (CYP51)
  amino acid sequence alignment, 145, 146t
  amino acid sequence alignment of binding site, 147
  Botrytis cinerea (BCCYP51), 147–149
  conformation analysis of metconazole, 147, 148f
  determination of binding site with fluconazole, 146
  docking of metconazole into binding site, 148f
  fungicidal activity of metconazole against plant pathogen, 145t
  homology modeling, 144
  multiple sequence alignment of binding site, 148f
  Mycobacterium tuberculosis (MTCYP51), 145–147
  proposed interaction mode for metconazole to, 144f
  substitution of amino acid residues in binding site, 148
  three-dimensional model of BCCYP51, 149

Cytochrome P450 monooxygenase, herbicide detoxification in plants, 19

D

Defenses, tick salivary molecules modulating, 356
Deforestation, vectorborne diseases, 318t
Demethylchlordimeform
  actions on *Bombyx mori* tyramine receptor, 187–189
  dose-response curve in inhibiting tyramine binding to B96Bom, 188f
  effects on forskolin-stimulated cAMP production, 189f
Dengue
  arboviral disease, 322
  control/elimination program, 324t
3-Desmethyl sulfentrazone (DMS), structure and nomenclature, 297
Dibenzoylhydrazines, insect-growth regulator, 274
Dihydrojasmone, synthesis of lactone analogs of, 270–271
Dihydropyridine miticide, lead optimization, 112–114
Dihydropyrimidines, 1,1-dicyano-2,2-bis(trifluoromethyl)ethylene (BTF) condensations, 113–114
Dipteran-specific insecticidal crystal proteins
  abbreviations, 345
  amino acid sequences of Cry11A fragments, 343
  association of 20-kDa and 45-kDa fragments, 340
  Bacillus thuringiensis subsp. *israelensis*, 335
  bioassay of mosquito larvicidal activity of Cry proteins, 336

brush border membrane vesicle (BBMV) preparation, 336
construction of fusion proteins with Cry11A fragments, 343, 344f, 345
construction of fusion proteins with Cry4A fragments, 338, 339f
coprecipitation experiments, 336
Cry11A insecticidal protein of 70-kDa, 342–343
immunohistochemical staining experiments, 336
ingestion of Cry4A crystal by larvae of *Culex pipiens*, 337–338
insecticidal activities of Cry11A fragments, 344f
insecticidal activities of Cry4A fragments, 341
insertion assay, 336
in vitro processing of Cry11A with trypsin and gut extracts from *C. pipiens*, 343f
in vitro processing of Cry proteins, 335
in vitro processing products from Cry4A with gut extracts of *C. pipiens*, 338f
in vivo binding experiments, 341–342
ligand blotting, 336
materials and methods, 335–336
predicted modes of membrane interaction of Cry4A, 342f
processing of Cry4A protoxin, 337
schematic of active form of Cry4A and subunits, 340f
Discovery
alternative pediculicides, 390–391
combinatorial chemistry, 110
designing combinatorial libraries for herbicide lead, 121–122
insecticides, 257
lead compound to pyridalyl, 258
natural products, 151–152
novel and actionable chemistry, 100
oxazolidine scaffold-based, library, 116–117
screening process, 133
Diseases
arboviral, 322–323
bacterial, 321
control strategies, 324
control tools, 325
parasitic, 319–320
See also Vectorborne diseases
5,6-Disubstituted nicotinic acid, preparation of libraries, 122–124
DNA microarray technology, inhibitor modes of action, 158
Dogs, vaccination against fleas, 362
Dopamine, dose-response curve in inhibiting tyramine binding to B96Bom, 187f
Drosophila melanogaster
amino acid sequences, 180t
cloning ecdysone receptor (EcR) and ultraspiracle (USP) for, 192
expression and functional analysis of EcR and USP proteins, 193–194
interactions of α subunits with neonicotinoids, 176–177
relationship of receptor-binding activities between insect orders, 196–198
See also Ecdysone agonists, non-steroidal

E

Ecdysone agonists, non-steroidal
binding activities, 195t
comparisons of primary sequences of ecdysone receptor (EcR) and ultraspiracle (USP) proteins, 194f

expression and functional analysis of EcR and USP proteins, 193–194
insecticidal activity, 193
insect molting, 192
relationship between molting hormonal activity and receptor-binding activity of, against *Chilo suppressalis*, 196*f*
relationship between molting insecticidal activity and receptor-binding activity of, against *C. suppressalis*, 196*f*
relationship of receptor-binding activities between *C. suppressalis* and *Drosophila melanogaster*, 197*f*
relationship of receptor-binding activities between insect orders, 196–198
relationships among receptor-binding, molting-hormonal and insecticidal activities against *C. suppressalis*, 194–196
structures of ecdysteroids and, 192*f*
Ecdysone receptor (EcR)
expression and functional analysis, 193–194
See also Ecdysone agonists, nonsteroidal
Echinochloa oryzicola
effects of pyrazolate and norflurazon on pigment contents, 166, 167*t*
herbicides treatment, 164
phytoene content, 168, 169*f*
pigments extraction and determination, 164
watergrass, 163
See also Pyrazole herbicides
Ectoparasitic arthropods
host immune defenses, 350
See also Anti-ectoparasite vaccines
Effective environmental concentration (EEC), toxicity threshold, 51
ELISA (enzyme-linked immunosorbent assay)
comparison with liquid chromatography/mass spectrometry/MS (LC/MS/MS), 301–303
method for immunogens, 298–299
Endocrine disruptors (EDs)
absorption and accumulation in plant species, 41–42
absorption of ^{14}C-labeled 17β-estradiol and nonylphenol, 46*f*
aryl hydrocarbon receptor (AhR) and estrogen receptor (ER) binding, 41–42
contamination of environment and agricultural products, 41
dose-dependent expression of green fluorescence protein (GFP) gene in transgenic *Arabidopsis*, 46*f*
dose-dependent GUS (β-glucuronidase) activity in transgenic tobacco plants, 43*f*
environmental water contamination in Japan, 41*t*
expression plasmid for AhR-dependence inducible expression system for GUS reporter gene, 43*f*
expression plasmids for ER-dependent inducible expression system, 45*f*
expression plasmids for GFP reporter gene, 44–46
histochemical staining of GUS activity in transgenic tobacco plants, 44*f*
transgenic *Arabidopsis* plants carrying recombinant ER gene, 44–46
transgenic tobacco plants carrying mouse AhR gene, 42–43
δ-Endotoxins
Bacillus thuringiensis, 335

See also Dipteran-specific insecticidal crystal proteins
Environmental contamination, endocrine disruptors (EDs), 41–42
Environmental impacts, green pesticides, 285–286
Environmentally benign synthesis. *See* Green synthesis; Insect-growth regulators
Environmental Protection Agency (EPA), pesticide testing, 14–15
Enzymes. *See* Cytochrome P450 14α-demethylase (CYP51)
Esters. *See* Ring-fused 2-pyridinone esters
Estrogen receptor (ER) gene, transgenic *Arabidopsis* plants, 44–46
Expressed sequence tag (EST), blood feeding arthropods, 367–368

F

Factors X and Xa, ticks blocking host coagulation, 352–353
Fairy rings, mushrooms, 63
Famoxate
  quinol oxidation, 139–140
  site topography, 138–140
  structure, 138*f*
Farmers, number of people fed by, 10
Farming practices, public acceptance, 15*t*
Field studies, pyridalyl, 265
Fleas, vaccine control, 362–363
Fluacrypyrim, trifluoromethyl containing acaricide, 305
Fluconazole, binding site, 146
Flufenacet, trifluoromethyl containing herbicide, 305
Fluorine-containing compounds agrochemicals, 304

See also 2-Substituted-3-trifluoromethylquinoxalines (3-TFQ)
Food affordability, United States, 8, 9*f*
Food surpluses, productivity, 4
Formamidines
  actions on *Bombyx mori* tyramine receptor, 187–189
  effects of demethylchlordimeform on forskolin-stimulated cAMP production, 189*f*
  structures, 184*f*
Fructose-1,6-bisphosphate aldolase (FBPase), inhibition, 152, 153*f*
Fruits and vegetables
  obesity epidemic, 11–12
  protective role against cancer, 13
Fungi
  metabolites for nematicides, 216
  See also Nematicidal compounds
Fungicides
  azoles, 143–144
  biochemical approach, 137–138
  growth and inhibition of *Saccharomyces cerevisiae*, 134–137
  overexpression lines, 137
  peptides, 80
  world market, 3
  See also Ipconazole; Metconazole; Site of action
2(5H)-Furanones, synthesis, 271–272

G

Gas chromatography–mass spectrometry (GC–MS), sex pheromone structure identification, 229–231
Gas chromatography with electroantennogram detection (GC–EAD), sex pheromone identification, 227–228

Gene expression profile libraries, whole genome DNA microarrays, 158–159
Genomics, blood feeding arthropods, 367–368
Germination, transgenic rice plants, 20, 22–23
Glyphosate, Roundup®, 29
Green fluorescence protein (GFP) reporter gene, transgenic *Arabidopsis* plants, 44–46
Green pesticides
  Bacillus thuringiensis, 287–288
  driving forces behind biopesticide development, 286–287
  environmental impacts, 285–286
  green chemistry philosophy, 284
  hazardous substances, 286
  human health, 285
  Messenger®, 290–291
  safety, 284
  Serenade®, 291–293
  spinosad, 289–290
  waste reduction, 286
Green synthesis
  E/Z-civetone, 268–270
  Z-civetone, 268, 269
  Claisen condensations and aldol reactions, 267–268
  2(5H)-furanones, 271–272
  lactone analogs of dihydrojasmone and *cis*-jasmone, 270–271
  R-mintlactone and R-menthofuran, 271–272
  R-muscone, 270
Groundwater samples. *See* Sulfentrazone
Guinea pigs
  anti-ectoparasite vaccine research, 359
  anti-tick immunity, 360–361
GUS (β-glucuronidase) reporter gene, transgenic tobacco plants, 42–43

H

Haematobia irritans, vaccine control, 364
Hazardous substances, green pesticides, 286
Head louse
  mechanisms of resistance to pyrethroids, 387–390
  resistance to permethrin, 384
  See also Pediculosis
Hematophagous insects, linking pathogen transmission to saliva molecules, 357–358
Herbicidal activity, screening library compounds, 106, 107t
Herbicides
  carotenoid biosynthesis, 162
  designing combinatorial libraries for lead discovery, 121–122
  detoxification by cytochrome P450 monooxygenase (P450), 19
  metabolism in higher plants, 19
  metabolism in transgenic rice plants, 23, 24f
  natural products, 151
  phloem mobility, 121
  phytotoxicity of, toward CYP1A1 and CYP2B6 transgenic rice plants, 21f
  tolerance of transgenic rice plants, in germination tests, 22t
  value in U.S. crop production, 8f
  world market, 3
  See also Pyrazole herbicides; Transgenic rice plants
Hereford cattle, vaccination, 361
Heterocyclic compounds, insect-growth regulators, 274
High performance liquid chromatography (HPLC), chiral, for enantiomers of type II pheromones, 231–232
Histamine, mediator in ticks, 351–352
Hit Generation, screening process, 133

Homogentisate
  conversion of 4-hydroxyphenylpyruvate to, 162
  effect on phytoene accumulation, 168, 169f
Homology modeling, cytochrome P450 14α-demethylase (CYP51), 144
Hormones. See Plant hormones
Horn flies, vaccine control, 364
Host immune defenses, arthropods and, 350
Human health, pesticides, 4, 285
Human lives saved, pesticides, 13, 14f
Human P450 genes. See Transgenic rice plants
Human pediculosis. See Pediculosis
Human population, disease outbreaks, 315, 318
Human vector control. See Vectorborne diseases
3-Hydroxymethyl sulfentrazone, structure and nomenclature, 297
4-Hydroxyphenylpyruvate dioxygenase (HPPD)
  concentration of metabolite in pyrazole herbicides solution in HPPD assay, 168, 170f
  effect of pyrazolate, pyrazoxyfen and metabolite on HPPD activity, 168, 170f
  natural products inhibiting, 156–158
  preparation and assay, 165
  target site, 162
  See also Pyrazole herbicides
Hypoderma species, vaccine control of myiasis, 366
Hypodermosis, vaccination against, 366

I

Imidacloprid
  agonist actions, 177f
  contour plots for electrostatic potential, 175f
  interactions of α subunits with neonicotinoids, 176–177
  neonicotinoids, 174
  structure, 173f
  See also Neonicotinoid insecticides
Immune defenses
  proteins binding mediators, 354
  tick salivary molecules modulating, 356
Immunization
  mosquitoes, 364–365
  See also Anti-ectoparasite vaccines
Infectious diseases
  increasing, 315
  large outbreaks, 316f
  unexpected outbreaks, 317f
  See also Vectorborne diseases
Insect-growth regulators
  amino pyridazinones, 279
  antifeedent activities against Asiatic corn borer, 281
  aromatic heterocycles, 274
  benzoylphenylureas, 274
  2,5-bis(2,4-dichlorophenyl)-1,3,4-oxadiazole (DCPO) and analogs, 274–275
  chronic growth inhibitory activity of oxadiazolyl pyridazinones and analogs, 280t
  core structures for benzoylphenylurea, dibenzoylhydrazine, and oxadiazole, 274
  dibenzoylhydrazines, 274
  eco-friendly, 274
  mode of action, 280
  novel oxadiazoles, 275–277
  oxadiazolyl pyridazinones and analogs, 277–281
  structures and bioactivities of oxadiazoles against larvae of *Pseudaletia separata*, 277
  symmetrical 2F-DCPO, 275–276

synthesis of oxadiazolyl pyridazinones, 278–279
target structures of oxadiazolyl pyridazinones, 278
Insecticidal activity
Chilo suppressalis, 194–196
fusion proteins of Cry11A fragments, 344*f*
fusion proteins of Cry4A fragments, 341
lead compound to pyridalyl, 259*t*
method for 2-substituted-3-trifluoromethylquinoxalines, 306
non-steroidal ecdysone agonists, 193
pyridalyl, 264*t*
See also Dipteran-specific insecticidal crystal proteins; 2-Substituted-3-trifluoromethylquinoxalines (3-TFQ)
Insecticidal nets. *See* Long lasting insecticidal net (LLIN); Malaria
Insecticides
discovery, 257
world market, 3
See also Neonicotinoid insecticides; Pyridalyl
Insect Management (IM). *See* Nicotinic acid
Insect molting, regulation, 192
Ipconazole
azole fungicide, 143–144
structure, 143*f*
Itch, ticks, 351–352
Ixodes scapularis
antigens by screening cDNA library, 356
blocking alternate pathway of activation, 355
saliva with IL-2 binding protein, 354–355
See also Ticks

J

Japan, contamination by endocrine disruptors (EDs), 41*t*
cis-Jasmone, synthesis of lactone analogs of, 270–271
Jasmonic acid and analogs. *See* Plant hormones

L

Lead discovery
designing combinatorial libraries for herbicide, 121–122
See also Discovery; Parallel synthesis technologies; Site of action
Leishmaniasis, parasitic disease, 320
Lepidoptera. *See* Sex pheromones
Lepidopterous pests
insecticidal activity of pyridalyl against, 264*t*
See also Pyridalyl
Lettuce seedling
efficacy of growth, 67
growth inhibition by cultivated mushrooms, 70*t*
growth inhibition by wild mushrooms, 68*t*, 69*t*
mushrooms inhibiting growth, 67–68
Libraries
defining quality, 105
design strategy, 101–102
gene expression profile, 158–159
optimization goal, 106
oxazolidine scaffold-based, 116–117
preparation of 5,6-disubstituted nicotinic acid, 122–124
ring-fused 2-pyridinone scaffold, 100
screening activity, 106, 107*t*

tick antigens by screening cDNA, 356
weed management, 100
See also Nicotinic acid
Lice
vaccine control, 363
See also Pediculosis
Lipopolysaccharide (LPS), lipid A, 92, 95
Liquid chromatography/mass spectrometry/MS (LC/MS/MS)
comparison to enzyme-linked immunosorbent assay (ELISA), 301–303
method for sulfentrazone, 299–300
Long lasting insecticidal net (LLIN)
durability to washing, 328
duration of Olyset under practical use, 330, 332
generation of efficacy after washing, 329
impact on malaria transmission, 329–330
malaria control, 327, 332–333
See also Malaria
Lowest-observable-effect-concentration (LOEC), toxicity, 54, 55f
Lucilia cuprina
myiasis vaccine, 366–367
vaccine control of myiasis, 366
Lyme disease, bacterial, 321
Lymphatic filariasis, parasitic disease, 319

M

Macrophase migration inhibitory factor (MIF), tick salivary glands, 355
Maize
compositional analysis plan, 30–31
composition evaluation, 33–34
event NK603 test line, 30
nutritional evaluation plan, 31–32
poultry performance evaluation, 34–35
rat safety assurance evaluation, 35–37
rat safety assurance study, 32t, 33
See also Biotech crops
Malaria
control/elimination program, 324t
durability of nets to washing, 328
long lasting insecticidal nets (LLIN), 327, 332–333
mesh sizes of nets, 327–328
mortality of mosquitoes after contact with nets, 328t
observation of female *Culex pipiens pallens* passing through nets, 328t
Olyset curation under practical use conditions, 330, 332
Olyset net characteristics, 327–332
Olyset net impact on transmission of, in endemic areas, 329–330
parasitic disease, 320
percentage permethrin remaining after seven years use, 331t
percentages of knock down and mortality of mosquitoes after net contact, 329t, 332t
regeneration of efficacy after washing nets, 329
Roll Back Malaria (RBM) by World Health Organization (WHO), 320, 327, 332–333
slide positive rate by mass blood survey, 330t
Malathion, resistance to pediculicides, 385–386
Market overview, agricultural chemicals, 3
Mating
Ascotis selenaria cretacea ratio in tea garden, 234t
disruption by type II pheromones, 234–235

disruption overview, 232–234
representative disruptants in world, 233*t*
See also Sex pheromones
Mechanisms
head louse resistance to pyrethroids, 387–390
resistance to pediculicides, 385–386
Media coverage, pesticide risks, 12
Meldrum's acid adduct (MAA)
library inputs, 105*f*
preparation, 103
R-Menthofuran, synthesis, 271–272
Messenger®, production, 290–291
Metabolism, herbicides in higher plants, 19
Metalloproteases, platelet aggregation, 353
Metconazole
azole fungicide, 143–144
conformation analysis, 147, 148*f*
docking in binding site, 148*f*
fungicidal activity against plant pathogens, 145*t*
interaction mode for *Botrytis cinerea* CYP51, 149*f*
proposed interaction mode to target enzyme, 144*f*
structure, 143*f*
See also Cytochrome P450 14α-demethylase (CYP51)
Method recoveries, sulfentrazone and metabolites in water, 300–301
Methoxyacrylate stilbene (MOAS)
quinol oxidation, 139–140
site topography, 138–140
structure, 138*f*
Metolachlor
analysis of residual, in CYP2B6 rice plants, 24–26
degradation in three-month-old rice plants, 26*t*
Milkweeds
density of stands, 53–54

distribution within and around cornfields, 56
food source for monarch butterfly, 51
host for monarch offspring, 55–56
See also Monarch butterfly
R-Mintlactone, synthesis, 271–272
Mix-and-split methodology, libraries, 123
Mixture strategy, scytalone dehydratase (SD) inhibitors, 110–111
Modeling. *See* Cytochrome P450 14α-demethylase (CYP51)
Mode of action
2,5-anhydro-D-glucitol (AhG), 153*f*
2,5-bis(2,4-dichlorophenyl)-1,3,4-oxadiazole (DCPO) and analogs, 274–275
ceramide synthetase inhibitors in plants, 154
cinmethylin proposed, 155
insect-growth regulators, 280
pyrazolate, 166, 168
pyridalyl, 265
Serenade®, 291
See also Site of action
Molecular biology, target site determination, 158–159
Molecular cloning. *See* B96Bom from silkworm
Molecular genetic approach, fungal lead, 137–138
Molecular modeling. *See* Cytochrome P450 14α-demethylase (CYP51)
Molting-hormonal activity, *Chilo suppressalis*, 194–196
Monarch butterfly
Bacillus thuringiensis (*Bt*) corn pollen, 48–49, 51
components of risk assessment, 51
density of milkweed stands, 53–54
dispersal of corn pollen, 54

distribution of milkweeds by cornfields, 56
estimates of exposure factors, 57
exposure of larvae under field conditions, 52
exposure to *Bt* corn pollen, 53
laboratory bioassays of pollen, 52
materials and methods, 51–52
milkweeds as host for offspring, 55–56
mortality, 52
overall risk, 57–58
percent growth inhibition of larvae, 55*f*
probabilities of toxicity for events, 54–55
probability of exposure to *Bt* corn pollen, 56–57, 59*t*
risk as function of exposure and effect, 51–52
risk conclusions, 58, 60
risk of impact to monarch populations, 57
risks to larvae, 54
survival and growth of neonates on milkweed, 53
susceptible larvae, 53
weight-of-evidence approach, 50–51
Mosquitoes
arbovirus transmission by saliva, 357
immunization, 364–365
lymphatic filariasis, 319
malaria, 320
West Nile virus, 323
See also Dipteran-specific insecticidal crystal proteins; Malaria
Motility, complex mixtures affecting zoospore, 209
R-Muscone, formal synthesis, 270
Mushrooms
allelopathic activity in, 64–65
description, 63

efficacy of lettuce seedling growth, 67
estimation of allelopathic activity, 66–67
fairy rings, 63
lettuce seedling growth inhibition, 67–68
percent inhibition of lettuce seedling growth, 68*t*, 69*t*, 70*t*
plant growth regulators in, 63–64
procedure and allelopathy assay for, litters, 67*f*
proposed release routes of allelochemicals from, 66*f*
safety of allelochemicals, 69
Mycobacterium tuberculosis, cytochrome P450 14α-demethylase (CYP51), 145–147
Myiasis
anti-*Lucilia cuprina* vaccine, 366–367
warble flies and sleep blowfly, 366

N

Natural compounds
activity of *p*-hydroxyphenylpyruvate dioxygenase (HPPD), 156–158
2,5-anhydro-D-glucitol (AhG), 152, 153*f*
asparagine synthetase (AS), 154–155
ceramide synthase biosynthesis, 152–154
1,4-cineole inhibiting AS, 154–155
discovery efforts, 151–152
doses and time for sampling, 158
fructose-1,6-bisphosphate aldolase (FBPase), 152
gene expression profile libraries, 158–159
herbicides, 151
mode of action of AhG, 153*f*

new methods for target site
determination, 158–159
new target sites, 152–155
old target sites, 156–158
Natural products
sorgoleone, 156, 157f
sulcotrione, 156, 157f
usnic acid, 156–158
See also Aldol reactions; Claisen
condensations; Green pesticides;
Insect-growth regulators
Nematicidal compounds
activities towards *Pratylenchus
penetrans*, 220t, 222t, 224t
bioassay for growth of lettuce
seedlings, 217–218
bioassay for growth of rice
seedlings, 218
bioassay for nematicidal activity,
217
bioassay for nematicidal activity
toward free-living nematode,
217
extraction and isolation method,
218
fermentation, 218
fungal metabolites, 216
materials and methods, 216–218
Penicillium cf. *simplicissimum*
(Oudemans) Thom, 218–220
Penicillium bilaiae Chalabuda,
222–224
unidentified *Aspergillus* sp., 221–222
Neonicotinoid insecticides
agonist actions of imidacloprid,
177f, 179f
amino acid sequences in loop D of
vertebrate and insect nAChR
subunits, 180t
chemical structures, 173f
contour plots for electrostatic
potential of imidacloprid, 175f
interactions of α subunits with,
176–177
interactions of loop D with, 178–179
interactions of loop F with, 178
nicotinic acetylcholine receptors
(nAChRs), 173
nithiazine, 173–174
schematic of nAChRs, 176f
structural and physicochemical
properties, 174–175
Nets, insecticidal. *See* Long lasting
insecticidal net (LLIN); Malaria
Nicotinic acetylcholine receptors
(nAChRs)
pesticide targets, 173
schematic, 176f
Nicotinic acid
attachment to Wang resin, 124
Crop Disease (CDM) and Insect
Management (IM), 127,
128t
library construction, 123f
pass rate of acid libraries vs.
random acids and chemistry,
127f
preparation of 5,6-disubstituted,
libraries, 122–124
screening acid libraries in weed
management, 126t
screening for three targeted
therapeutic areas, 128t
screening results, 125–127
yield and purity of 25-membered
rehearsal library of 5,6-
disubstituted, 125t
See also Libraries
Nithiazine, insecticidal activity, 173–174
Nitrile, library inputs, 105f
Norflurazon
effect on pigment contents in
watergrass, 166, 167t
phytoene accumulation, 166,
167t
See also Pyrazole herbicides
Nutrition, pesticide use, 4

O

Obesity, fruits and vegetables, 11–12
Octopamine
  biogenic amine, 184
  dose-response curve in inhibiting tyramine binding to B96Bom, 187f
  dose-response curves in cAMP production, 186f
  See also B96Bom from silkworm; Biogenic amines
Oligosaccharides, synthesis, 90–91, 92
Olyset net. *See* Malaria
Onchocerciases
  control/elimination program, 324t
  parasitic disease, 319
Oomycete phytopathogens
  complex mixtures affecting zoospore motility, 209
  differentiation of *Aphanomyces cochlioides* zoospores, 211
  disease by, 203
  diverse activity of secondary metabolites, 206–207
  host-specific signal substances, 205–206
  life cycle disruptions, 203
  motiles zoospores, 203
  motility inhibitory and cell lytic substances against *Aphanomyces* zoospores, 208–209
  putative signal transduction pathways, 211–212
  regulation of developmental transition by *Amaranthus gangeticus* metabolites, 210–211
  repellant activity of estrogenic compounds towards *Aphanomyces* zoospores, 207–208
  specificity in chemotaxis, 206
  stimuli triggering characteristic behavior and morphological changes, 204t
  zoospore development and regeneration, 211
Optimization
  N-azoyl phenoxypyrimidine herbicide, 114–115
  dihydropyridine miticide, 112–114
  goal of library, 100, 106
  lead compound to pyridalyl, 258, 260
  pyridalyl, 260–263
  See also Parallel synthesis technologies
Orgyia postica, type II pheromones for tussock moth, 230–231
Overexpression lines, fungal lead, 137
Oxadiazoles
  biological properties against *Pseudaletia separata* Walker, 276
  2,5-bis(2,4-dichlorophenyl)-1,3,4-oxadiazole (DCPO) and analogs, 274–275
  structure-activity relationship, 277
  symmetrical 2F-DCPO, 275–276
  target structures, 275
  unusual synthesis of symmetrical 2F-DCPO, 276
  See also Insect-growth regulators
Oxadiazolyl pyridazinones and analogs
  amino pyridazinones, 279
  chronic growth inhibitory activity, 280t
  structure, 277–278
  synthesis, 278–279
  See also Insect-growth regulators
Oxazolidine, scaffold-based discovery library, 116–117

P

Pain, ticks, 351–352
Parallel synthesis technologies

N-azoyl phenoxypyrimidine
herbicide lead optimization,
114–115
1,1-dicyano-2,2-
bis(trifluoromethyl)ethylene
(BTF), 112
dihydropyridine miticide lead
optimization, 112–114
dihydropyrimidines via BTF
condensations, 113f
DuPont Crop Protection, 110
DuPont's scytalone dehydratase
(SD) structure design program,
110–111
fungicide lead compound, 116f
insecticide field candidates, 114f
oxazaspirodecane library synthesis
scheme, 116f
oxazolidine scaffold-based
discovery library, 116–117
SD inhibitors via focused mixture
strategy, 110–111
selected SD inhibitors, 110f
selected SD inhibitors from mixture
strategy, 111f
solid phase synthesis method,
115f
synthesis of lead dihydropyridine,
112f
Parasitic diseases
leishmaniasis, 320
lymphatic filariasis, 319
malaria, 320
onchocerciasis, 319
river blindness, 319
trypanosomiasis, 319–320
See also Vectorborne diseases
Pathogen transmission
block by immunity to arthropod
vector, 358–359
saliva molecules, 357–358
Pediculosis
adulticidal and ovicidal activities of
compounds from clove oil,
391t

deduced amino acid sequences of
partial sodium channel α-
subunit genes, 387f
detection of knockdown resistance
(kdr) mutations for monitoring
resistance, 390
discovery of new alternative
pediculicides, 390–391
extent and mechanism of resistance
to pediculicides, 385–386
filter paper diffusion bioassay, 391
functional analysis of mutations,
288
head louse resistance to permethrin,
384
linkage analysis of mutations to
resistant phenotype, 388–389
malathion resistance, 385–386
molecular analysis of knockdown
resistance (kdr) in permethrin-
resistant head lice, 387–388
molecular mechanisms of head
louse resistance to pyrethroids,
387–390
monitoring tools for resistance
management, 387–390
mortality-survival time regression
lines, 389f
parasitic infestation of humans, 384
relationship between phenotype
resistant level and percent of
resistant homozygotes, 389f
resistance ratio of head lice from
FL, CA, and TX, 386t
serial invasive signal amplification
reaction (SISAR) vs. sequencing
results, 390t
susceptibility to insecticides, 384–
385
Penicillium sp.
compounds from *Penicillium* cf.
simplicissimum (Oudemans)
Thom, 218–220
compounds from *Penicillium
bilaiae* Chalabuda, 222–224

nematicidal activity of compounds toward *Pratylenchus penetrans*, 220*t*, 224*t*
See also Nematicidal compounds
Peptidomimetics, geometrically defined
advantage of cyclopropylideneacetates, 80
aliphatic amines in last steps, 82–83
bicyclic, using chiral amines, 84*t*
from cyclopropylideneacetates, 80*f*
kinetic resolution during ring closure, 82*f*
optically active piperazinones and lactams by kinetic resolution, 82*t*
piperazinone formation under modified Schotten–Baumann reaction, 83*f*
sequence for construction of bicyclic, 83*f*
structural analogue search, 80
structural flexibility in Michael addition of 2-chloro-2-cyclopropylideneacetate and cyclizations, 81*t*
synthesis of monocyclic peptidomimetic, 81*f*
Permethrin
head louse resistance to, 384
Olyset and polyethylene nets, 328
percent remaining in nets after seven years, 331*t*
See also Long lasting insecticidal net (LLIN)
Peronosporomycetes
new classification for Oomycetes, 203
See also Oomycete phytopathogens
Pesticides
activists exaggerating, 11–12
affordability of food in United States, 8, 9*f*
arable land in production, 4, 6*f*
cancer and consumption of fruits and vegetables, 13
consumers lacking appreciation for, 10, 11*t*, 15*t*
food surpluses, 4
fruit and vegetable consumption, 11–12
high technology agriculture, 10
human health and nutrition, 4
media coverage, 12
natural chemicals, 10, 11*t*
number of people fed per farmer, 10*f*
productivity, 4, 7*f*, 8
risk analysis, 13
testing requirements, 14–15
value of herbicides in U.S. crop production, 8*f*
See also Green pesticides
Pharmacologically active molecules
saliva of blood-feeding arthropods, 350–357
See also Ticks
Phase tagging. See Synthesis based on affinity separation (SAS)
Pheromones. See Sex pheromones
Phloem mobility, finding leads, 121
Phytoene
accumulation, 162
norflurazon and, accumulation, 166
See also Pyrazole herbicides
Plant essential oils, alternative pediculicides, 390–391
Plant hormones
abscisic acid (ABA) and jasmonic acid (JA), 246
analogs of ABA, 247–248
JA analogs (MJA and MepiJA), 249
JA analogs, antimetabolites, 250–252
plant growth inhibitory activity of ABA analogs, 248*t*
plant growth regulatory activity of JA analogs, 250*t*, 251*t*

practical synthesis, 252–253
roles, 246
synthesis of 11- or 12-fluoro methyl jasmonate (MJA) analogs, 250
synthesis of ABA and β-ionylideneacetic acid (β-IAA), 252
synthesis of methyl epijasmonate (MepiJA), 249
synthesis of methyl tuberonate (MTA), 251–252
synthesis of MJA and MTA, 253
synthetic schemes of ABA analogs, 248
tuberonic acid (TA), 251
Plant protection. See Sex pheromones
Plaque, bacterial disease, 321
Platelet aggregation, ticks, 353
Plipastatins
Serenade®, 292
See also Green pesticides
Plutella xylostella, insecticidal activity of pyridalyl, 264t
Pollen, corn. See Monarch butterfly
Population, mushrooming and disease outbreaks, 315, 318
Potato scab. See Streptomyces species
Poultry
maize performance evaluation, 34–35
nutritional evaluation for maize event NK603, 31–32
See also Biotech crops
Pratylenchus penetrans
nematicidal activity against, 220t, 222t, 224t
See also Nematicidal compounds
Production, arable land, 4, 6f
Productivity
America's farmers, 8, 9f
pesticide use and, 4, 7f, 8
Prostaglandin E_2, ticks, 353
Proteins. See Dipteran-specific insecticidal crystal proteins

Proteomics, blood feeding arthropods, 367–368
Pseudaletia separata
chronic growth inhibitory activity of oxadiazolyl pyridazinones and analogs, 280t
structures and bioactivities of oxadiazoles against, 277
See also Insect-growth regulators
Public acceptance, farming practices, 15t
Pyrazolate. See Pyrazole herbicides
Pyrazole herbicides
cell growth condition, 164
chemical structures of pyrazolate, pyrazoxyfen, and metabolite, 163f
concentration of metabolite in solution for 4-hydroxyphenylpyruvate dioxygenase (HPPD), assay, 168, 170f
determination of metabolite in aqueous solution, 165
Echinochloa oryzicola watergrass, 163
effect of, on HPPD activity, 168, 170f
effect of homogentisate on phytoene content, 168, 169f
effect of metabolite on carrot cell growth, 168, 169f
effects of pyrazolate and norflurazon on pigment contents in watergrass, 166, 167t
herbicides treatment, 164
HPPD preparation and assay, 162, 165
materials and methods, 163–165
mode of action of pyrazolate, 166, 168
phytoene accumulation and norflurazon, 166, 167t
phytoene content in third leaves of watergrass, 168, 169f

pigments extraction and determination, 164
plant materials, 163
pyrazolate and pyrazoxyfen, 162–163
Pyrazoxyfen. *See* Pyrazole herbicides
Pyrethroids, mechanisms of head louse resistance, 387–390
Pyridalyl
  beneficial arthropods at 100 ppm, 265*t*
  biological properties, 263–265
  discovery of lead compound, 258
  effect of linkage moiety between two rings on *Spodoptera litura* activity, 263*t*
  effect of phenyl ring substituents on *S. litura* activity, 261*t*
  effect of pyridyl ring substituents on *S. litura* activity, 262*t*
  evolution to find second lead compound, 258, 260
  field studies, 265
  insecticidal activity against insecticide resistant strain of *Plutella xylostella*, 264*t*
  insecticidal activity against lepidopterous pests, 264*t*
  insecticidal activity of lead compound against *S. litura*, 259*t*
  insecticide, 257
  laboratory studies, 263–264
  mode of action, 265
  optimization of linkage moiety between phenyl and pyridyl rings, 262–263
  optimization of substituents on phenyl ring, 260–261
  optimization of substituents on pyridyl ring, 261–262
  selection, 263
  structure-activity relationships for compound in artificial diet assay, 259*f*
  structure and nomenclature, 257*f*

Pyridine carboxylic acids, combinatorial libraries, 122–123
Pyridine-containing acids, asymmetry, 122
2-Pyridinone esters. *See* Ring-fused 2-pyridinone esters
Pythia search engine, combinatorial chemistry, 76, 77*f*, 84

Q

Quinol oxidation, famoxate vs. methoxyacrylate stilbene, 139–140
Quinoxalines. *See* 2-Substituted-3-trifluoromethylquinoxalines (3-TFQ)

R

Rats
  plan for evaluating maize event NK603, 32*t*, 33
  safety assurance evaluation, 35–37
  See also Biotech crops
Receptor-binding
  Chilo suppressalis, 194–196
  relationship between insect orders, 196–198
Regulators, insect-growth. *See* Insect-growth regulators
Resistance, pediculicides, 385–386
Resistance management. *See* Pediculosis
Rhodopeptin, structural information, 80
Rift Valley fever, arboviral disease, 322
Ring-fused 2-pyridinone esters
  broad leaf and grass herbicidal activity, 107*t*
  calculated log P vs. molecular weight for candidate virtual library, 104*f*

candidate library screening, 104
defining quality library, 105
design strategy, 101–102
herbicide property ranges for commercial products and, library, 102t
lead generation effort, 100–101
library Meldrum's acid adduct inputs, 105f
library nitrile inputs, 105f
library scaffold, 100
screening for activity, 106
synthesis, 103, 104f
Risk analysis
monarch butterfly and corn pollen, 57–58, 60
pesticides, 13, 14f
See also Monarch butterfly
River blindness, parasitic disease, 319
Roll Back Malaria (RBM), World Health Organization (WHO) campaign, 320, 327, 332–333
Roundup®, glyphosate, 29
Roundup Ready®, maize event NK603, 29
Russet scab toxin
analysis method, 241
isolation and identification, 242–243
isolation by *Streptomyces cheloniumii*, 241
necrosis-inducing activity, 244
phytotoxin, 240, 244
purification procedure, 243f
structure, 240f
See also *Streptomyces* species

S

Saccharomyces cerevisiae
gene expression profile libraries, 158–159
growth curve, 134f
growth inhibition, 134–135

Safety, green pesticides, 284
Safety concerns
food and pesticides, 15
See also Biotech crops
Saliva
pathogen transmission, 357–358
See also Ticks
Sample management facility (SMF), screening process, 133
Sampling, doses and time points, 158
Sand flies, modulating host immune defenses, 357
Sanger reagent, benzotriazole synthesis, 76–77
Scaffold-based discovery library, oxazolidine, 116–117
Screening
hit generation, 133
process in discovery, 133
See also Nicotinic acid
Scytalone dehydratase (SD)
inhibitors via focused mixture strategy, 110–111
selected inhibitors, 110f
selected inhibitors from mixture strategy, 111f
Serenade®
Bacillus subtilis QST-713, 291
effect of lipopeptides on pathogen spore germination, 292–293
lipopeptides, 291
mode of action, 291
molecular structure, 291–292
production, 293
structures of agrastatins, 292f
surfactins, 292
See also Green pesticides
Sex pheromones
biosynthesis and perception, 235–236
biosynthetic pathways, 235f
chiral high performance liquid chromatography (HPLC) analysis, 231–232

disruption by type II compounds, 234–235
foresight of research for identification, 232
gas chromatography with electroantennogram (EAG) detector (GC–EAD), 227–228
identification of natural, 227–232
Lepidoptera, 227
mating disruption overview, 232–234
Orgyia postica tussock moth, 230–231
representative lepidopteran, and abbreviations, 227f
representative mating disruptants, 233t
type II by highly evolved species, 229–231
type I with conjugated diene system, 228–229
work at Chemical Ecology Laboratory in Tokyo, 227
Signal transduction pathway, zoospore, 211–212
Silkworm. *See* B96Bom from silkworm
Site of Action
chemistry to gene (C2G) approach, 133–134, 138
discovery, 133
effects of lead 1 on transcript levels of specific gene clusters, 136f
effects of various supplement pools on yeast growth, 135f
fungal lead 1, 134–137
fungal lead 2, 137
fungal lead 3, 137–138
future directions, 140
growth curve of *Saccharomyces cerevisiae*, 134f
inhibitors famoxate and methoxyacrylate stilbene, 138–140

site topography, 138–140
target identification, 133–134
Site topography, famoxate vs. methoxyacrylate stilbene, 138–140
Sorgoleone, activity of p-hydroxyphenylpyruvate dioxygenase (HPPD), 156, 157f
Spinosad, green pesticides, 289–290
Spinosyns A and D, green pesticides, 289–290
Spodoptera litura
  insecticidal activity of lead compound to pyridalyl, 259t
  See also Pyridalyl
Streptomyces species
  analysis for phytotoxins, 241
  bacterial strains, 241
  concanamycin A and B, 240
  isolation and identification of russet scab phytotoxin, 242–243
  isolation of russet scab toxin by *S. cheloniumii*, 241
  materials and methods, 241
  necrosis-inducing activity of russet scab toxin, 244
  pathogenicity assay and phytotoxin production by, 242t
  pathogenicity assay in greenhouse, 241
  phytotoxin thaxtomin A, 239–240
  production of russet scab toxin by other strains, 244
  production of thaxtomin A and concanamycins, 242
  purification procedure for russet scab toxin, 243f
  russet scab toxin, 240, 244
  structures of phytotoxins, 240f
  symptoms by pathogenicity assay, 241–242
Structure-activity relationships. *See* 2-Substituted-3-trifluoromethylquinoxalines (3-TFQ)

2-Substituted-3-
trifluoromethylquinoxalines (3-
TFQ)
  biological testing method, 306
  effect of R^1 substitution of, on
    acaricidal activity, 308, 309t
  effect of R^2 substitution at 2-
    position of, on acaricidal
    activity, 307t
  effect of R^2 substitution of 6,7-Cl_2-
    3-TFQ on acaricidal activity,
    308, 309t
  effect of R^2 substitution of 6,7-Cl_2-
    3-TFQ on acaricidal and
    insecticidal activity, 311t
  effect of R^2 substitution of 6,7-Cl_2-
    3-TFQ on insecticidal activity,
    310t
  effect of R substitution at 3-
    position of, on acaricidal
    activity, 307, 308t
  structure activity relationships,
    307–310
  synthesis, 305–306
Sulcotrione, activity of p-
  hydroxyphenylpyruvate
  dioxygenase (HPPD), 156, 157f
Sulfentrazone
  comparing immunogen and
    reactivity, 298t
  3-desmethyl sulfentrazone (DMS)
    structure and nomenclature,
    297f
  ELISA (enzyme-linked
    immunosorbent assay) kit for,
    and sulfentrazone-3-carboxylic
    acid (SCA) residues, 296–297
  ELISA method, 298–299
  ELISA vs. LC/MS/MS comparison,
    301–303
  herbicide, 296
  3-hydroxymethyl sulfentrazone
    (HMS) structure and
    nomenclature, 297f

  liquid chromatography/mass
    spectrometry/MS (LC/MS/MS)
    method, 299–300
  metabolites HMS and DMS, 297
  method recoveries of, and SCA in
    water, 300–301
  N–C–C–O bonds in HMS and
    SCA, 298f
  positive and negative residues by
    ELISA and LC/MS/MS, 302t
  SCA structure and nomenclature,
    296f
  structure and nomenclature, 296f
Surfactins
  Serenade®, 292
  See also Green pesticides
Syntheses. See Parallel synthesis
  technologies
Synthesis based on affinity separation
  (SAS)
  diarylpiperazine preparation, 89–90
  interaction between crown ether
    (32-crown-10) and ammonium
    ion, 88–91
  interaction of barbituric acid and
    artificial receptor, 91–97
  phase tagging, 87
  schematic, 88f
  synthesis of oligosaccharides, 90–
    91, 92
  Triton X-100 as tag, 90–91

T

Target enzymes. See Cytochrome
  P450 14α-demethylase (CYP51)
Target identification, chemistry to
  gene (C2G), 133–134, 138
Target sites
  asparagine synthetase (AS), 154–
    155
  ceramide synthase biosynthesis,
    152–154

fructose-1,6-bisphosphate aldolase
    (FBPase), 152
4-hydroxyphenylpyruvate
    dioxygenase (HPPD), 156–158,
    162
    new methods for determination,
        158–159
    nicotinic acetylcholine receptors
        (nAChRs), 173
    See also Natural compounds
Thaxtomin A
    analysis method, 241
    pathogenicity assay and production,
        242*t*
    production, 242
    Streptomyces scabies, 239–240
    structure, 240*f*
    See also Streptomyces species
Thifluzamide, trifluoromethyl
    containing fungicide, 305
Thrombin inhibitors, ticks, 353
Ticks
    antigens by screening cDNA
        library, 356
    blood coagulation, 352
    bradykin, 351
    calreticulin, 354
    expression of macrophage
        migration inhibitory factor
        (MIF), 355
    histamine, 351–352
    host immune interactions, 350
    IL-2 binding protein of *Ixodes
        scapularis*, 354–355
    inhibiting alternative pathway of
        complement activation, 355
    inhibitors of factor Xa and X, 352–
        353
    linking pathogen transmission to
        saliva molecules, 357–358
    Lyme disease, 321
    modulating immune defenses, 356
    pain/itch, 351–352
    pharmacologically active
        molecules, 356–357

prostaglandin E_2 as vasodilator,
    353
saliva, 351
salivary gland molecules, 354
thrombin inhibitors, 353
transmitting infectious agents, 349
vaccine research, 360–362
Tobacco plants, transgenic, containing
    aryl hydrocarbon receptor (AhR)
    gene, 42–43
Topography of site, famoxate vs.
    methoxyacrylate stilbene, 138–140
Transgenic rice plants
    analysis of residual metolachlor in
        CYP2B6, 24–26
    CYP1A1, 20, 22
    CYP2B6, 22–23
    cytochrome P450 monooxygenase
        (P450), 19
    degradation of metolachlor in
        three-month-old, 26*t*
    germination tests, 20, 22–23
    metabolism of herbicides in,
        expressing CYP1A1 and
        CYP2B6 gene, 23, 24*f*
    phototoxicity of herbicides toward
        CYP1A1 and CYP2B6, 21*f*
    production of, expressing human
        P450 genes, 19–20
    tolerance of, CYP1A1 and
        CYP2B6 to herbicides in
        germination tests, 22*t*
Transgenic tobacco plants
    aryl hydrocarbon receptor (AhR)
        gene, 42–43
    GUS (β-glucuronidase) reporter
        gene, 42–43
    histochemical staining of GUS
        activity, 44*f*
Trifluoromethyl-containing
    compounds
    agrochemicals, 305
    See also 2-Substituted-3-
        trifluoromethylquinoxalines (3-
        TFQ)

Triton X-100, tag in synthesis, 90–91
Trypanosomiasis, parasitic disease, 319–320
Tussock moth, type II pheromones, 230–231
Tyramine
  actions of formamidine insecticide on *Bombyx mori*, receptor, 187–189
  dose-response curve in inhibiting tyramine binding to B96Bom, 187f
  dose-response curves in cAMP production, 186f
  functional properties of B96Bom receptor in cell line, 185–186
  See also B96Bom from silkworm; Biogenic amines

U

Ultraspiracle (USP)
  expression and functional analysis, 193–194
  See also Ecdysone agonists, non-steroidal
United States, affordability of food, 8, 9f
Urbanization, vectorborne diseases, 318t
Usnic acid, activity of *p*-hydroxyphenylpyruvate dioxygenase (HPPD), 156–158

V

Vaccines
  advantages of anti-arthropod, 349–350
  challenges for anti-arthropod, 368–369
  vector-blocking, 365–366
  See also Anti-ectoparasite vaccines

Vector-blocking vaccine
  concept, 365–366
  mosquito molecular biology, 364
Vectorborne diseases
  arboviral diseases, 322–323
  bacterial diseases, 321
  contributing factors, 315, 318
  control tools, 325
  dengue, 322
  disease control strategies, 324
  epidemics during 1994–99, 317f
  increasing, 315
  influences of emergent/resurgent, 318t
  leishmaniasis, 320
  Lyme disease, 321
  lymphatic filariasis, 319
  malaria, 320
  mobility of human populations, 318
  mushrooming population, 315, 318
  onchocerciases (river blindness), 319
  parasitic diseases, 319–320
  plague, 321
  relationship between population growth and ease of travel, 318f
  trypanosomiasis, 319–320
  twenty year history of large outbreaks, 316f
  unexpected outbreaks, 317f
  urbanization, 318
  vector control/elimination programs, 324t
  West Nile virus, 323
Vector suppression, control of arthropod-borne infectious agents, 349
Vegetables. *See* Fruit and vegetables

W

Waste reduction, green pesticides, 286
Weed management, library target, 100–101

Weed Management (WM). *See* Nicotinic acid
West Nile virus, arboviral disease, 323
World market
   agricultural chemicals, 3
   industry consolidation, 4, 5*f*

Y

Yellow fever, arboviral disease, 322, 324*t*

Z

Zoospores
   accumulation at infection sites, 203
   Aphanomyces cochlioides zoospores, 211
   complex mixtures affecting motility, 209
   development and regeneration, 211
   motility inhibitory and cell lytic substances against, 208–209
   repellent activity of estrogenic compounds, 207–208
   signal transduction pathway, 211–212
   specificity in chemotaxis, 206
   specificity of chemotaxis, 206
   stimuli triggering characteristic behavior and morphological changes, 204*t*
   See also Oomycete phytopathogens